Simplified Classical Mechanics, Volume 2 (Second Edition)

Gravity and the conservation laws

Online at: https://doi.org/10.1088/978-0-7503-6402-7

Simplified Classical Mechanics, Volume 2 (Second Edition)

Gravity and the conservation laws

Gregory A DiLisi

John Carroll University, University Heights, OH, USA

IOP Publishing, Bristol, UK

ISBN   978-0-7503-6402-7 (ebook)
ISBN   978-0-7503-6398-3 (print)
ISBN   978-0-7503-6399-0 (myPrint)
ISBN   978-0-7503-6401-0 (mobi)

DOI   10.1088/978-0-7503-6402-7

Version: 20250901

IOP ebooks

British Library Cataloguing-in-Publication Data: A catalogue record for this book is available from the British Library.

Published by IOP Publishing, wholly owned by The Institute of Physics, London

IOP Publishing, No.2 The Distillery, Glassfields, Avon Street, Bristol, BS2 0GR, UK

US Office: IOP Publishing, Inc., 190 North Independence Mall West, Suite 601, Philadelphia, PA 19106, USA

This book is dedicated to my family:

to my grandparents, Tommaso and Carmela Frate,

to my parents, Richard and Mary DiLisi,

to my siblings, Rick DiLisi, Carla Solomon, and Jennifer Newton,

to my wife, Linda,

to my daughter, Carmela,

and

to the wonderful creatures who inhabit our home.

Contents

Preface

We derive the word 'physics' from the Greek word 'physika,' which translates into 'pertaining to natural things.' Therefore, as we begin our study of 'physics,' we quite literally begin our study of 'natural things.' When I was in high school, my physics teacher once asked me: 'Who is history's greatest physicist? Newton? Einstein? Galileo?' I thought for a moment and gave him an answer that absolutely floored him. 'Batman,' I said, 'because his entire life is dedicated to figuring-out how and why things move. All of his equipment helps him solve motion ... grapplings, cables, levers, pulleys, projectile-launchers, etc.'

Michael Keaton as *Batman,* from Louis Tussaud's Waxworks in Ontario, Canada.

Image Credit: Author.

Etymology:
   The word 'physics' comes from the Greek word 'physika' that translates into 'pertaining to natural things.'

In short, that's what this book is about—describing how and why things *move*. Instead of viewing physics as the study of 'natural things' (from its literal Greek translation), an excellent, alternate way to understand physics is simply to view it as the study of motion. Albert Einstein once said: 'Nothing happens until something moves.' Einstein's quotation is used in a number of contexts. Most often, it is applied in the area of self-help. The implication is that in order for something to happen in your life, you must be proactive and make something happen. However, when interpreted literally, his quotation is equally profound ... nothing *does* happen until something moves. Think about it ... you cannot even generate a thought without some electrical signal creating the movement of charged particles somewhere in your brain. Even your emotions involve the complex movement of chemicals and electrical signals throughout your body. Pay careful attention and you will notice that every problem you encounter in this series on *Simplified Classical Mechanics* involves something moving. Therefore, the goal of this textbook is simple: to develop problem—solving techniques that handle different types of motion.. The situations you encounter in everyday life vary considerably, so our problem-solving techniques must vary accordingly. If Batman truly is history's greatest physicist, the we will discover that the techniques that defeat the Joker may not work well on the Riddler or Penguin. Consequently, we need to develop an entire arsenal of problem-solving tactics—all of which are meant to help us understand and describe objects *in motion!*

Before developing these problem-solving techniques, I will describe how I designed this series on *Simplified Classical Mechanics* and how it came to be created. First, I will describe the structure and topics covered in a typical introductory physics sequence of courses. This overarching structure will provide the *context* for which the chapters in the volumes were designed. Next, I will describe the *format* of how each topic is presented. The format will demonstrate how each section is written and how each topic in the volumes is arranged. Finally, I will more thoroughly describe the overall *purpose* of the series.

1. **Structure of the typical introductory physics sequence of classes:** An introductory physics sequence of classes is typically divided into three distinct courses. Believe it or not, the three courses can be nicely understood by the special properties, observed by ancient scientists, associated with four stones. This series focuses on only the first course, 'Classical Mechanics.' However, the distinctions among these three courses are useful to describe because they frame the context under which 'Classical Mechanics' is housed. The distinctions among these three courses are briefly described below:

    ❶ *'Classical Mechanics'*—Consider the sandstone shown below:

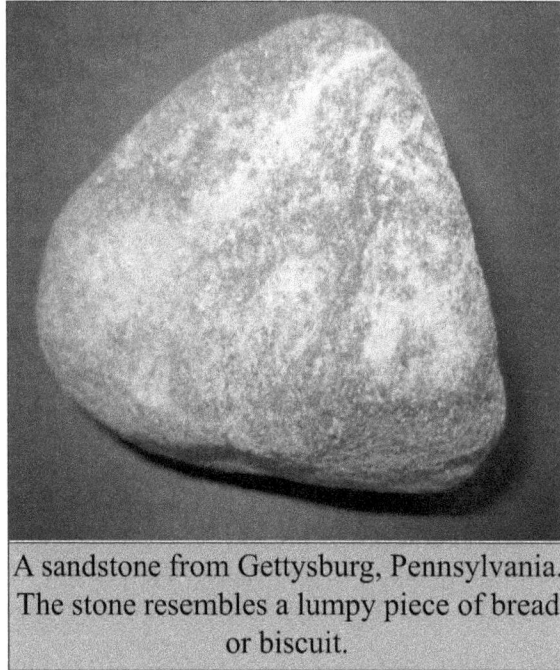

A sandstone from Gettysburg, Pennsylvania.
The stone resembles a lumpy piece of bread
or biscuit.

Image Credit: Author.

Ancient scientists observed that when held above the Earth's surface
and released, it had the wonderous power of falling toward the ground. (In
my opinion, this is indeed a wonderous power ... an object moves towards
the Earth even though nothing touches it. Can you explain how or why
this happens?) The object was said to possess the Greek word 'maza' that
gave it the ability to interact with the Earth through 'gravitas.'

Etymology:
The word 'mass' comes from the Greek word 'maza' that translates into 'heavy
stone' or 'lumpy bread' while the word 'gravity' comes from the Latin word 'gravitas'
that translates into 'heaviness' or 'seriousness.'

Film critics often say that an actor may not have the 'gravitas' to play a certain role,
meaning he or she may not have the needed seriousness, or on-screen presence, to pull
off a certain performance.

A course in 'Classical Mechanics' focuses on the motion or behavior
of objects with **mass** and whose dimensions and/or speeds are familiar
to us. Your intuition from everyday experiences will be generally

correct because, like it or not, you have mass (maybe more than you would like) and are used to interacting with the Earth through the force of 'gravity.' If you use the etymological origins of the words associated with this introductory physics course, in a 'Classical Mechanics' course, you are literally studying pieces of lumpy bread (mass) that have heaviness (interact through gravity).

❷ *'Classical Electricity and Magnetism'*—Consider the stones of fossilized tree resin, known as amber, shown below:

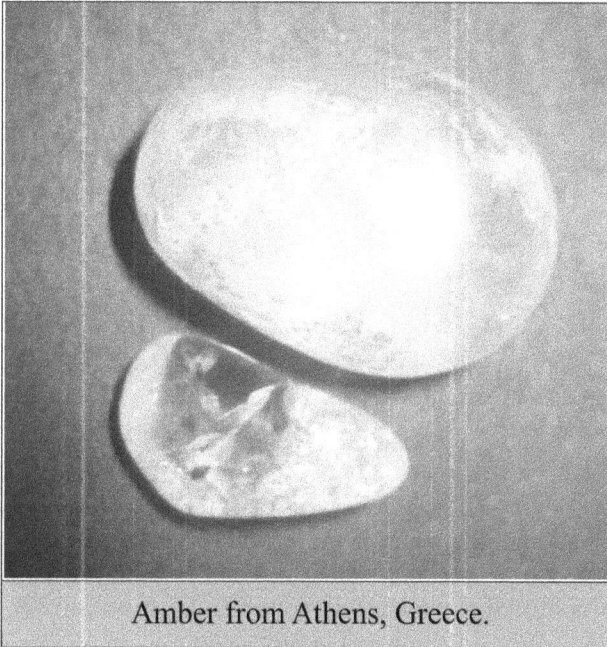

Amber from Athens, Greece.

Image Credit: Author.

Ancient scientists observed that when rubbed with animal fur, these stones had the wonderous power of lifting grass clippings (or hay) simply by passing the stones over the grass clippings. (In my opinion, this is indeed a wonderous power … a stone can lift another object without even touching it. The stone can pull on the grass clippings and overcome gravity to lift them off the ground. Can you explain how or why this happens?) Such a stone, known in Greek as an 'ēlektron,' was truly special because of its unique ability to lift certain objects after being rubbed.

Etymology:
The word 'electricity' comes from the Greek word 'ēlektron' that translates into 'amber.'

Also consider these metallic stones from Volos, Greece shown below:

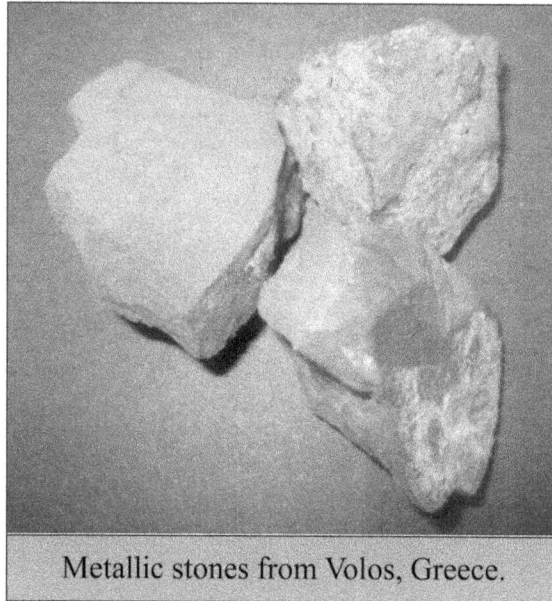

Metallic stones from Volos, Greece.

Image Credit: Author.

Ancient scientists observed that these stones, *without even being rubbed*, had the wonderous power of lifting other metallic objects simply by passing the stone over the metallic object. (In my opinion, this is indeed a wonderous power ... a stone can lift another object without even touching it. The stone can pull on other metal objects and overcome gravity to lift them off the ground. Can you explain how or why this happens?) You will be fascinated to know that Volos, Greece was once known as the province of Magnesia, so these stones were known in Greek as 'Magnesia,' or simply 'stones from Magnesia.'

Etymology:
The word 'magnetism' comes from the Greek word 'Magnesia' that translates into 'stones from Magnesia.'

A course in 'Classical Electricity and Magnetism' focuses on the motion or behavior of objects with **charge** and whose dimensions and/or speeds are familiar to us. Your intuition from everyday experiences will be somewhat challenged because you typically do not carry huge amounts of static or moving charge with you, so are therefore not used to interacting with other objects through the forces of 'electricity' or 'magnetism.' If you use the etymological origins of the words associated with the title of this introductory physics course, in an 'Electricity and Magnetism' course, you are literally studying pieces of fossilized tree resin (amber or electricity) and stones from Magnesia (magnets).

❸ *'Modern Physics'* or *'Quantum Mechanics'*—Consider the pebble shown below:

A pebble from Olmsted Falls, Ohio.

Image Credit: Author.

Ancient scientists observed that a pebble had the wonderous power of doing things that larger stones could not. For example, it could be made to fit into tighter spaces into which the larger stones simply could not fit. (In my opinion, this is indeed a wonderous power ... a pebble could now find its way into spaces which were once *restricted* to larger stones. Can you explain how or why this happens?) The pebble, which we can label by the Latin word, 'quantum,' might be the smallest sized stone that can exist.

Etymology:

The word 'quantum' comes from the Latin word 'quantum' (the plural is 'quanta') that translates into 'amount' or 'how much.' The implication here is that a 'quantum' represents the minimum amount of a quantity that can exist ... the smallest possible amount of something. Therefore, the word 'quantum' can be applied to many different quantities. For example, a quantum of light is called the 'photon,' a quantum of electricity is called the 'electron,' etc.

A course in 'Modern Physics' or 'Quantum Mechanics' focuses on the motion or behavior of objects whose dimensions and/or speeds are very unfamiliar (i.e., extremely small sizes and extremely fast speeds) to us. Your intuition from everyday experiences will tell you nothing about a particular problem because you are neither extremely small nor extremely fast. If you use the etymological origins of the words associated with the title of this introductory physics course, in a 'Quantum Physics' course, you are literally studying the smallest possible pieces of things (quanta).

The distinction of these courses is summarized in the following diagram:

INTRODUCTORY PHYSICS COURSES

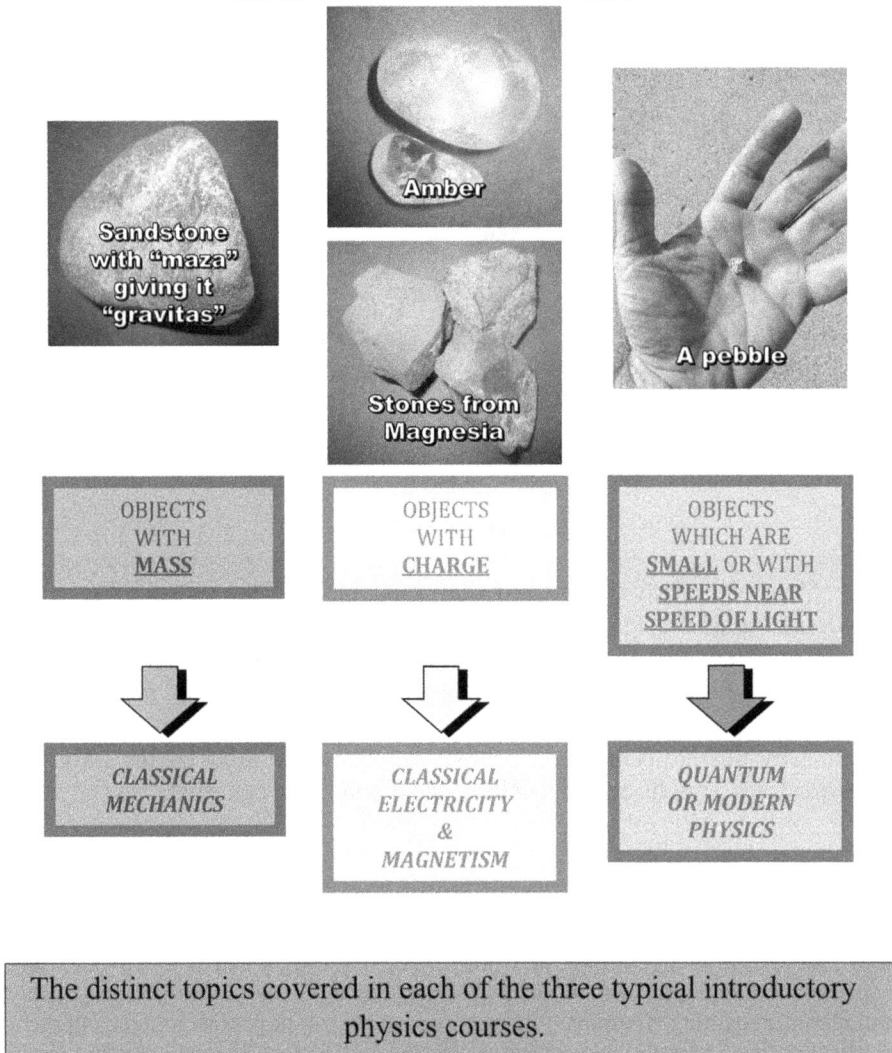

The distinct topics covered in each of the three typical introductory physics courses.

Image Credit: Author.

2. **Format of this series:** This series was generated from the set of notes that I have used to teach my introductory physics courses on *Simplified Classical Mechanics.* After decades of teaching these courses, I finally decided to convert these notes into a series of textbooks. This conversion from notes to textbooks was simply a natural progression as a next step along the development of my *Simplified Classical Mechanics* courses. I only assume that the reader understands mathematics at the level of high school *Algebra.* The reader need not be familiar with high school physics, nor Calculus. What makes this series of books unique (indeed distinctive from typical Algebra-based introductory physics books) is that it combines my experience as a physicist with my experience as a science educator. In this light, I have merged my expertise and content knowledge of physics with my pedagogical expertise as a science educator. Thus, the physics content is presented according to well-established pedagogical methods that are grounded in research. This research has demonstrated that no matter what the teacher's style, grade level (Primary, Middle Childhood, or Adolescent/Young Adult), subject matter, or economic background of the students, a properly taught lesson contained 10 elements that enhanced and maximized learning. These 10 elements form the structure of each chapter in this text:

GETTING READY:

❶ **Anticipatory Set:** This is an activity designed to deliberately focus a student's attention and to develop mental *and* physical readiness in the student for ensuing instruction.

❷ **Objective:** This occurs when the teacher communicates to the student **WHAT** the student will be able to do by the end of instruction.

❸ **Purpose:** This occurs when the teacher communicates to the student **WHY** the students need to learn the upcoming material and what the student will be able to do with the instruction.

GIVING INFORMATION:

❹ **Instructional Input:** This section involves the determination by the teacher of what information is needed by the student to accomplish the current objective.

❺ **Modeling:** This is the teacher showing in graphic form, or demonstrating, what the finished product will looks like. The student *sees* an example of an acceptable finished product or process.

❻ **Checking for Understanding:** This step occurs when the teacher uses a variety of questioning techniques to determine *'Got it, yet?'* and to pace the lesson.

KEEPING INFORMATION:

❼ **Guided Practice:** This step ensures that the instruction has been correctly processed by the students before letting them loose to practice independently. This step also ensures that the students' initial attempts in new learning are carefully guided so they are accurate and successful.

❽ **Closure:** This is a step to add new meaning to a lesson and for the teacher to check, one last time, that the new information has been processed correctly.

❾ **Independent Practice:** This step is just a fancy expression for homework; however, homework should be done only when the student can perform the desired task without major errors, discomforts, or confusion ... only then can students be given written/verbal assignments to practice the new skill/process with no teacher influence.

❿ **Peer Teaching:** Although students may be able to 'do' certain skills, optimum learning has not taken place until students can process the steps to perform the skill; can teach these steps to someone else; and can recognize errors and self-correct these errors.

3. **Purpose of this series:** On one hand, the purpose of this series of books is to teach you *how* to solve physics problems; in other words, how to put math and physics together to obtain a numerical or algebraic result and then interpret these results physically. These skills are certainly important and will be needed in more advanced science and engineering courses. However, more important than developing problem-solving skills and physical-interpretation skills, the purpose of this series is to survey the basic concepts, or what I consider to be the 'grand themes,' of *Simplified Classical Mechanics.* Most practicing physicists and engineers state that the primary skill on which they most often rely in the practice of their profession is their knowledge of basic physics principles. The primary purpose of this series is to provide the reader with a solid understanding of the foundational content knowledge of *Classical Mechanics.* Therefore, the topics in this series have been arranged around the basic concepts, or 'grand themes,' of *Simplified Classical Mechanics.*

Volume 1 is titled, 'Foundations of Motion':

❶ **Volume 1, chapter 1—Tools and vectors:** Chapter 1 is simply about transmitting information! The conventions used to transmit certain types of numerical information are crucial concepts that must be addressed at the outset of any series on *Classical Mechanics.* These conventions will be addressed when discussing 'Scalars' versus 'Vectors.'

❷ **Volume 1, chapter 2—Kinematics:** Chapter 2 focuses on the difference between asking, 'How does an object move?' versus 'Why does an object move?' This distinction requires a paradigm shift in the mind of the reader. Therefore, the reader must train himself or herself to clarify, 'Am I trying to describe *how* the object moves ... or *why* the object moves?' This distinction will be addressed in chapter 2 where we focus our discussion on the question of: '*How* does an object move?'

❸ **Volume 1, chapter 3—Uniformly accelerated motion:** Chapter 3 develops the first of many problem-solving recipes and applies it to situations of idealized accelerations, short-term-accelerations, and free-fall. The recipe is then extended to two-dimensional motion.

❹ **Volume 1, chapter 4— Newton's laws of motion:** Chapter 4 focuses on the question of: '*Why* does an object move?' To answer that question, we turn to Isaac Newton. The hallmark of any good introductory physics series is its treatment of Newton's laws of motion. These laws are difficult concepts for most readers for a number of reasons: they have a *reputation* as being difficult concepts; they require the mastery of multiple sub-skills; and problems involving these laws can be cast in a variety of formats.

❺ **Volume 1, chapter 5—Problem-solving with Newton's laws of motion:** We can view chapter 4 as learning how to build a car and chapter 5 as learning how to drive a car. In chapter 5, we practice … practice … practice … solving problems using Newton's laws of motion.

❻ **Volume 1, chapter 6—Uniform circular motion:** In chapter 6, we literally go in circles. Every problem we encounter will involve motion in a circular path, or partial circular path. To analyze this type pf motion, we will once again apply Newton's laws of motion, but employ a problem-solving formula for the special case of circular motion.

Volume 2 is titled, 'Gravity and the Conservation Laws':

❶ **Volume 2, chapter 1—The universal law of gravitation:** The notion that forces act through their associated fields is first introduced when discussing Newton's Universal Law of Gravitation. A huge conceptual leap is needed by the reader—an object can cause another object to move without even touching it! This is a difficult concept to reconcile with our everyday experiences but it makes perfect sense when we realize that is exactly how the Earth acts on us. Gravity is able to pull on us even though we are not in direct contact with the Earth. Also, the concept of 'super-position' (and when it is applicable) is introduced in chapter 1. Super-position is so crucial to the development of problem-solving skills that I consider it central to any good introductory physics course. This concept will be illustrated in a number of example problems.

❷ **Volume 2, chapter 2—The conservation laws:** The concept of conservation laws is first addressed in a typical *Classical Mechanics* course. This concept is also crucial for developing problem-solving techniques in almost all physics courses, regardless of the content or level. Conservation laws are used in every type of physics course: mechanics, electricity, magnetism, thermodynamics, electronics, quantum mechanics, nuclear and particle physics, etc. The first two of these conservation laws, 'The Conservation of Energy' and 'The Conservation of Momentum' will be addressed in this chapter.

❸ **Volume 2, chapter 3—Rotational motion:** Finally, this chapter introduces the concept of center-of-mass and rotational motion. Instead of limiting our motion to a straight line, the objects we examine can now rotate, spin, and tumble. The motion we discuss can get fairly sophisticated because both linear and rotational motion are often coupled. In other words, an object may move along a linear path while simultaneously spinning about some axis of rotation.

❹ **Volume 2, chapter 4—Transition to classical electricity and magnetism:** Finally, the series concludes by examining the natural transition from 'Classical Mechanics' to 'Classical Electricity and Magnetism.' How does one transition from the study of objects with mass to the study of objects with charge? We will make this transition and take a brief look at some of the interesting phenomena associated with charge.

Acknowledgments

Several individuals significantly strengthened the presentation of this material. Working 'behind the scenes,' their efforts merit every possible recognition. Therefore, I would like to thank:

My high school English and Latin teacher, **Richard Grejtak**. A gifted teacher and scholar, Mr Grejtak taught me how to write a sentence. Decades after enrolling in his courses, I still strive to emulate the excellence and excitement that he brought to each class. As a teacher, I could not ask for a better role-model.

My undergraduate and graduate teachers, **George Livesay**, **Carl Sagan**, and **Ken Kowalski**. I couldn't ask for better mentors in the fields of STEM, the art of teaching, and the subtleties of advising.

Colleagues, **Kathleen Manning**, **Gerald Jorgenson**, and **Joseph Trivisonno**. Kathleen, Gerry, and Joe helped me acclimate to university life and classrooms. Much of the pedagogical expertise appearing throughout this book was handed down from these gifted educators.

My support team at home: **Linda DiLisi, MD** and **Carmela DiLisi**. Over the years, Linda and Carmela have served as volunteers for many of the activities developed in this book. For example, Linda and Carmela appear in the chapter on kinematics while Carmela appears in the demonstrations on rotational motion. Additionally, Carmela was the artist for many of the graphics appearing throughout the book.

My graduate assistants over the past several years: **Brittany Bosowski**, **Alison Chaney**, and **Stella McLean**. Brittany, Alison, and Stella handled much of the proof-reading of the work appearing in the upcoming pages. Theirs was time-consuming and tedious work … and they all handled it superbly!

My able and supportive team of publishers: **Ashley Gasque** and **Bethany Hext**, from Institute of Physics Publishing, who turned the idea of this book into reality.

Author biography

Gregory Anthony DiLisi

Gregory DiLisi earned his Bachelor of Science degree, with distinction, from Cornell University in Applied and Engineering Physics. He then earned his Master of Science and Doctor of Philosophy degrees in Condensed Matter Physics from Case Western Reserve University. Since then, he has taught a wide range of physics courses at the high school, undergraduate, and graduate levels. He is currently a professor at John Carroll University, where he has held appointments in two departments —physics and education. As a faculty member, he developed courses on topics including: computational physics, experimental physics, instructional technology, interdisciplinary science, physics for engineers, problem-solving, science and society, and science methods. As an experimental physicist, he specialized in liquid crystals and complex fluids with publications appearing in peer-reviewed journals such as: *Journal de Physique II*, *Liquid Crystals*, *Microgravity Science and Technology*, and *Physical Review A*. His research focused on the viscoelastic properties and surface interactions of oligomeric liquid crystals as well as the stability of liquid bridges as they shift from micro- to hyper-gravity environments. In this area, he authored a book, *An Introduction to Liquid Crystals*, which explored the various building blocks of liquid crystalline systems as well as the experimental techniques used to probe them. In the area of science education, his research initially focused on developing problem-solving strategies and team-building skills in undergraduate engineering and science students. However, his current research focuses on using case studies as a pedagogical approach to teaching physics. In these areas, he has publications appearing in peer-reviewed journals such as: *The Journal of College Science Teaching*, *The Journal of STEM Education: Innovation and Research*, and *The Physics Teacher*. He authored a five-volume series on *Classical Mechanics* and his most recent book, titled *Case Studies in Forensic Physics*, examines case studies which illustrate how scientists no longer adopt a strictly passive approach to analyzing historical events—instead, they bring sophisticated analytical tools to scrutinize why certain events happened. He has been the Principal Investigator of externally-sponsored research through several grants from agencies such as: The American Association of Physics Teachers, The National Aeronautics and Space Administration, The National Science Foundation, and the American Institute of Physics. He was chosen to be the Ohio Educator Fellow for both of NASA's *Stardust* and *Cassini* space probes and serves as a consultant to numerous educational outreach initiatives. He has authored over forty peer-reviewed journal articles and is an international speaker, having presented at numerous scientific and educational conferences of various professional societies.

IOP Publishing

Simplified Classical Mechanics, Volume 2 (Second Edition)
Gravity and the conservation laws
Gregory A DiLisi

Chapter 1

The universal law of gravitation

The chapter starts by asking the reader to look up to the sky and observe some patterns of celestial objects. Some of these patterns can be seen during the day, during the night, or while viewing the Moon. Understanding these patterns formed the backbone of the history of modern astronomy and culminated in Isaac Newton's Universal Law of Gravitation, the law that states that every mass in the Universe attracts every other mass in the Universe. Furthermore, the notion that forces act through their associated '*fields*' is first introduced when discussing Newton's Universal Law of Gravitation. Here, a huge conceptual leap is needed by the reader—an object can attract another object without even touching it! The '*fields*' is a difficult concept to reconcile with everyday experiences but it makes perfect sense—the Earth can pull on objects even though it is not in direct contact with those objects.

To know the mighty works of God; to comprehend His wisdom and majesty and power; to appreciate, in degree, the wonderful working of His laws, surely all this must be a pleasing and acceptable mode of worship to the Most High to whom ignorance cannot be more grateful than knowledge.
—Nicolaus Copernicus

Finally we shall place the Sun himself at the center of the Universe. All this is suggested by the systematic procession of events and the harmony of the whole Universe, if only we face the facts, as they say, 'with both eyes open.'
—Nicolaus Copernicus

…for until that God who rules all the region of the sky … has freed you from the fetters of your body, you cannot gain admission here. Men were created with the understanding that they were to look after that sphere called Earth, which you see in the middle of the temple. Minds have been given to them out of the eternal fires you call fixed stars and planets, those

doi:10.1088/978-0-7503-6402-7ch1

spherical solids which, quickened with divine minds, journey through their circuits and orbits with amazing speed...

—Marcus Tullius Cicero

You teach your daughters the diameters of the planets and wonder when you are done that they do not delight in your company.

—Samuel Johnson

When I investigate and when I discover that the forces of the heavens and the planets are within ourselves, then truly I seem to be living among the gods.

—Leon Battista Alberti

It may be that our role on this planet is not to worship God, but to create him.

—Arthur C Clarke

To those who think that the law of gravity interferes with their freedom, there is nothing to say.

—Lionel Tiger

Someday, after mastering winds, waves, tides and gravity, we shall harness the energy of love; and for the second time in the history of the world, man will have discovered fire.

—Pierre Teilhard de Chardin

1.1 Motivation

In volume 1, chapter 1, we developed some mathematical tools. The focus of that chapter was to develop our conventions for scalar notation and vector notation.

In volume 1, chapter 2, we defined four kinematic quantities (speed, velocity, acceleration, and jerk) that will be used throughout this text to describe how objects move.

In volume 1, chapter 3, we remained focused on describing **how** objects move and developed the UAM technique for solving problems. We saw that this technique involved three vector equations and is most applicable to 'idealized situations,' short-term accelerations, and free-falling objects (i.e., projectile motion). This was our first technique allowing for numerical analysis of motion.

In volume 1, chapters 4 and 5, we tackled the question of **why** objects move. Thanks to Isaac Newton and his three laws of motion, we saw that objects move because a net external force exists on them. This was our second technique allowing for numerical analysis of motion.

In volume 1, chapter 6, we adapted Newton's laws of motion to the special situation of *'Uniform Circular Motion'* or *'UCM.'* In a sense, UCM is a sub-strategy of Newton's laws of motion. In other words, the focus of volume 1, chapters 4 and 5 was the general motion of objects when they are subjected to various common forces while the focus of volume 1, chapter 6 was a specific case of objects moving in a circle.

To kick off volume 2, recall that when discussing Newton's laws of motion, we promised to revisit the concept of field forces and the mysterious recurrence of the value $9.8 \, \mathrm{m \, s^{-2}}$. Therefore, volume 2, chapter 1 focuses more on the historical development of gravity and the concept of field forces rather than on a new problem-solving technique.

1.2 Getting ready

1.2.1 Anticipatory set

All of us have looked up at the night sky and wondered exactly what we were we seeing—we have some vague notion that what we are looking at are countless bright and dim points of light that have names like 'planets,' 'stars,' 'nebulae,' 'galaxies' and 'star clusters.' After staring at the night sky for a while, you might even begin to notice some patterns in the way these points of light move. For example, one obvious pattern you may have noticed is that the entire sky spins 360° about a point in a 24 h period. For our anticipatory set, let us focus on describing these 'patterns in the night sky' and then see if later in the chapter, we can make sense of them.

Imagine on a clear night, you look up at the sky and fix your stare on a bright point of light. This point of light can be any celestial object you can imagine. Your chosen object might be a planet, a star, or a galaxy. For our discussion, let us imagine you happen to pick the planet Mars.

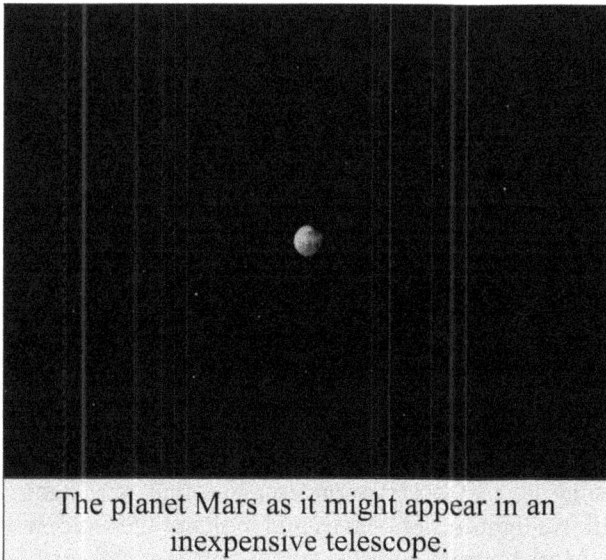

The planet Mars as it might appear in an inexpensive telescope.

Mars Image Credit: NASA/JPL-Caltech.

Over the course of the night, all of us would agree that Mars, and all celestial objects, rise somewhere in the east, travel across the night sky then set in the west. Since the Earth rotates once about its axis in a day, 360° in 24 h, we know that objects travel across the sky at a rate of 15° every hour. To get a feel for this rate, extend your arm as far forward as it will go. Bend you wrist upward so that it is perpendicular to your forearm. The distance from your wrist to the tips of your fingers roughly spans a 15° arc of the sky.

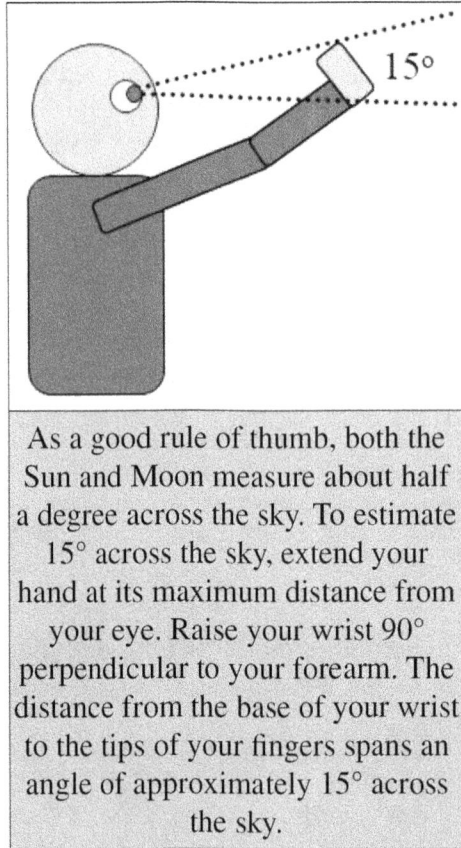

As a good rule of thumb, both the Sun and Moon measure about half a degree across the sky. To estimate 15° across the sky, extend your hand at its maximum distance from your eye. Raise your wrist 90° perpendicular to your forearm. The distance from the base of your wrist to the tips of your fingers spans an angle of approximately 15° across the sky.

We've just established a pattern; namely, on a given night, all the objects in the sky travel from east to west at a rate of $15°\ \text{h}^{-1}$. Now let us try a more difficult, but far more interesting type of observation. Imagine that one night, you sit in your family room, with the lights off of course, and look out the window at Mars. On this particular night, Mars happens to be so bright that you can easily see it even from inside your family room. You now sit in your favorite comfortable chair with you head resting motionless again the back cushion ... or perhaps you lean against the windowsill … and grab a magic marker. You take careful note of the location of Mars as you look outside your window. If the window were closed, you could carefully draw a small dot or arrow on your window to coincide with the location of Mars. You look at your watch and note that the time is exactly 11:54 pm.

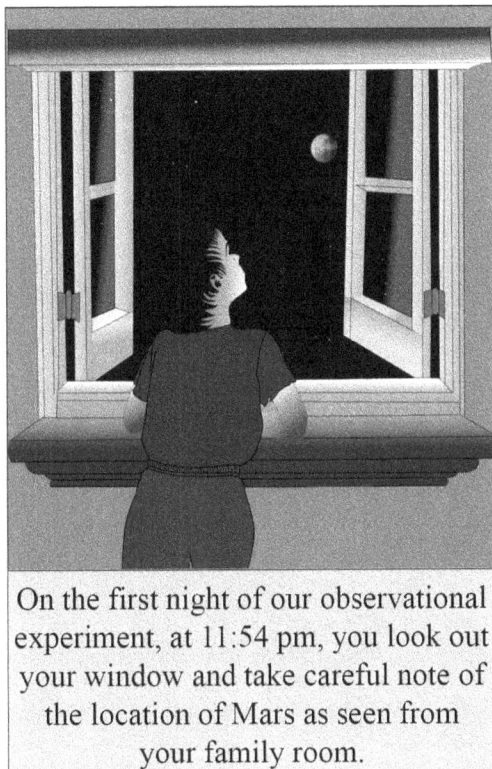

On the first night of our observational experiment, at 11:54 pm, you look out your window and take careful note of the location of Mars as seen from your family room.

Child Image Credit: Pixabay by ANINDYA_2022.
Mars Image Credit: NASA/JPL-Caltech.

As we mentioned earlier, over the course of **one** night, Mars will move from east to west as the Earth spins about its axis. However, what would happen if you returned the next night at exactly 11:54 pm, sat in the exact same chair or stood in the exact same position, with your head in the exact same location, and recorded the position of Mars? *Would Mars be at the same location at the same time of night or would it have somehow moved?* If you did the experiment, you would find two very interesting patterns emerge:

1. First, from **night to night at a specific time**, the locations of all the stars would remain almost constant but the locations of the planets would move west to east against the background of the fixed stars. You would think that the planets 'wander' from night to night, like restless points of light that can't seem to sit still. In fact, we get the word 'planet' from the Greek word for 'wanderer.'

Etymology:

The word 'planet' comes from the Greek word 'planetes' that translates into 'wanderer.'

The following series of pictures demonstrates the 'wandering' of the planet Mars at 11:54 pm over the course of four successive nights:

This series of drawings depicts your family room window as seen at 11:54 pm on four successive nights. The bright point of light that you have identified as Mars, moves slightly more to the east every night. The scales of the size of Mars and its motion are exaggerated in this drawing. You can see why ancient astronomers labeled these points of light as "planets" since these celestial objects appear to "wander" across the sky from night to night.

Child Image Credit: Pixabay by ANINDYA_2022.
Mars Image Credit: NASA/JPL-Caltech.

2. Second, just when you get used to the planets moving to the east from night to night, another strange pattern emerges. Occasionally, some of the planets' motions appear to reverse direction, and the planets, for a short time, move from east to west against the background of the fixed stars. We call the backwards motion of a planet, from east to west, 'retrograde motion.' Eventually, the retrograde motions stop and the planets begin to move toward the east again. On your family room window, over the course of year, you would see Mars trace out a path of recurring loops indicating one retrograde motion after another.

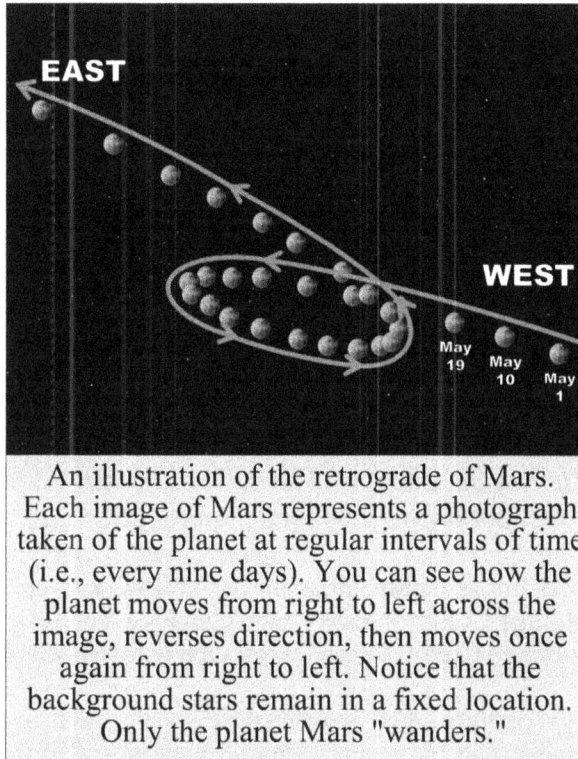

An illustration of the retrograde of Mars. Each image of Mars represents a photograph taken of the planet at regular intervals of time (i.e., every nine days). You can see how the planet moves from right to left across the image, reverses direction, then moves once again from right to left. Notice that the background stars remain in a fixed location. Only the planet Mars "wanders."

Mars Image Credit: NASA/JPL-Caltech.

Now let us try another observational experiment that is only slightly different from the one we just performed. If you thought the previous experiment produced strange results, wait until you see what happens this time! In this experiment, you once again sit down in your favorite chair in your family room, fasten your head to a fixed location, except now we perform our experiment at noon. You look at your watch, note that the time is exactly 12:00 pm and place a dot or arrow on the window at the exact location of the Sun. Finally, just like in the previous experiment, you repeat your observation day after day, always at exactly 12:00 pm. *Is the Sun always at the same location at 12:00 pm or, like the planets, does it wander from day to day?* The series of drawings below depicts the Sun at 12:00 pm on four successive days:

June 21: Summer Solstice in Northern Hemisphere	September 21: Autumn Equinox in Northern Hemisphere	December 21: Winter Solstice in Northern Hemisphere	March 21: Vernal Equinox in Northern Hemisphere

Contrary to popular intuition, the Sun is not always in the same position in the sky at 12:00 pm. The scale of the size of othe Sun and its motion are exaggerated in this drawing.

Child Image Credit: Pixabay by ANINDYA_2022.
Cloud Image Credit: Pixabay by Clker-Free-Vector-Images.
Sun Image Credit: Pixababy by OpenClipart-Vectors.

If you were to perform this experiment over the course of a year, you would actually see the Sun trace out a figure-8 pattern, called the 'analemma' of the Sun.

Etymology:

The word 'analemma' comes from the Greek words 'ana' and 'lambanein' which translate into 'up' and 'to take.' When the two words are combined, they translate into 'to take up' or 'support,' such as a sling for a broken arm or a tabulated scale, both of which had the shape of a figure-8. Thus, the figure-8 became known as an 'analemma.'

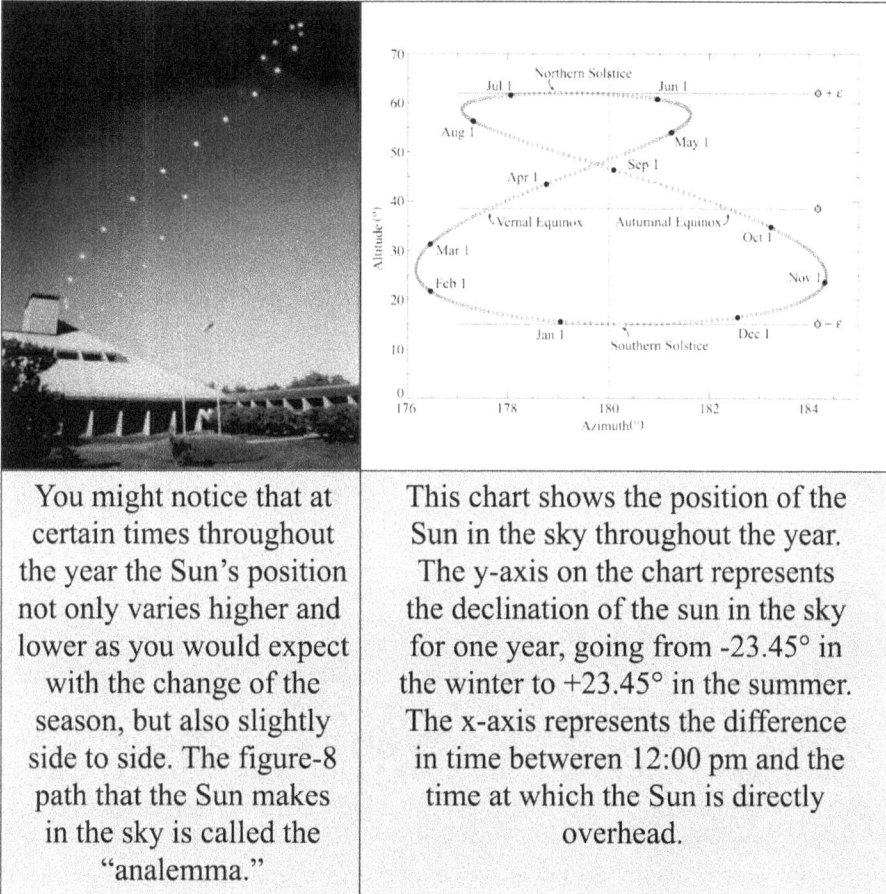

You might notice that at certain times throughout the year the Sun's position not only varies higher and lower as you would expect with the change of the season, but also slightly side to side. The figure-8 path that the Sun makes in the sky is called the "analemma."	This chart shows the position of the Sun in the sky throughout the year. The y-axis on the chart represents the declination of the sun in the sky for one year, going from -23.45° in the winter to +23.45° in the summer. The x-axis represents the difference in time between 12:00 pm and the time at which the Sun is directly overhead.

This Analemma image (Photo taken in 1998-99 of analemma from office window of Bell Labs, Murray Hill, NJ) has been obtained by the author from the Wikimedia website where it was made available under a CC BY-SA 3.0 licence. It is included within this article on that basis. It is attributed to Jfishburn. Right image: Image Credit: NASA/JPL-Caltech.

Ancient astronomers had a difficult time explaining the analemma of the Sun and the retrograde of certain planets. In this chapter, we will see how the quest to understand these patterns, and other 'patterns of the night sky,' led scientists to one of the most fundamental force laws in physics. This quest, undertaken by many scientists, from many countries, over hundreds of years, culminated in Isaac Newton's formulation of the Universal Law of Gravitation.

1.2.2 Objective

By the end of this chapter, you will be able to:
- State the Universal Law of Gravitation in words and represent the statement in equation form.
- Explain each symbol in the mathematical representation of the Universal Law of Gravitation.
- Use the Universal Law of Gravitation to compute the acceleration due to gravity at different locations on the surface of the Earth and at different locations on the surface of any planet, moon, asteroid, or other celestial body.
- Qualitatively describe the concept of a field and how a force can act through a field.
- Quantitatively compute a gravitational field.
- Describe the general approach to calculating any field.
- Trace the historical development of the Universal Law of Gravitation and the concept of field forces.

1.2.3 Purpose

This information is needed:
- To describe the origins of Earth's gravitational force and why the acceleration due to gravity is assigned the value of $g = 9.8$ m s^{-2}.
- To realize that the acceleration due to gravity on the surface of the Earth is not constant at 9.8 m s^{-2} but actually varies depending on location. However, you must also realize that this variation of g is so minor that g can in fact very accurately be considered a constant.
- To allow you to compute the acceleration due to gravity at any location in the Universe.
- To understand how the Earth, and other objects with mass, are able to pull on other objects with mass through the force of gravity. The concept of a gravitational field will allow you to understand how two objects can exert a force on one another even though the two objects are not in contact with one another.

1.3 Giving information

1.3.1 Instructional input

1.3.1.1 History of modern astronomy

The hallmark of this chapter is Isaac Newton's Universal Law of Gravitation. In the next section, we'll state this law in words and systematically transform those words into a concise mathematical equation. However, to state and verify the Universal Law of Gravitation, Newton relied on the work of several other physicists, astronomers, and mathematicians living in many different countries. Over the course of hundreds of years, these scientists often made their crucial conceptual and experimental contributions at great personal cost. Perhaps you are familiar with a famous quotation attributed to Isaac Newton as it appeared in a letter he wrote to fellow English scientist Robert Hooke that was dated February 5 in either 1675 or 1676: 'If I have seen further, it is by standing on the shoulders of giants.' With this quotation, Newton was himself paying tribute to these scientists as the individuals who laid the foundation of our understanding of gravity and celestial mechanics. In this section, we'll take a brief look at the contributions of five important scientists.

❶ **Ptolemy of Alexandria**
Born: unknown, approximately 100 AD in Egypt
Died: approximately 168 AD in Alexandria, Egypt

Ptolemy Image Credit: aipsidtr/Shutterstock.com

Ptolemy Image Credit: German Vizulis/ Shutterstock.com

Although we know that Ptolemy was one of the most influential astronomers and geographers of his time, we know very little about Ptolemy's life. Ptolemy made astronomical observations from Alexandria, in Egypt, during the years 127 −41. In fact, historians have determined that Ptolemy made his first observation on March 26, 127 and made his last on February 2, 141. Ptolemy also used observations made by 'Theon, the mathematician,' who almost certainly was **Theon of Smyrna**, his teacher. Many of Ptolemy's early works are dedicated to 'Syrus' who may have also been one of his teachers in Alexandria, but nothing is known of 'Syrus.'

Ptolemy's most famous works are his eight-volume manuscript titled, *Geography* and his thirteen-volume work titled, *Almagest.* Interestingly, in *Geography*, Ptolemy established the modern-day terms 'longitude' and 'latitude.' In *Almagest*, Ptolemy's theories on the motions of the five known planets, Sun, and Moon are his most important original work and represent his most enduring contribution to astronomy. The planetary theory that Ptolemy developed is a sophisticated mathematical model that very accurately fits observational data. The model he produced, although complicated and flawed by today's standards, represents the motions of the planets fairly well; thus Ptolemy's model cannot be so easily dismissed since it does fit experimental data quite nicely. Ptolemy's model is the following: the Earth is fixed at the center of the Solar System. Surrounding the Earth is the sphere of stars that rotates once around the Earth every day. Also surrounding the Earth, but inside the sphere of stars, are spheres that carry upon them the Sun, Moon, and the five known planets. Because of the complicated motions of the planets over time, some planets are actually mounted to complicated systems of spheres, mounted to other spheres, which are mounted to other spheres. These planet-carrying spheres rotate around the Earth at certain rates to produce the planetary motions we see in the night sky. A possible representation of Ptolemy's model of the Solar System is shown below:

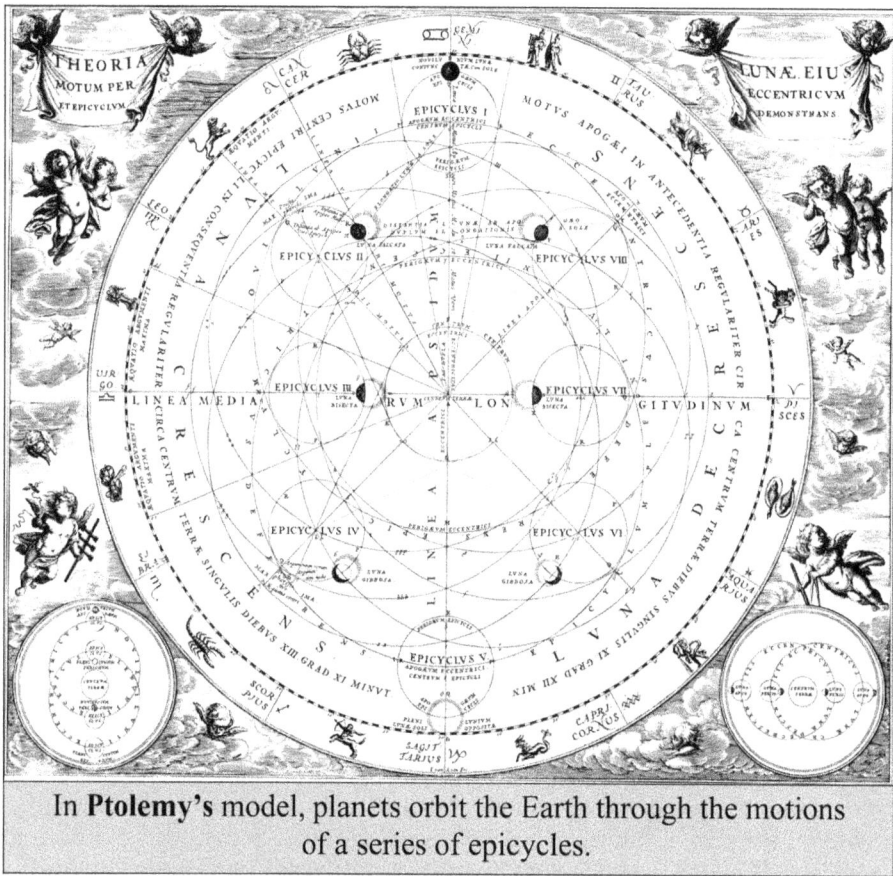

In **Ptolemy's** model, planets orbit the Earth through the motions of a series of epicycles.

Ptolemy's Model Image Credit: Pixabay by GDJ.

We call Ptolemy's model of the Solar System, as represented by combinations of spherical motions, the 'epicycle model' ('epi' translates into 'upon' or 'around,' and 'cycle' translates into 'circle'—thus the 'epicycle model' is one of 'circles around circles'). Ptolemy's model of an 'Earth-centered' or 'geocentric' Solar System dominated astronomy for over 1400 years.

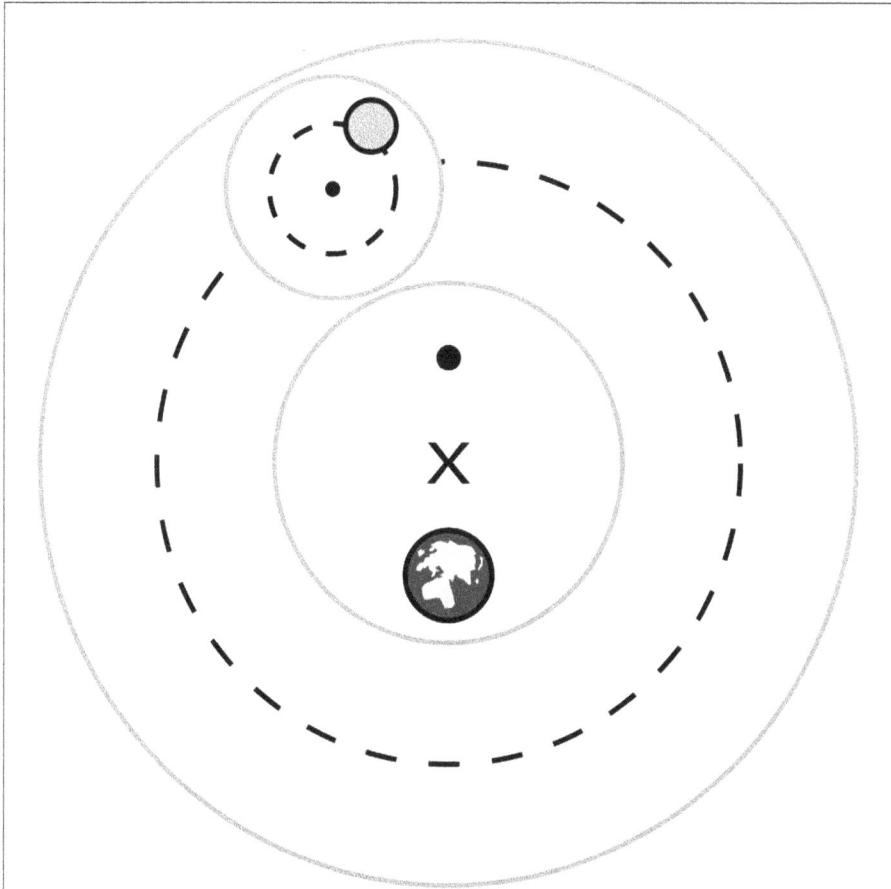

Here we see a schematic representation of Mars orbiting Earth according to **Ptolemy's** model of the Solar System. With Earth fixed at the center, Mars rotates about a small disk, which in turn is mounted to a larger disk that rotates about the Earth. When viewed from the Earth, such a system would indeed account for the retrograde motion of Mars in the sky.

This Ptolemaic elements image has been obtained by the author from the Wikimedia website, where it is stated to have been released into the public domain. It is included within this article on that basis.

To better visualize the 'epicycle model' of the Solar System, physicists often call Ptolemy's model the 'Spinning Teacup' model. Most amusement parks have a version of the 'Spinning Teacup' ride in which people sit in giant teacups that are free to spin about their center axis. However, the teacups themselves are mounted to a giant floor that is free to spin. Thus, the 'Spinning Teacup' amusement park ride is a set of rotating teacups, mounted to a rotating floor. If we think of the Solar System as the 'Spinning Teacup' amusement park ride, the Earth would be located at the center of the rotating floor. Imagine yourself sitting atop the Earth and the planets to be the people sitting in the various teacups. When the ride starts and both the floor and teacups begin to spin, the motion of the people about you would indeed mimic the motion of the planets as seen from an observer on the Earth.

The "Spinning Teacup" amusement park ride–the ride is essentially an epicycle model of the Solar system.

Image Credit: Author.

As we mentioned before, Ptolemy's model does accurately explain the patterns we observe in the night sky. An observer on the Earth looks up to the night sky and locates a planet. As the planet moves about its epicycles, it would trace out a path in the sky similar to the retrograde pattern we observed in our anticipatory set. The retrograde of a planet is therefore just the looping path that a point mounted to an epicycle traces out against a fixed backdrop.

> **The Ptolemaic model of the Solar System preserves the importance of the Earth as the center of the Solar System; maintains that the Earth is stationary (if it were spinning about an axis, wouldn't we fly off it?); and most importantly, it accurately describes the observational data.**

Someone would have to present a lot of evidence and convince us of the 'unimportance of the Earth' in the structure of the Solar System before we would abandon Ptolemy's model.

❷ **Nicolaus Copernicus**
Born: February 19, 1473
(Thorn, Poland)
Died: May 24, 1543
(Frauenburg, Poland)

Nicolaus Copernicus
Mounment in Warsaw,
Poland

Left image: Copernicus Image Credit: Morphart Creation/Shutterstock.com. Right image: Nicolaus Copernicus Statue Image Credit: Pixabay.

Nicolaus Copernicus (his Polish name was Mikolaj Kopernik) belonged to a wealthy merchant family of German origin. In the early 1490s, Copernicus studied canon law in the church of Frauenburg and maintained a position there his entire life although he was never officially ordained as a priest. Early in his career, Copernicus studied economics, mathematics, and medicine at Kracow, Bologna, Rome, Padua, and Ferrara. He practiced medicine for a few years and gained a considerable reputation as a 'skilled and cautious physician.'

Copernicus' interest turned to astronomy in 1510 but during his lifetime he made few astronomical observations and relied instead on data that was published in various literatures. Sometime between 1510 and 1514, he published an early description of his 'heliocentric' or 'Sun-centered' model of the Solar System in his manuscript 'Commentariolus.' In this manuscript, Copernicus outlined his theory that the Earth is not the center of the Universe, but only the center of the lunar orbit. Copernicus firmly believed his heliocentric theory was an actual representation of Nature and did not regard it as simply a mathematical hypothesis even though after Copernicus' death, many of his contemporaries and subsequent generations generally claimed that he viewed his model as only a mathematical simplification.

One advantage of the Copernican model of the Solar System is that the model required fewer basic assumptions than the Ptolemaic idea. The Copernican model only assumed the planets revolved about the Sun and the Earth rotated about a tilted axis. However, Copernicus' model failed to gain popular acceptance for two reasons: (1) His model demoted the Earth from the center of the Solar System to the third planet orbiting the Sun. (2) Copernicus' model placed the planets in perfectly circular orbits about the Sun. With a heliocentric structure and circular planetary orbits, Copernicus' model actually made less accurate astronomical predictions than Ptolemy's model. Thus, Copernicus' idea was generally ignored and viewed more as a mathematical simplification than an actual physical model of the Solar System.

In 1539, Copernicus took on 'Rheticus' (Georg Joachim von Lauchen) as his only student and asked him to write a popularization of the heliocentric theory, which Rheticus did in 1540 with the publication of *Narratio Prima*. In 1543, Rheticus convinced Copernicus to publish the manuscript, *De Revolutionibus Orbium Coelestium*, which contained Copernicus' full mathematical analysis of a heliocentric Solar System. The heliocentric model of the Solar System was not new (similar theories had been proposed by Aristarchus and Nicholas of Cusa), but Copernicus was the first to describe the model in full mathematical detail. He was the first person in history to create a complete and general model of the Solar System that combined mathematics, physics, and astronomical data. Unfortunately, Rheticus handed over the final publication to a Lutheran minister who added a disclaimer to the manuscript, stating that the heliocentric model was not an actual physical model but rather a mathematical model that simplifies calculations. Thus began the idea that Copernicus did not actually believe in his heliocentric view of the Solar System.

> **The Copernican model of the Solar System is a "heliocentric" model that assumes circular planetary orbits. Because of the assumption of circular planetary orbits, Copernicus' model does not explain astronomical observations as well as the Ptolemaic Model. This, in addition to demoting importance of the Earth in the structure of the Solar System, contributed to Copernicus' model being ignored.**

❸ **Galileo Galilei**
<u>Born:</u> February 18, 1564
(Pisa, Italy)
<u>Died:</u> January 8, 1642
(Florence, Italy)

Galileo Galilei etching
on the Italian lira
banknote

Galileo Galilei Statue
in Florence, Italy

Left image: Galileo Image Credit: vkilikov/Shutterstock.com.
Right image: Galileo Statue Image Credit: Pixabay by wgbieber.

We presented a brief biographical sketch of Galileo in volume 1, chapter 3. However, when discussing the history of modern astronomy, we should elaborate on the impact of Galileo's contributions to our understanding of the Solar System and the universality of physical principles. In volume 1, chapter 3, we discussed how Galileo grew interested in the heavens and built his own a telescope in 1609 after the discovery of lenses was reported from Holland. Galileo used his telescope to discover sunspots, lunar craters, the phases of Mercury and Venus, the rings of Saturn, and the moons of Jupiter—the four large moons of Jupiter are now called the 'Galilean Moons' in his honor.

The four "Galilean Moons" of Jupiter as seen in a photographic montage. From left to right, the moons are: **IO** (pronounced *"Ē-O"*), **Europa, Ganymede,** and **Callisto**

Galilean Moons Image Credit: SN VFX/Shutterstock.com.

The phases of Venus—Notice that only a crescent Venus is visible.

Phases of Venus Image Credit: Pixabay by WikiImages.

Before Galileo, the Copernican model of the Solar System had no advocate. Certainly, no experimental proof existed to support Copernicus' heliocentric model. However, Galileo's observations demonstrated that the Copernican theory was correct. For example, the phases of Venus would only be observed if the orbit of Venus about the Sun lies inside the orbit of the Earth about the Sun. Such an observation supports Copernicus' view of the planets orbiting the Sun with the Earth having the third most distant orbit. What about the importance of the Earth? The Ptolemaic model placed the Earth at the center of the Solar System and surely the Earth deserves a place of such high prominence and importance. On the other hand, the Copernican model demoted the Earth to the humbler role of just one of several planets orbiting the Sun. Galileo's observations of Jupiter's four moons, **Io**, **Europa**, **Ganymede**, and **Callisto** clearly supported the idea that not every celestial object has to orbit the Earth. Jupiter is the center of the orbits of these four moons, not the Earth!

Finally, Galileo should be credited as the first physicist and astronomer to understand the concept of the 'universality of physical principles.' Prior to Galileo, scientists viewed celestial objects as 'the heavens.' In their status as 'heavenly,' celestial objects were thought to obey their own sets of rules. For instance, the motions of the planets and Moon could not be understood by studying the motion of a falling apple on the Earth. 'The heavens' had their set of rules while Earth-bound objects had a different set. By observing Io, Europa, Ganymede, and Callisto orbiting Jupiter, Galileo realized that the laws governing the motions of these four moons could also be the laws that govern the motion of the Earth about the Sun. When studying Jupiter's moon system, Galileo realized he was looking at a miniature Solar System. Whatever he discovered about this miniature Solar System could also be applied to the actual Solar System. Thus, Galileo made the huge conceptual leap of realizing that by studying the motions of these moons around Jupiter, we can learn how the planets orbit the Sun. The notion that all objects, whether in the heavens or on Earth, obey the same physical laws is called the 'universality of physical principles.'

> Galileo contributed experimental observations that supported the heliocentric model of the Solar System. His observations legitimized the demotion of the Earth from the center of the Solar System to the third most distant of several planets orbiting the Sun. Finally, perhaps more important than his astronomical observations, Galileo contributed the concept of the universality of physical principles.

❹ Tycho Brahe
Born: December 14, 1546
(Knudstrup, Denmark)
Died: October 24, 1601
(Prague, Denmark)

Left image: Tycho Brahe Image Credit: Ntguilty/Shutterstock.com.
Right image: Jacques de Gheyn II/ MFAH.

Tycho Brahe (his Danish name was Tyge Brahe) was a Danish astronomer who set out to make the most accurate observations of his time in order to determine whether the Ptolemaic or Copernican model of the Solar System was correct. Tycho was born a twin but the other child was stillborn. At age one, Tycho was stolen from his crib by his uncle Jorgen and raised by his uncle's family. The reason for the abduction was that Tycho's aunt was infertile and Tycho's father Otto, promised that his first male child would be given to Jorgen. The families reconciled over Tycho's abduction because Tycho's mother, Becite Bille, provided Otto with many other male children. Tycho went on to have eight children of his own.

In December 1566, the 20 year-old Tycho attended a betrothal dance and argued with his student, Manderup Parsbjerg, over who was the better mathematician. Their argument ended without incident. However, at another party, the argument continued and the two men dueled resulting in the amputation of Tycho's nose. He devised a gold and silver nose-plug that he wore to conceal his disfigurement

King Frederick II of Denmark sponsored Tycho's astronomical observations. Frederick built Tycho the world's first observatory, a castle called Uraniborg on the island of Hveen (now Ven) between Denmark and Sweden. At Uraniborg, Tycho and his wife, Kristine Barbara, spent over 35 years engaged in the rigorous observation and recording of planetary and stellar positions with the highest possible accuracy. To make these observations, Tycho and Kristine used Tycho's invention, the 'Giant Quadrant' to sight the planets and stars. His large, innovative instruments yielded measurements that were accurate to within four minutes of arc. He was a superb technologist who stretched his instruments to the limits of their accuracy. In today's jargon, we would think of Tycho as possessing 'state-of-the-art' astronomical instruments. He carefully compiled his data and even developed his own 'geoheliocentric theory' of planetary motions, in which the Sun orbited the Earth and the other planets orbited the Sun. However, despite his best efforts, Tycho was unable to turn his observations into a coherent theory of the Solar System. For that, he needed to turn to someone else.

In 1600, Tycho took on the brilliant 23 year-old German mathematician and physicist, Johannes Kepler, as his student. Tycho assigned Kepler the task of explaining the orbit of Mars. Tycho was jealous of Kepler's superior mathematical skills and was hesitant to share all of his data with Kepler. Tycho was not about to turn over his life's work to a rival! A great way to explain the dynamics between Tycho and Kepler can be best described by my college astronomy professor, planetary scientist and best-selling author, Carl Sagan. In 1986, I took an undergraduate senior-level astronomy course with Sagan at Cornell University. During class, he recounted an episode of his popular television series *Cosmos*, in which Tycho and Kepler grapple with their attitudes toward one another. I can still hear Sagan creating the scene with his famous, distinctive voice: *'The birth of modern science, which is the fusion of observation and theory, teetered on the precipice of their mutual distrust.'*

Your friendly neighborhood author (left) and Carl Sagan (right) at Cornell University in the fall of 1986. After the enormous success of his Cosmos series, Sagan returned to Cornell and taught an undergraduate senior-level seminar on astronomy and critical-thinking in science. Sagan's course was the highlight of my experiences at Cornell.

Image Credit: Author.

In 1601, Tycho supposedly died from kidney stones that, coupled with excessive over-eating and drinking, caused his bladder to burst. Tycho's last words, spoken in a delirium to Kepler, were: 'Let me not seem to have lived in vain … let me not seem to have lived in vain.' Tycho is reported to have written his own epitaph: 'He lived like a sage and died like a fool.' Following Tycho's death, Kepler obtained possession of Tycho's observations, and devoted his life to analyzing them.

Immediately after Tycho's death, rumors began to circulate that he had not died of an infected bladder caused by kidney stones, but that he was in fact a victim of either intentional or unintentional poisoning. Since Tycho exhibited symptoms of heavy metal poisoning or poisoning by certain plants, the rumors of the strange circumstances surrounding his death continued for centuries. In the 1990s, a series of investigations were conducted to determine the actual cause of Tycho's death. If kidney stones were the culprit, they should still be present in Tycho's remains. If poisoning was the culprit, an analysis should reveal trace amounts of the ingested toxin. The results of the investigations are summarized below in an article from the *Planetarian, 30* (December 2001):

During a banquet at the Bohemian count of Rosenberg, Tycho was too courteous to obey the calls of nature during the hour-long dinner and finally his bladder burst, which led to his death. Or so the story goes. But is this the real cause of Tycho's death? Is it at all possible to die from a burst bladder?

In 1996 it was possible to carry out another analysis using the PIXE-method (Particle Induced x-ray Emission), this time on hair from Tycho with the root preserved. The result was that the mercury was not from an outside source but actually had been digested. Using the growth rate of hair it was concluded that Tycho was poisoned by mercury one day before his death.

These forensic investigations show Tycho Brahe died of mercury poisoning. Even though it cannot be excluded, it is not likely that Tycho was murdered, but most likely he conducted his own death by using his own mercury-rich medicines the day before his death. This was done to help cure his disorder of the urinary system (prostatic hypertrophy or less likely bladder stones, since no stones were to be found in the coffin). It was not a burst bladder caused by his courteousness, but mercury in his own medicines that led to the uremia of which he died.

Tycho Brahe was an astronomical technologist who, for 35 years, recorded planetary and stellar positions to the highest possible accuracy of his time. His compilation of astronomical data would make experimental verification of subsequent theories possible.

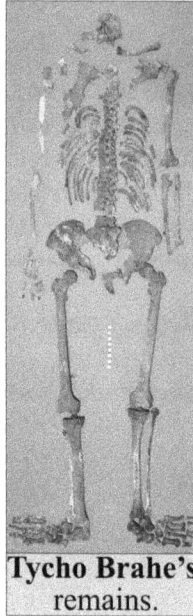

Tycho Brahe's remains.

Reproduced from Kacki S *et al* 2018 Rich table but short life: Diffuse idiopathic skeletal hyperostosis in Danish astronomer Tycho Brahe (1546-1601) and its possible consequences *PLoS ONE* 13 e0195920. CC BY 4.0.

❺ Johannes Kepler
Born: December 27, 1571 (Wiel der Stadt, Swabia)
Died: November 15, 1630 (Regensburg, Bavaria)

Tycho Brahe and Johannes Kepler Mounment in Pragure, Czech Republic.

Left image: This Johannes Kepler image has been obtained by the author from the Wikimedia website, where it is stated to have been released into the public domain. It is included within this article on that basis. Right image: Kepler Monument Image Credit: Pixabay by Karl_Napp.

Johannes Kepler's life, though filled with extraordinary scientific achievement, would be one of great personal tragedy. His revolutionary theories on planetary motion and his single-minded determination to understand physical principles and the patterns of the night sky are a testimony to human endurance and perseverance. Kepler was born to a thin garrulous woman and a mercenary soldier who eventually abandoned the family. Kepler described his childhood as only having two pleasant memories: seeing a comet from the top of a hill and observing an eclipse of the Moon.

As an adult, Kepler had a poor self-image and described himself as a 'mangy dog.' Physically, he was described as short, frail, nearsighted, plagued by fevers and stomach ailments, yet was driven to succeed. He had a ready wit and a modest manner that won him many friends. Interestingly, though he revolutionized astronomy, he had poor eyesight.

Kepler obtained his first teaching position as an emergency substitute when a death left a vacancy at the Lutheran school in Graz. He was a horrible teacher and during lectures, often lapsed into strange mathematical digressions. While lecturing, he mumbled, digressed, and was utterly incomprehensible. When he returned for his second year of teaching, no students enrolled in his classes.

Kepler's wife and children had little understanding of or compassion for Kepler. His wife absolutely despised Kepler's profession and he described her as stupid, sulking, lonely, and melancholy. Theirs was not a happy marriage. Kepler often worked in deplorable conditions and was continually exiled form one location to another yet he believed that the Universe could be described by precise mathematical formulas. He described God as 'the mathematician supreme' and once wrote: 'Geometry is God himself.'

Kepler began to work for Tycho Brahe in 1600 at the age of 28. Kepler had taken the position of an assistant to Brahe in order to get access to Brahe's planetary data. Immediately, Tycho assigned Kepler the task of explaining the orbit of Mars but Tycho, cautious about turning over his life's work to a potential rival, only gave Kepler limited scraps of data. Upon Tycho's death, Kepler was able to analyze the vast amounts of Tycho's data and he was determined to use the data to prove the validity of the Copernican theory. Kepler so strongly believed in the Copernican model of the Solar System that he went about explaining the orbit of Mars in terms of a perfectly circular orbit. After studying Tycho's data, Kepler described the idea of planetary *circular* orbits as 'a cartful of dung' and in 1605, realized that an ellipse explained the orbit beautifully. He published his results in a 70-chapter book called *Astronomia Nova* (i.e., *The New Astronomy*) that received mixed reviews. In his manuscript, Kepler described three laws of planetary motion which were able to predict planetary positions

100 times more accurately than those of Copernicus. For the formulation of these laws, Kepler is considered the founder of physical astronomy.

❶ **Law #1 (published in 1609):** Planets move in elliptical orbits about the Sun with the Sun located at one focus.

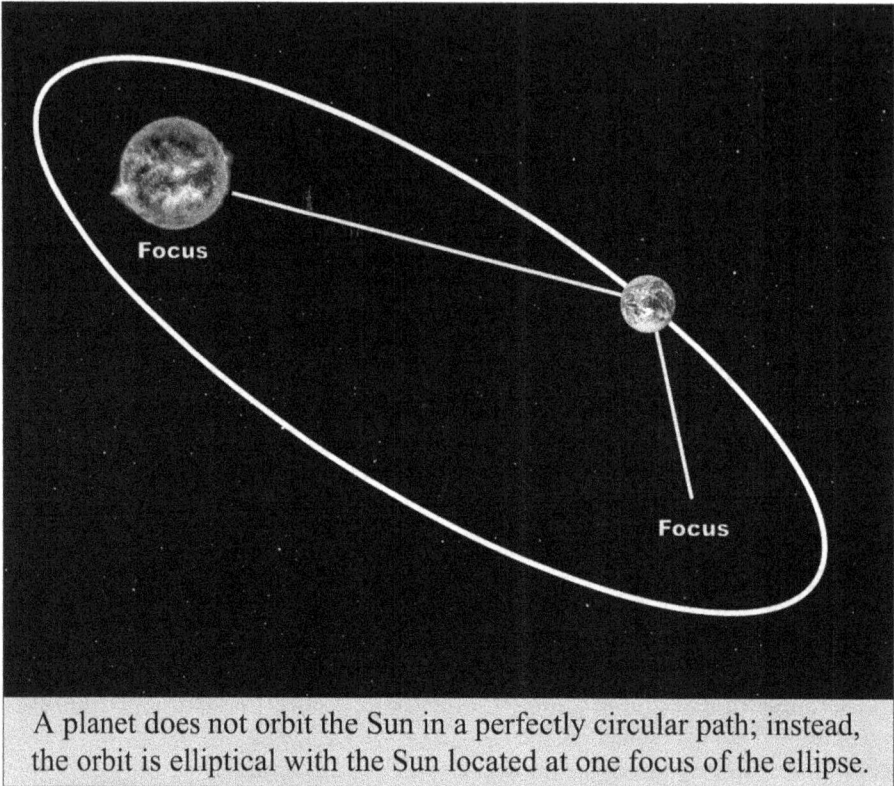

A planet does not orbit the Sun in a perfectly circular path; instead, the orbit is elliptical with the Sun located at one focus of the ellipse.

Earth Image Credit: NASA. Sun Image Credit: NASA.

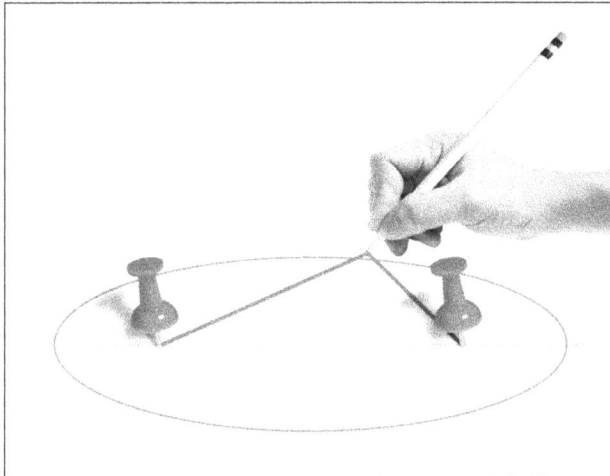

You can draw a perfect ellipse by pushing two thumbtacks into a piece of paper. The thumbtacks will serve as the foci of the ellipse. Now connect a string between the two points. Pull on the string with the tip of your pencil until the string is taut to both thumbtacks. Finally, let your pencil trace-out a set of points while keeping the string constantly taut. The resulting shape is an ellipse. Mathematically, a circle is the set of points whose distance from a **single point** (called the "*center*" of the circle) is constant. Likewise, an ellipse is the set of points whose combined distance from **two points** (called the "*foci*" of the ellipse) is constant.

Hand Image Credit: Pixabay by franciscofranklin. Thumbtack Image Credit: Pixabay by OpenClipart-Vectors.

Etymology:

The word 'ellipse' comes from the Greek word 'elleipsis' that translates into 'to fall short.'

As shown below, one way to generate an ellipse is to cut through a cone at an angle with a plane. The intersecting points would trace out the shape of an ellipse. Thus, an ellipse is the shape resulting from a plane 'falling short' of cutting through the base of a cone.

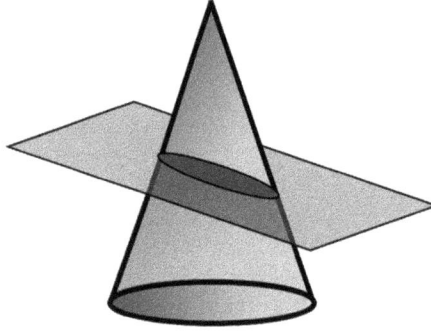

A planet's elliptical orbit is described by its 'eccentricity,' a parameter symbolized by ε. The more elliptical a planet's orbit, the higher its eccentricity. For example, a perfect circle has an eccentricity of $\varepsilon = 0$; the orbit of the Venus (the most *circular* of all the planetary orbits) is $\varepsilon = 0.0068$; and finally, the orbit of the Pluto (the most *elliptical* of all the planetary orbits) is $\varepsilon = 0.25$.

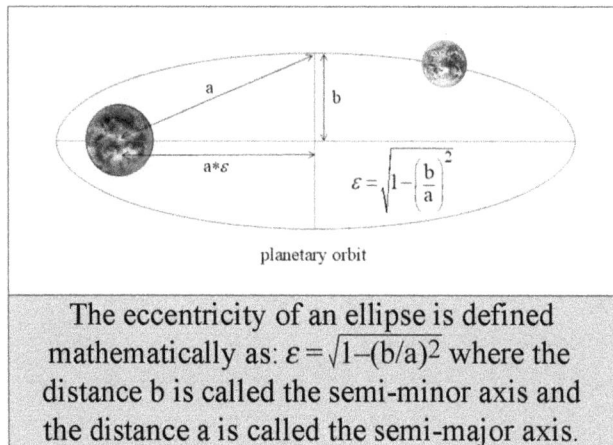

a

b

a*ε

$$\varepsilon = \sqrt{1 - \left(\frac{b}{a}\right)^2}$$

planetary orbit

The eccentricity of an ellipse is defined mathematically as: $\varepsilon = \sqrt{1 - (b/a)^2}$ where the distance b is called the semi-minor axis and the distance a is called the semi-major axis.

Earth Image Credit: NASA. Sun Image Credit: NASA.

❷ **Law #2 (published in 1609):** An imaginary line connecting a planet to the Sun sweeps out equal areas in equal time intervals.

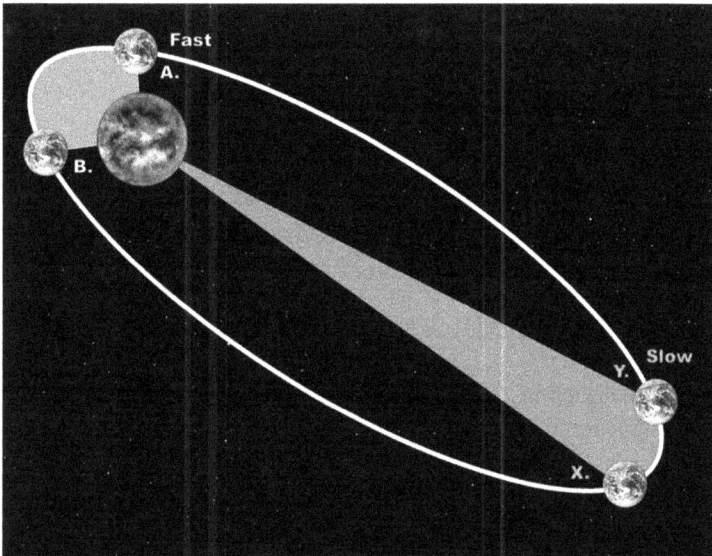

Imagine we take four snapshots of a planet revolving about the Sun. Photograph (B) is taken exactly one week after photograph (A). Likewise, photograph (Y) is taken exactly one week after photograph (X). We don't care about the interval of time between photographs (B) and (X)–it can be a month, a year, a decade, etc. The line between the Sun and the planet sweeps out equal areas in equal time intervals. Kepler's Second Law is essentially stating that a planet's orbital velocity must decrease as it moves further from the Sun. According the Kepler's second law, the green area equals the yellow area since they were traced out in the same time interval.

Earth Image Credit: NASA. Sun Image Credit: NASA.

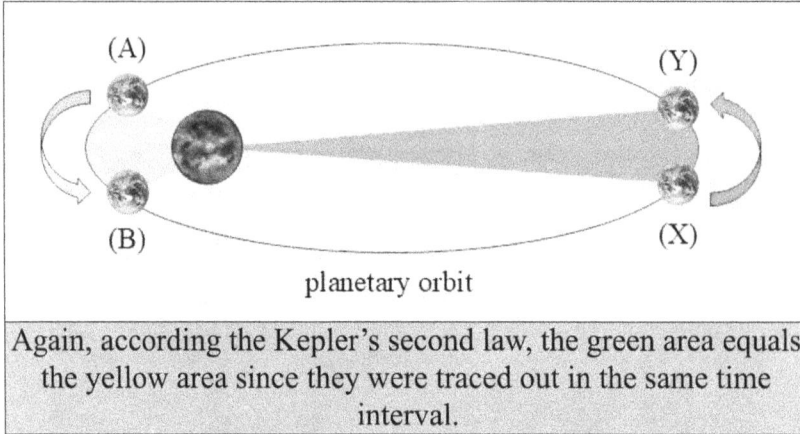

planetary orbit

Again, according the Kepler's second law, the green area equals the yellow area since they were traced out in the same time interval.

Earth Image Credit: NASA. Sun Image Credit: NASA.

❸ **Law #3 (published in 1619):** The time for a planet to revolve once around the Sun is proportional to its mean distance from the Sun raised to the 3/2 power or:

$$T_{\text{one orbit}} \propto R^{3/2}.$$

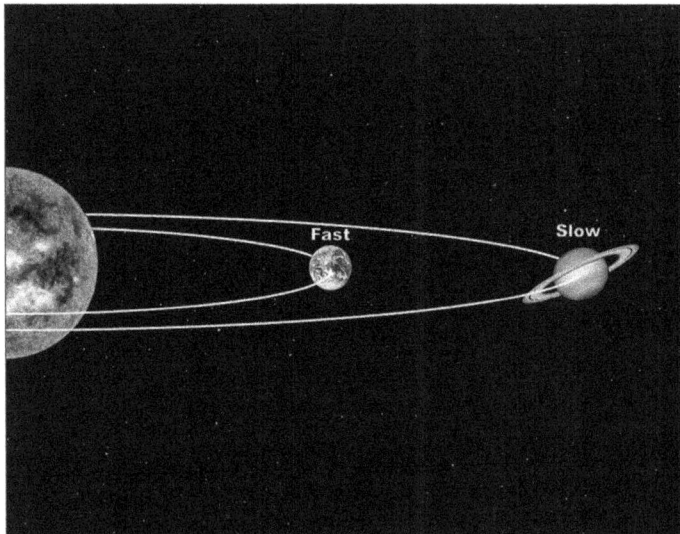

The further a planet is from the Sun, the slower it revolves around the Sun; however, the orbital speed isn't just slower, it is precisely related to the distance from the Sun according to Kepler's third law of planetary motion. We will explore this precise relationship in the Checking for Understanding section of this chapter.

Earth Image Credit: NASA. Saturn Image Credit: NASA/JPL-Caltech. Sun Image Credit: NASA.

PLANET	R = MEAN DISTANCE *(in millions of kilometers)*	T = PERIOD *(in Years)*	$T/R^{3/2}$
Mercury	57.9	0.241	0.000547
Venus	108.2	0.615	0.000546
Earth	149.6	1	0.000547
Mars	227.9	1.88	0.000546
Jupiter	778.3	11.86	0.000546
Saturn	1,427	29.5	0.000547
Uranus	2,870	84	0.000546
Neptune	4,497	165	0.000547
Pluto	5,900	248	0.000547

If you were to plot T versus $R^{3/2}$ for each planet in our Solar System, you would generate a straight line. In other words, $T/R^{3/2}$ is the same value for every planet in our Solar System.

Continuing with Kepler's biography, Kepler's mother, *Katharina Kepler*, was a cantankerous old woman who sold drugs and herbal concoctions to soldiers. When several soldiers became ill, she was accused of being a witch and was removed from her home in a laundry basket and imprisoned. Kepler ultimately freed her after seven years of unremitting effort.

In 1619, eight days after publishing his Third Law of Planetary Motion, the Thirty Years War began leaving Kepler penniless and pleading for money. His wife and son died from an epidemic spread by the soldiers and he was ultimately excommunicated from the Lutheran Church. After Kepler's death, the war completely obliterated his grave.

> **Johannes Kepler was a mathematical workhorse who was the first scientist to correctly understand the structure of the Solar System as heliocentric, with the planets orbiting the Sun at different speeds along elliptical paths. Kepler's three laws of planetary motion paved the way for Isaac Newton's Universal Law of Gravity.**

1.3.1.2 The universal law of gravitation

With the revolutionary heliocentric model of **Copernicus**, the conceptual break-throughs of **Galileo**, the accurate observational work of **Brahe**, and the exhaustive mathematics of **Kepler**, the table was set for **Isaac Newton** to encapsulate all the laws governing celestial mechanics into a single unifying theory and equation. In volume 1, chapter 4, we recounted how in 1687, Isaac Newton published a manuscript known today simply as the *Principia*. In addition to stating the three laws of motion and laying the foundations of light and optics, Newton discovered the 'Universal Law of Gravitation,' which we will abbreviate as the 'ULG.' Unlike his laws of motion, Newton did not provide us with a simple, all-encompassing Latin phrase in the *Principia* to summarize his understanding of the ULG. Therefore, rather than presenting a Latin statement and providing a translation, as we did in volume 1, chapter 4, we will summarize the ULG in our own words:

Newton's Universal Law of Gravitation:

Every particle of matter in the Universe attracts every other particle with a force that is directly proportional to the product of the masses of the particles and inversely proportional to the square of the distance between them.

As we've seen before, Newton's seemingly simply findings actually pack quite a wallop! The ULG is no exception. To understand Newton's ULG and to use it as a problem-solving instrument, we must translate Newton's discovery into an equation and carefully define what each word means. Since mathematics is just a concise symbolic language, we must ensure that the relations described by our syntax are embodied within our equation. Let us review our statement phrase-by-phrase so that Newton's discovery may guide us in the formation of our ULG equation. After every few words, we'll be able to add another symbol to our evolving equation:

We start the ULG with the phrase: *'Every particle of matter in the universe attracts every other particle with a force ...'*

- First, we need to define exactly what the ULG gives us. In other words, 'What will the ULG compute for us when we use it? ' Newton's ULG determines **a force**. In volume 1, chapter 4, we saw that Newton devoted an entire law of motion, law #3, to precisely define what a force is: 'A force is any interaction between two bodies.' This is a crucial concept to understanding the ULG; namely, it calculates the force between *only* two objects with mass. If we want to calculate the interactions between more than two objects, we must use the ULG between pairs of objects, then sum the results as vectors (i.e., we must use the principle of superposition).

- Also from volume 1, chapter 4, we know force is a vector and therefore two bits of numeric information are required to completely specify a force. We will write the force in vector notation, $\vec{F} = (F_x, F_y)$, and add or subtract our results according to our rules of vector addition and subtraction. Combining these first two interpretations we obtain:

$$\textbf{ULG: } \vec{F_g} = ?$$

Etymology:

The word 'gravity' comes from the Latin word 'gravis' that translates into 'heavy' or 'weighty.'

Notice that we include the small arrow in our evolving equation to symbolize the vector nature of a force.

- We label this force as 'gravitational force' or the 'force of gravity' because it happens to describe the interaction that each of us has with the Earth, or what we would call our 'weight.' The Latin word for 'weight' is 'gravis,' and so our interaction with the Earth (the constant pull we feel toward the Earth) has come to be known as the 'force of gravity.' To update our evolving ULG equation, let us add a subscript 'g' to our symbol for force to denote 'gravity' or 'gravitational':

$$\textbf{ULG: } \vec{F_g} = ?$$

Notation:

We have just introduced a new notation to our lexicon of mathematical symbols. Scientists and engineers often use the '$\wedge$' (called a 'hat') to symbolize a pure direction. Traditionally, the '$\wedge$' is called a 'caret' or 'circumflex' and can be typed from the SHIFT character above the number-6 on a standard keyboard. In general, a hat only serves to specify the direction of a vector.

For example, $\vec{A} = 3\hat{x}$ miles (spoken as '3 x-hat miles'), translates into '3 miles along the x-axis' or, using our vector notation, $\vec{A} = (3, 0)$ miles.

As another example, $\vec{B} = 4\hat{x} + 5\hat{y}$ Newtons (spoken as '4 x-hat and 5 y-hat Newtons') translates into '4 Newtons along the x-axis and 5 Newtons along the y-axis,' or $\vec{B} = (4, 5)$ Newtons.

When you read our ULG equation, the $\hat{r}$ simply translates to the following words: 'attractive along the direction of the line joining the two particles.'

- Newton's law is 'universal' and that's where we get the 'U' in 'ULG.' No particle of matter can escape the ULG. Although the ULG describes how each of us interacts with the Earth, the ULG more generally applies to EVERY particle of matter in the Universe. Also, not only is every particle of matter in the Universe subjected to the ULG, but **every particle of matter in the Universe interacts with every other particle of matter in the Universe!** Gravity is a force between almost any two objects you can imagine: you and the Earth, you and this book, you and your dog, your dog and this book, your dog and your cat, etc. Students tend to think that the phrases 'gravity,' 'gravitational' or the 'force of gravity' only describe how *the Earth* pulls on all objects, causing them to fall. The interaction between you and the Earth may in fact be the largest of these forces, but a gravitational force does still exist between you and every other mass in the Universe. Thus, the ULG applies to each particle of matter and determines that particle's gravitational interaction with every other particle of matter.
- The gravitational force is always attractive, never repulsive. Masses pull each other together; they never push each other apart. You always fall toward the Earth, never away. Let us update our evolving ULG equation by adding a symbol that reminds us that gravity is always attractive. The symbol that physicists use to specify the attractive nature of gravity is $\hat{r}$ and is spoken as '*r hat.*'

$$\textbf{ULG: } \vec{F}_g = ? \; \hat{r}$$

Let us continue to dissect our rendition of the ULG that continues with the phrase: '*Every particle of matter in the Universe attracts every other particle with a force ... that is directly proportional to the product of the masses of the particles and ...*'

- Now the ULG begins to take on some computational meat. The value or magnitude of the force, according to Newton's discovery, is proportional to the product of the two masses:

$$\textbf{ULG: } \vec{F}_g \propto (m_1 \times m_2)\hat{r}.$$

At this point, you would verbalize this proportion as: 'The force of gravity is proportional to the product of m_1 and m_2, and attractive along the direction of the line joining the two particles.'

We conclude our rendition of the ULG with the phrase: '*Every particle of matter in the Universe attracts every other particle with a force ... that is directly proportional to the product of the masses of the particles and ... inversely proportional to the square of the distance between them.*'

- We continue the translation of Newton's words by including a symbol to represent the inverse square relationship of the force of gravity with respect to separation distance. For instance, if you INCREASE the separation distance between two masses by a factor of 4, the force of gravity DECREASES by a factor of 4^2, or 16. The relationship is inverse (increasing distance, decreases force) and squared (changing distance by 4 changes force by 4^2 or 16):

$$\textbf{ULG: } \vec{F_g} \propto \frac{(m_1 \times m_2)}{r^2} \hat{r} \, .$$

You would verbalize this proportion as: 'The force of gravity is proportional to the product of m_1 and m_2, inversely proportional to the square of the distance between them, and directed along the line joining the two particles.'

Also at this time, we emphasize a critical notational convention used in the ULG. The r appearing in the denominator of the ULG denotes the distance between the **centers** of masses m_1 and m_2. We conceptualize any two masses as 'point masses,' meaning that we can collapse their shapes to that of a point and treat the two masses as if all of their mass were concentrated at their centers. In effect, replace any object with a particle of equal mass located at its center. For example, if we separate two bowling balls each of radius 0.5 m, by 4 m, the actual distance between their centers is: $4 + 0.5 + 0.5 = 5$ m. The situation is shown below:

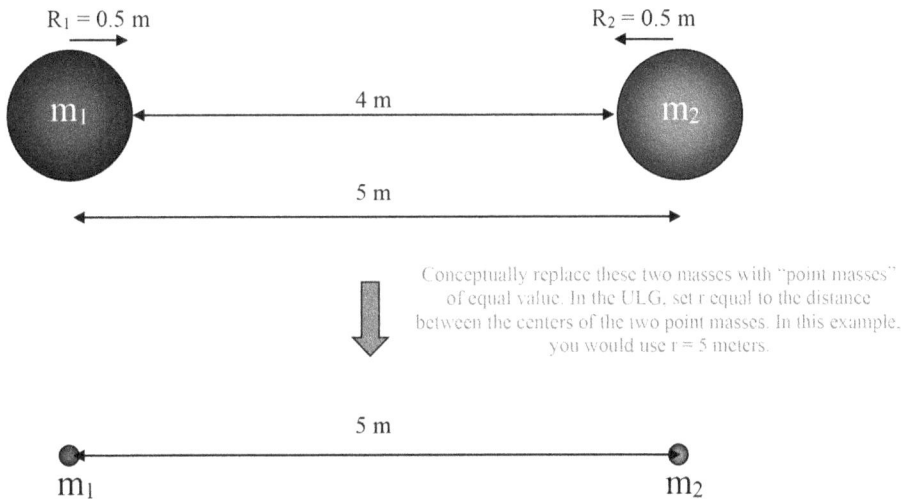

Conceptually replace these two masses with "point masses" of equal value. In the ULG, set r equal to the distance between the centers of the two point masses. In this example, you would use r = 5 meters.

- The final piece to write the ULG as an equation is to convert the proportionality into an equality. In general, any proportion can be converted into an equation by simply including a constant of proportionality—a fixed number that appropriately scales the proportionality to make the units work.

In our case, the constant of proportionality is symbolized by 'G' and is called the 'Universal Gravitation Constant.' The value of 'G' and how it was numerically determined is described in the next section.

Putting everything together, we get the final mathematical form of Newton's ULG. You would verbalize the ULG as: 'The force of gravity equals some constant (i.e., G) times the product of m_1 and m_2, divided by the square of the distance between m_1 and m_2, and is directed along the line joining the two particles.'

$$\boxed{\text{ULG: } \vec{F_g} = \frac{G \times m_1 \times m_2}{r^2}\hat{r}}$$

1.3.1.3 Determining the universal gravitation constant, G

Often in the development of scientific and mathematical concepts, proportionalities must be converted into equalities. As mentioned above, physicists and mathematicians make this conversion by including a constant of proportionality in the equation. Such is the case when we incorporate G into the ULG. Let us take a closer look at why this conversion is necessary and how we might experimentally determine a numeric value for G.

For the moment, let us pretend that the ULG did not contain the Universal Gravitation Constant and that instead, we wrote the proportionality as a straight equality:

$$\vec{F_g} = \frac{m_1 \times m_2}{r^2}\hat{r}.$$

If we write the ULG as shown above, we have created a dilemma in our definition of force because we now have two equations that can be used to define the basic **m.k.s.** unit of force, the 'Newton':

- The first equation is Newton's second law, $\vec{F} = m \times \vec{a}$. According to this law, 1 kg, accelerated at 1 m s^{-2}, produces a force of 1 Newton. We would therefore define a Newton as the force of 1 kg accelerating at a rate of 1 m s^{-2}. Imagine setting up this situation with an experiment—perhaps you take a 1 kg marble and very carefully push it across a smooth Teflon tabletop or perhaps you find a 1 kg wooden block and accurately accelerate it across an air-track. To get a feel for what a Newton would be in these scenarios, think of a Newton as roughly equivalent to the weight of a good-sized apple or about a quarter of a pound.
- The other equation of course is the ULG, which we are pretending has the form, $\vec{F_g} = (m_1 \times m_2/r^2)\hat{r}$. According to this law, two 1 kg masses, separated by 1 m, would exert a gravitational force of 1 Newton on one another. We would therefore define a Newton as the gravitational force between two 1 kg

masses, separated by a distance of 1 m. Imagine setting up this situation in a laboratory—perhaps you find two 1 kg marbles, separate them by 1 m then use a spring to measure the gravitational force of one on the other. To get a feel for what a Newton would be in this scenario, imagine a weight so incredibly small, you essentially couldn't even feel it. Only a very sensitive balance would be able to detect the force of a Newton under these conditions.

So without the inclusion of the Universal Gravitation Constant G, Newton's second law and the ULG produce conflicting definitions of the fundamental unit of force, the Newton. Physically, we cannot change Nature—if we make Newton's second law our defining equation of force, then 1 Newton is roughly the weight of an apple. However, if we make the ULG our defining equation of force, then 1 Newton is an extremely small value. To reconcile this discrepancy, we have to choose one of our equations to serve as our definition of a Newton then mathematically scale the other equation to jive with that definition. In other words, we could use Newton's second law as our definition of force so that 1 Newton is roughly the weight of a hefty-sized apple, but then include a 'decreasing scaling factor' in the ULG that guarantees two 1 kg masses separated by 1 m results in a very small force; or we can use the ULG as our definition of force so that 1 Newton is a very small value, but then include an 'increasing scaling factor' in Newton's second law that guarantees 1 kg accelerating at 1 m s^{-2} results in a force of about a quarter of a pound. Either approach is valid, but we must make a choice and stick with it. In case you haven't guessed, physicists use Newton's second law as the definitional equation of force and include a scaling factor in the ULG so that it jives with that definition.

Operational Definition:

Newton's second law is chosen to be the operational definition of a force and the basic **m.k.s.** unit of force, the 'Newton.' Therefore, a Newton is **defined** to be the force of 1 kg accelerating at a rate of 1 m s^{-2}.

Thus, every subsequent equation (more will follow in later courses in physics) involving a force must include a factor, a constant of proportionality, to scale that equation to agree with the operational definition of force.

Provided G is a very small number, we must include it in our ULG so that mathematically the ULG results in a very small force between two 1 kg masses, separated by 1 m. Let us now turn our attention to determining a precise value for G, anticipating that this value will be a very small number.

Historically, the value of G was (and still is) experimentally determined using a device known as a 'torsion balance.' Although the torsion balance has been used historically in a number of different experiments, in 1798, Sir Henry Cavendish (1731−810) was the first to use it to determine the value of G.

Biographical Information:
SIR HENRY CAVENDISH

Picture and signature of Henry Cavendish

Image Credit: Wikimedia Commons
Public Domain

BORN: October 10, 1731 (Nice, Kingdom of Sardinia)
DIED: February 24, 1810 (London, England)

Sir Henry Cavendish was born into high aristocratic lineage and became immensely wealthy through a number of inheritances. He discovered Hydrogen and named the element. Throughout his life, he shunned people and lived a life of isolation with few friends. He was extremely uncomfortable in society and avoided it when he could. In fact, stories claim he was only able to speak to one person at a time and when someone sought out his opinion, they might receive a mumbled reply, but more often than not they would hear a peeved squeak (his voice seems to have been high-pitched) and turn "to find an actual vacancy and the sight of Cavendish fleeing to find a more peaceful corner." He devoted himself entirely to science and experimented with all disciplines of science. He wrote a vast number of papers but published fewer than twenty and therefore many of his original theories were credited elsewhere because of his unwillingness to publish. In the late nineteenth century, long after Cavendish's death, James Clerk Maxwell looked through Cavendish's papers and found observations and results for which others had been given credit (*i.e.*, Ohm's Law, Dalton's Law, Charles Law of Gases, etc.) Cavendish possessed the experimental and theoretical talents comparable only to Newton and was without scientific peer during the 1800s. Cavendish died as one of the wealthiest men in Britain.

This Cavendish Henry signature image has been obtained by the author from the Wikimedia website, where it is stated to have been released into the public domain. It is included within this article on that basis.

The torsion balance is made of a light rigid rod shaped like an 'inverted T.' This rod is suspended from the ceiling by a 'torsion fiber' (usually a very thin quartz rod or stiff wire) to which a flat mirror is mounted. Two small spherical masses, each of mass m_1, are attached to the ends of the inverted T and a light source (today, we would use a laser) is directed at the mirror. The reflection of the light source off of the mirror strikes a ruler or curved scale. We place a mark on the ruler to indicate the initial position of the light beam. The initial configuration of a torsion balance is shown below:

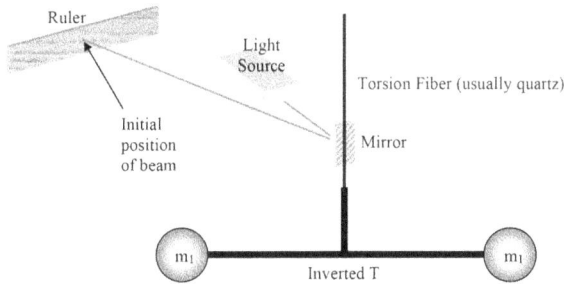

To operate the torsion balance, we carefully bring a mass m_2 near one of the masses mounted to the inverted T. The attractive gravitational force between m_1 and m_2 twists the torsion fiber and hence turns the mirror through a small angle θ. The angle θ can be accurately measured on the ruler since the reflection of the light source will move with the mirror:

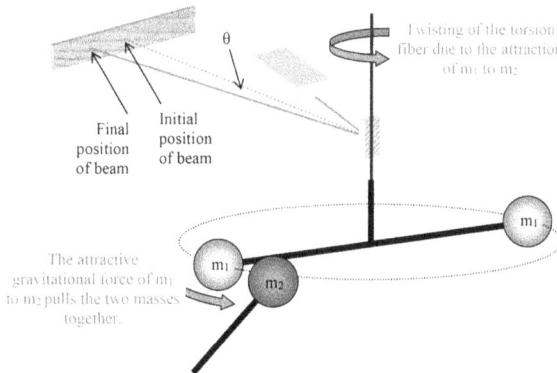

By carefully calculating the angle through which the torsion balance moves, the values of masses m_1 and m_2, the separation distance of masses m_1 and m_2, and the force required to twist the quartz fiber, a precise value of G can be determined. As described above, the principle behind the torsion balance is that the gravitational force between two masses causes the quartz fiber 'to twist.' The apparatus is labeled as a 'torsion balance' since it is a scale or balance that utilizes the twisting of the quartz fiber to determine its measurements.

Etymology:

The word 'torsion' comes from the Latin word 'torquere' that translates into 'to twist.'
    The word 'balance' comes from the Latin word 'bilancia' that translates into 'two scales.'

Obviously, to determine an accurate value of G, measurements must be repeated many times while the experimenter pays careful attention to laboratory conditions. A good experimentalist, like Cavendish, must remove any air drafts and the effects of extra masses like the walls of the laboratory, a nearby table, and even the experimenter himself. Air drafts and extra masses will rotate the torsion balance ever so slightly and thus impact the experimental determination of G. Torsion balances have come a long way since the days of Cavendish, however, the basic principle still applies: measure the deflection of an instrument as the gravitational force between two masses pulls the masses together.

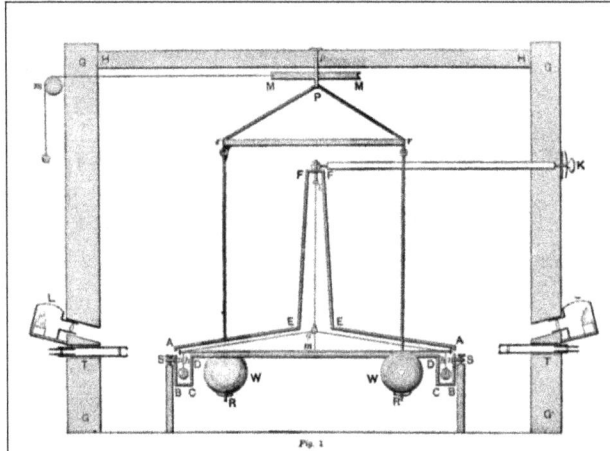

The torsion balance experiment of **Sir Henry Cavendish** who in 1797-98 was the first to experimentally measure the Universal Gravitational Constant G. Notice that the value of G was not experimentally determined (in 1798) until nearly a century after the publication of Newton's *Principia* (published in 1687).

This Cavendish Experiment image has been obtained by the author from the Wikimedia website, where it is stated to have been released into the public domain. It is included within this article on that basis.

Today, even satellites are in the business of determining G. A satellite emits a laser beam that reflects off of the Earth and back to the satellite. Using this reflected light, physicists measure the relative distance between the satellite and the Earth. Knowing the exact separation distance and the gravitational force exerted on the satellite, these physicists determine the value of G. The currently accepted value for G is:

$$G = 6.673 \times 10^{-11}\,\text{N} \cdot \text{m}^2\,\text{kg}^{-2}.$$

We can now re-state the Universal Law of Gravitation according to Newton's *Principia* and translate it into a concise equation:

"Every particle of matter in the universe attracts every other particle with a force that is directly proportional to the product of the masses of the particles and inversely proportional to the square of the distance between them."

⇩

$$\text{ULG: } \vec{F_g} = \frac{G \times m_1 \times m_2}{r^2}\hat{r} = \frac{(6.673 \times 10^{-11}) \times m_1 \times m_2}{r^2}\hat{r}$$

Armed with a mathematical representation of Newton's Universal Law of Gravitation, we are now in a position to add it to our arsenal of problem-solving strategies.

1.3.2 Modeling

❶ **EX 1:** A point-sized mass $m_1 = 3$ kg is placed at the origin. Another point-sized mass $m_2 = 4$ kg is placed at the location (8,0) m. No other masses are near these two masses, not even the Earth. To visualize this problem, imagine that it takes place in some remote location in outer space where we can assume all other masses in the Universe are infinitely far away. Calculate the gravitational force of m_2 on m_1 and the gravitational force of m_1 on m_2.

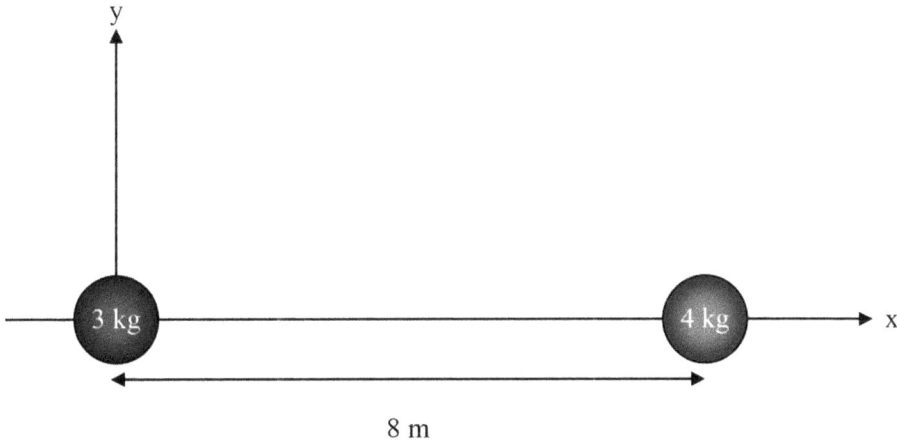

8 m

☑ **ANSWER:** The ULG equation computes a force. Since all masses are given in kilograms and all distances in meters, we can simply plug the numbers into the ULG and interpret the results:

$$\vec{F}_g = \frac{G * m_1 * m_2}{r^2} \hat{r}$$

$$= \frac{\left(6.673 \times 10^{-11}\right) * 3 * 4}{8^2} \hat{r}$$

$$= 1.25 \times 10^{-11} \, \hat{r} \text{ Newtons.}$$

Plug in the numbers but keep $\hat{r}$ separate—$\hat{r}$ does not enter into any of the computations; instead, it is only a reminder that the direction of the gravitational force is attractive.

However, we've been a bit cavalier about interpreting the 'directional reminder' $\hat{r}$ and writing our answer as a vector. Recall as we translated our rendition of the ULG into equation form, we included the $\hat{r}$ to remind us that the force of gravity is *attractive along the direction of the line joining the two particles.* In this example, the *x*-axis is the line that joins the two particles. Since we know gravity is **always attractive**, the 3 kg mass is *attracted toward* the 4 kg mass (i.e., along the **positive** *x*-axis) while the 4 kg mass is *attracted toward* the 3 kg mass (i.e., along the **negative** *x*-axis)—this is an action/reaction pair according to Newton's third law of motion. Therefore, we must write the force on the 3 kg mass as 1.25×10^{-11} Newtons **to the right** and the force on the 4 kg mass as 1.25×10^{-11} Newtons **to the left**. Thankfully, we have vector notation that allows us to mathematically keep track of directions.

$\vec{F}_{g \text{ (on 3 kg)} \atop \text{from 4 kg}} = \left(1.25 \times 10^{-11}, 0\right)$ Newtons

$\vec{F}_{g \text{ (on 4 kg)} \atop \text{from 3 kg}} = \left(-1.25 \times 10^{-11}, 0\right)$ Newtons

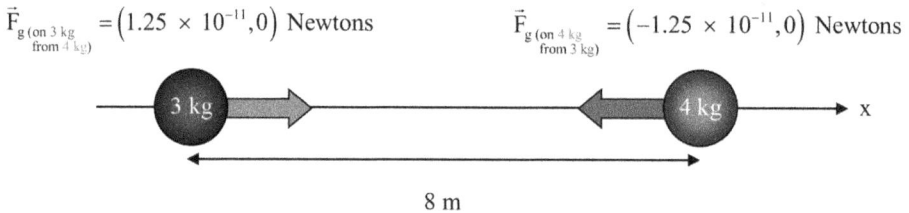

8 m

☺ **COMMENTS:** Notice that our answer for force is a vector and therefore must have two bits of numeric information:

$\vec{F}_{g \text{ (on 3 kg)} \atop \text{from 4 kg}} = \left(1.25 \times 10^{-11}, 0\right)$ Newtons

This answer contains 2 pieces of numeric information: (i.) a force directed to the right, and (ii.) no force directed up or down.

$\vec{F}_{g \text{ (on 4 kg)} \atop \text{from 3 kg}} = \left(-1.25 \times 10^{-11}, 0\right)$ Newtons

This answer contains 2 pieces of numeric information: (i.) a force directed to the left, and (ii.) no force directed up or down.

Also, notice the function of the directional reminder $\hat{r}$. At no point in our **numeric** calculation does $\hat{r}$ play a role. To compute the numeric value of the gravitational force, we only need to employ the formula: $F_g = (G \times m_1 \times m_2)/r^2$. However, $\hat{r}$

symbolically reminds us that we must write the gravitational force as a vector—and the direction of this vector is attractive along the line joining the two masses. Essentially, $\hat{r}$ is just a short-handed symbol that constantly screams at us: 'Hey, don't forget to write your answer in vector notation!'

❷ **EX 2:** A point-sized mass $m_1 = 3$ kg is placed at the origin. Another point-sized mass $m_2 = 4$ kg is placed at the location (8,0) m. Finally, a third point-sized mass $m_3 = 5$ kg is placed at the location (0,6) m. No other masses are near these two masses, not even the Earth. To visualize this problem, imagine that it takes place in some remote location in outer space where we can assume all other masses in the Universe are infinitely far away.
  a. Calculate the net gravitational force on the 3 kg mass.
  b. Calculate the net gravitational force on the 4 kg mass.
  c. Calculate the net gravitational force on the 5 kg mass.

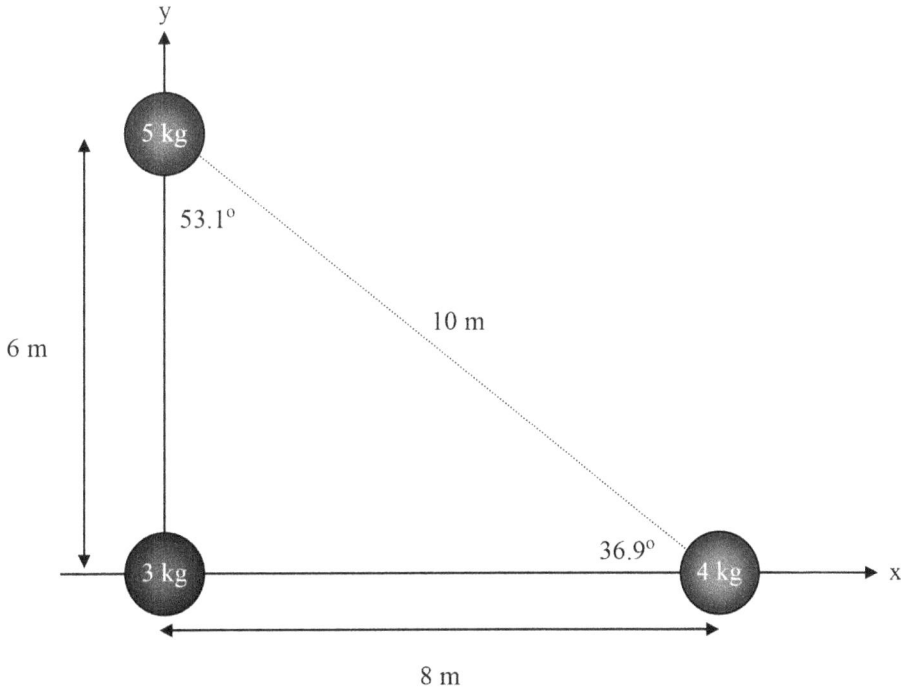

☑ **ANSWER:** Let us plug the numbers into the ULG formula, $F_g = (G \times m_1 \times m_2)/r^2$, then assign directions based on our knowledge of vector notation.

a. Calculate the net gravitational force on the 3 kg mass.

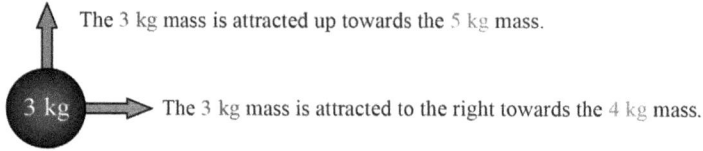

The 3 kg mass is attracted up towards the 5 kg mass.

3 kg — The 3 kg mass is attracted to the right towards the 4 kg mass.

$$\vec{F}_{g \text{ (on 3 kg from 4 kg)}} = \frac{\left(6.673 \times 10^{-11}\right) * 3 * 4}{8^2} \hat{r}$$

$= 1.25 \times 10^{-11} \hat{r} \text{ Newtons}$

$= \left(1.25 \times 10^{-11}, 0\right) \text{ Newtons}$

The 3 kg mass is attracted to the 4 kg mass along the positive x-axis.

$$\vec{F}_{g \text{ (on 3 kg from 5 kg)}} = \frac{\left(6.673 \times 10^{-11}\right) * 3 * 5}{6^2} \hat{r}$$

$= 2.78 \times 10^{-11} \hat{r} \text{ Newtons}$

$= \left(0, 2.78 \times 10^{-11}\right) \text{ Newtons}$

The 3 kg mass is attracted to the 5 kg mass along the positive y-axis.

Therefore, the net force is the vector sum: $\vec{F}_{g \text{ (total on 3 kg)}} =$ $(1.25 \times 10^{-11}, 2.78 \times 10^{-11})$ Newtons.

b. Calculate the net gravitational force on the 4 kg mass.

The 4 kg mass is attracted left and up (specifically at 36.9°) towards the 5 kg mass.

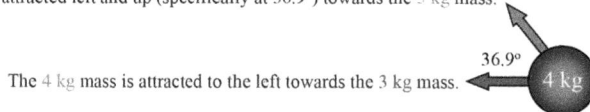

36.9°

The 4 kg mass is attracted to the left towards the 3 kg mass. ← 4 kg

$$\vec{F}_{g \text{ (on 4 kg from 3 kg)}} = \frac{\left(6.673 \times 10^{-11}\right) * 3 * 4}{8^2} \hat{r}$$

$= 1.25 \times 10^{-11} \hat{r} \text{ Newtons}$

$= \left(-1.25 \times 10^{-11}, 0\right) \text{ Newtons}$

The 4 kg mass is attracted to the 3 kg mass along the negative x-axis.

$$\vec{F}_{g \text{ (on 4 kg from 5 kg)}} = \frac{\left(6.673 \times 10^{-11}\right) * 4 * 5}{10^2} \hat{r}$$

$= 1.33 \times 10^{-11} \hat{r} \text{ Newtons}$

$= \left(-1.33 \times 10^{-11} * \cos\left(36.9°\right), 1.33 \times 10^{-11} * \sin\left(36.9°\right)\right) \text{ Newtons}$

$= \left(-1.06 \times 10^{-11}, 7.99 \times 10^{-12}\right) \text{ Newtons}$

The 4 kg mass is attracted to the 5 kg mass along the line at 36.9° up from the negative x-axis.

Therefore, the net force is the vector sum: $\overrightarrow{F}_{g\ (\text{total on } 4\ kg)}=$ $(-2.31 \times 10^{-11}, 7.99 \times 10^{-12})$ Newtons.

c. Calculate the net gravitational force on the 5 kg mass.

53.1° The 5 kg mass is attracted right and down (specifically at 53.1°) towards the 4 kg mass.

The 5 kg mass is attracted down towards the 3 kg mass.

$$\overrightarrow{F}_{g\ (\text{on } 5\ kg\ \text{from } 3\ kg)} = \frac{\left(6.673 \times 10^{-11}\right)*3*5}{6^2}\,\hat{r}$$

$= 2.78 \times 10^{-11}\,\hat{r}$ Newtons

$= \left(0, -2.78 \times 10^{-11}\right)$ Newtons

The 5 kg mass is attracted to the 3 kg mass along the negative y-axis.

$$\overrightarrow{F}_{g\ (\text{on } 5\ kg\ \text{from } 4\ kg)} = \frac{\left(6.673 \times 10^{-11}\right)*4*5}{10^2}\,\hat{r}$$

$= 1.33 \times 10^{-11}\,\hat{r}$ Newtons

$= \left(1.33 \times 10^{-11}*\cos\left(36.9°\right), -1.33 \times 10^{-11}*\sin\left(36.9°\right)\right)$ Newtons

$= \left(1.06 \times 10^{-11}, -7.99 \times 10^{-12}\right)$ Newtons

The 5 kg mass is attracted to the 4 kg mass along the line at 53.1° up from the negative y-axis.

Therefore, the net force is the vector sum: $\overrightarrow{F}_{g\ (\text{total on } 5\ kg)}=$ $(1.06 \times 10^{-11}, -3.58 \times 10^{-11})$ Newtons.

☺ **COMMENTS:** We now see the full complexity of the symbolic directional reminder $\hat{r}$. We must use the ULG formula $F_g = (G \times m_1 \times m_2)/r^2$ to compute *only* the magnitude of the gravitational force between two masses. After the magnitude is computed, $\hat{r}$ reminds us to write this gravitational force along a direction that is attractive along the line connecting the two masses. Often, writing the force as a vector requires us to calculate angles and decompose vectors into their *x*- and *y*- components.

❸ **EX 3:** A point-sized mass $m_1 = 3$ kg is placed at the origin. A point-sized mass $m_2 = 4$ kg is placed at the location (8,0) m. A third point-sized mass $m_3 = 5$ kg is placed at the location (0,6) m. Finally, a fourth point-sized mass $m_4 = 6$ kg is placed at the location (5,7) m. No other masses are near these masses, not even the Earth. To visualize this problem, imagine that it takes place in some remote location in outer space where we can assume all other masses in the Universe are infinitely far away. Calculate the net gravitational force only on the 3 kg mass.

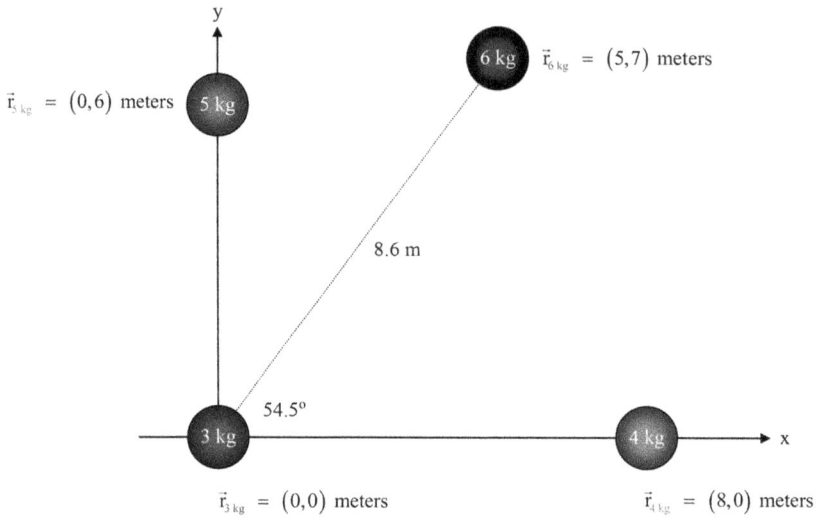

y

6 kg $\vec{r}_{6\,kg}$ = (5,7) meters

$\vec{r}_{5\,kg}$ = (0,6) meters   5 kg

8.6 m

54.5°

3 kg

$\vec{r}_{3\,kg}$ = (0,0) meters

4 kg   x

$\vec{r}_{4\,kg}$ = (8,0) meters

☑ **ANSWER:** The 3 kg mass is attracted separately to the other three masses:

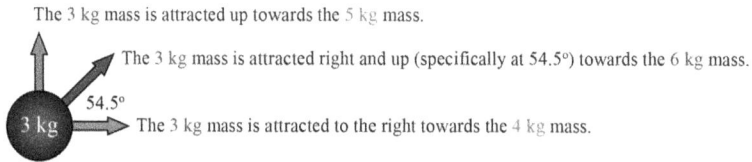

The 3 kg mass is attracted up towards the 5 kg mass.

The 3 kg mass is attracted right and up (specifically at 54.5°) towards the 6 kg mass.

54.5°

3 kg   The 3 kg mass is attracted to the right towards the 4 kg mass.

$$\vec{F}_{g\,(on\,3\,kg\,from\,4\,kg)} = \frac{\left(6.673 \times 10^{-11}\right) * 3 * 4}{8^2} \hat{r}$$

$$= 1.25 \times 10^{-11} \hat{r} \text{ Newtons}$$

$$= \left(1.25 \times 10^{-11}, 0\right) \text{ Newtons}$$

The 3 kg mass is attracted to the 4 kg mass along the positive x-axis.

$$\vec{F}_{g\,(on\,3\,kg\,from\,5\,kg)} = \frac{\left(6.673 \times 10^{-11}\right) * 3 * 5}{6^2} \hat{r}$$

$$= 2.78 \times 10^{-11} \hat{r} \text{ Newtons}$$

$$= \left(0, 2.78 \times 10^{-11}\right) \text{ Newtons}$$

The 3 kg mass is attracted to the 5 kg mass along the positive y-axis.

$$\vec{F}_{g\,(on\,3\,kg\,from\,6\,kg)} = \frac{\left(6.673 \times 10^{-11}\right) * 3 * 6}{\left(8.6\right)^2} \hat{r}$$

$$= 1.62 \times 10^{-11} \hat{r} \text{ Newtons}$$

The 3 kg mass is attracted to the 6 kg mass at 54.5° up from the positive x-axis.

$$= \left(1.62 \times 10^{-11} * \cos\left(54.5°\right), 1.62 \times 10^{-11} * \sin\left(54.5°\right)\right) \text{ Newtons}$$

$$= \left(9.43 \times 10^{-12}, 1.32 \times 10^{-11}\right) \text{ Newtons}$$

Therefore, the net force is the vector sum: $\vec{F}_{g \text{ (total on 3 kg)}}=$ $(2.19 \times 10^{-11}, 4.10 \times 10^{-11})$ Newtons.

1.3.3 Instructional input

1.3.3.1 Weight

The '**WEIGHT**' of an object is simply the total gravitational force exerted between that object and all other objects in the Universe ... and I mean ALL other objects in the Universe! Every object with mass exerts a gravitational force on every other object with mass; therefore the '**WEIGHT**' of an object, a fancy word for the sum total of all gravitational forces on an object, must be calculated by summing the gravitational force between that object and every other object in the Universe. If I ask you your weight, your modesty might prevent you from answering aloud, yet you literally would need the rest of the semester, if not your life, to give the precise answer. To answer the question, you would first take a snapshot of the Universe so that everything and everyone were held in a fixed location. Then you would have to calculate the gravitational force between you and your dog, you and your car, you and the planet Pluto, you and the tiniest asteroid, etc. You would need to continue making such calculations until every object in the Universe had been taken into account. When these calculations were complete, you would then need to sum your answers as vectors so that the final would be written with vector notation.

Fortunately, as we discussed in volume 1, chapter 4, such a precise calculation is, most of the time, not necessary. As you can imagine, after several calculations, you would begin to realize that most of the gravitational interactions between an object and another object are extremely small. In fact, the only significant number would come from the interaction of that object with the nearest, massive celestial body ... the Earth (or whatever celestial body you happen to be standing upon). So right off the bat, we make a significant time-saving approximation: The weight of an object is simply the gravitational force between that object and the Earth. Though this will serve as our definition of weight, keep in mind that it is an approximation. If, for example, an object is close to two or more celestial bodies (imagine a satellite halfway between the Earth and Moon) then the object's weight obviously must be calculated using the gravitational interaction between both celestial bodies. Let us look at two examples that illustrate the major concepts of calculating an object's weight.

1.3.4 Modeling

❶ **EX 1:** For this problem, you will need to know the international astronomical symbols for the Earth and the Sun. Every celestial object has an international astronomical symbol: The Earth is symbolized by a motherly sphere, represented by a circle with a cross inside it ($\oplus$) while the Sun is symbolized by the egg of creation, represented by a circle with a dot inside it ($\odot$). You have a mass of 75 kg (roughly 165 lbs) and are standing on the surface of the Earth. The mass of the Earth is $M_\oplus = 5.98 \times 10^{24}$ kg and its radius is $R_\oplus = 6.38 \times 10^6$ m. Exactly 2 m to your left is your friend Ted who has a mass of 85 kg. Exactly 1.5 m to your right is another friend

Fred who has a mass of 65 kg. Finally, you notice that the Sun is directly overhead and is a distance of 1.5×10^{11} m above you. The mass of the Sun is $M_\odot = 1.98 \times 10^{30}$ kg and its radius is $R_\odot = 6.995 \times 10^8$ m. Assuming no other objects exist in the Universe, calculate your weight. A drawing, not to scale, is shown below:

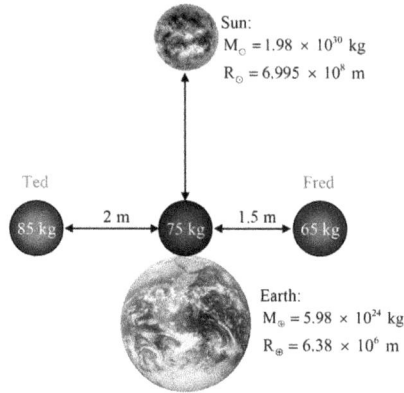

Sun:
$M_\odot = 1.98 \times 10^{30}$ kg
$R_\odot = 6.995 \times 10^8$ m

Ted   2 m   Fred
85 kg   75 kg   1.5 m   65 kg

Earth:
$M_\oplus = 5.98 \times 10^{24}$ kg
$R_\oplus = 6.38 \times 10^6$ m

"Earth Image Credit: The 'Blue Marble' from Apollo 17,
Credit: NASA. Sun Image Credit: NASA."

☑ **ANSWER:** Our strict definition of an object's weight is the total gravitational force acting on that object by all other objects in the Universe. In this problem, since the Universe only contains four objects in addition to you, we can actually do a vector summation of four forces.

- $\vec{F}_{g \text{ (on you from Ted)}} = \dfrac{\left(6.673 \times 10^{-11}\right)*75*85}{2^2}\hat{r}$

  $= 1.06 \times 10^{-7}\,\hat{r}$ Newtons

  $= \left(-1.06 \times 10^{-7}, 0\right)$ Newtons

  You are attracted to Ted along the negative x-axis.

- $\vec{F}_{g \text{ (on you from Fred)}} = \dfrac{\left(6.673 \times 10^{-11}\right)*75*65}{1.5^2}\hat{r}$

  $= 1.44 \times 10^{-7}\,\hat{r}$ Newtons

  $= \left(1.44 \times 10^{-7}, 0\right)$ Newtons

  You are attracted to Fred along the positive x-axis.

- $\vec{F}_{g \text{ (on you from Sun)}} = \dfrac{\left(6.673 \times 10^{-11}\right)*75*M_\odot}{\left(1.5 \times 10^{11} + R_\odot\right)^2}\hat{r}$

  Notice that we use the distance between you and the center of the Sun.

  $= \dfrac{\left(6.673 \times 10^{-11}\right)*75*1.98 \times 10^{30}}{\left(1.5 \times 10^{11} + 0.995 \times 10^9\right)^2}\hat{r}$

  $= 0.436\,\hat{r}$ Newtons

  $= \left(0, 0.436\right)$ Newtons

  You are attracted to the Sun along the positive y-axis.

- $\vec{F}_{g \text{ (on you from Earth)}} = \dfrac{\left(6.673 \times 10^{-11}\right)*75*M_\oplus}{\left(R_\oplus\right)^2}\hat{r}$

  Notice that we use the distance between you and the center of the Earth—that distance is just the radius of the Earth.

  $= \dfrac{\left(6.673 \times 10^{-11}\right)*75*5.98 \times 10^{24}}{\left(6.38 \times 10^6\right)^2}\hat{r}$

  $= 735\,\hat{r}$ Newtons

  $= \left(0, -735\right)$ Newtons

  You are attracted to the Earth along the negative y-axis.

Therefore, the net force is the vector sum: $\vec{F}_{g \text{ (total on you)}} \approx (0, -735)$ Newtons, which is just the gravitational force between you and the Earth. The other three forces are negligibly small in comparison.

☺ **COMMENTS:** Let us keep this comment short and simple—an object's weight is the sum of all gravitational forces on that object by all other objects in the Universe. However, since all other forces are negligibly small, we can very closely approximate an object's weight as only its gravitational attraction to the Earth.

❷ **EX 2:** You have a mass designated by M (M can be any value, but it's a known value) and travel to four locations on the surface of the Earth. The mass of the Earth is $M_\oplus = 5.98 \times 10^{24}$ kg and its radius is $R_\oplus = 6.38 \times 10^6$ m. Assuming that only the force of gravity between you and the Earth determines your weight, calculate your weight at the following four locations (the following diagram was drawn close to scale):

    a. Cleveland, Ohio (Cleveland's elevation is 177 m or 582 feet above sea level.)
    b. Denver, Colorado (Denver's elevation is 1610 m or 5280 feet above sea level.)
    c. Mount Everest (The tallest point above sea level is the summit of Mount Everest at 8850 m or 29 035 feet; however, because of the equatorial bulge, the farthest point from the center of the Earth is Chimborazo, Ecuador. Despite being 8481 feet lower in elevation than Everest, Chimborazo is 7096 feet farther than the summit of Everest from the Earth's center.)
    d. The Challenger Deep (The deepest point in any of the Earth's oceans is the 'Challenger Deep' a trench in the Pacific Ocean's Mariana Trench at 11 033 m or 36 201 feet below sea level.)

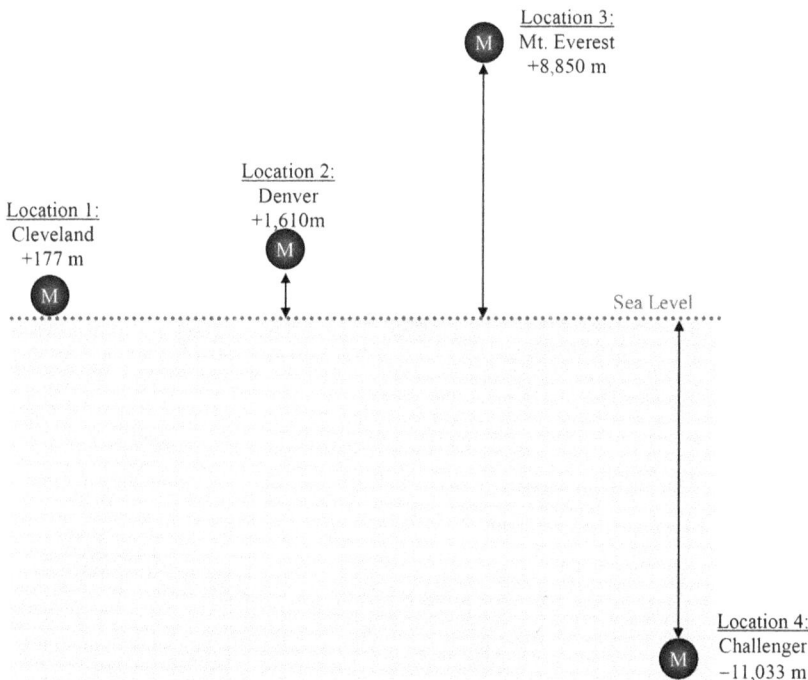

Location 3:
Mt. Everest
+8,850 m

Location 2:
Denver
+1,610m

Location 1:
Cleveland
+177 m

Sea Level

Location 4:
Challenger
−11,033 m

e. ☑ ANSWER: We'll use the ULG between the mass M and the Earth. The distance between the centers of the two masses will be the radius of the Earth plus or minus the distance of the location from sea level.

a. $\vec{F}_{g \, (on \, M \, from \, Earth \, at \, Cleveland)}$ $= \dfrac{\left(6.673 \times 10^{-11}\right) * M_{\oplus} * M}{\left(R_{\oplus} + \text{elevation of Cleveland}\right)^2} \hat{r}$ ⬅ Notice that we add the elevation to the radius of the Earth because Cleveland is further from the center of the Earth than sea level.

$= \dfrac{\left(6.673 \times 10^{-11}\right) * \left(5.98 \times 10^{24}\right) * M}{\left(6.38 \times 10^{6} + 177\right)^2} \hat{r}$

$= 9.80 * M \, \hat{r} \text{ Newtons}$

$= \left(0, -9.80 * M\right) \text{ Newtons}$

b. $\vec{F}_{g \, (on \, M \, from \, Earth \, at \, Denver)}$ $= \dfrac{\left(6.673 \times 10^{-11}\right) * M_{\oplus} * M}{\left(R_{\oplus} + \text{elevation of Denver}\right)^2} \hat{r}$ ⬅ Notice that we add the elevation to the radius of the Earth because Denver is further from the center of the Earth than sea level.

$= \dfrac{\left(6.673 \times 10^{-11}\right) * \left(5.98 \times 10^{24}\right) * M}{\left(6.38 \times 10^{6} + 5,280\right)^2} \hat{r}$

$= 9.79 * M \, \hat{r} \text{ Newtons}$

$= \left(0, -9.79 * M\right) \text{ Newtons}$

c. $\vec{F}_{g \, (on \, M \, from \, Earth \, at \, Mt. \, Everest)}$ $= \dfrac{\left(6.673 \times 10^{-11}\right) * M_{\oplus} * M}{\left(R_{\oplus} + \text{elevation of Mt. Everest}\right)^2} \hat{r}$ ⬅ Notice that we add the elevation to the radius of the Earth because Everest is further from the center of the Earth than sea level.

$= \dfrac{\left(6.673 \times 10^{-11}\right) * \left(5.98 \times 10^{24}\right) * M}{\left(6.38 \times 10^{6} + 8,850\right)^2} \hat{r}$

$= 9.78 * M \, \hat{r} \text{ Newtons}$

$= \left(0, -9.78 * M\right) \text{ Newtons}$

d. $\vec{F}_{g \, (on \, M \, from \, Earth \, at \, Challenger)}$ $= \dfrac{\left(6.673 \times 10^{-11}\right) * M_{\oplus} * M}{\left(R_{\oplus} - \text{depth of Challenger}\right)^2} \hat{r}$ ⬅ Notice that we subtract the depth from the radius of the Earth because Challenger is closer to the center of the Earth than sea level.

$= \dfrac{\left(6.673 \times 10^{-11}\right) * \left(5.98 \times 10^{24}\right) * M}{\left(6.38 \times 10^{6} - 11,033\right)^2} \hat{r}$

$= 9.84 * M \, \hat{r} \text{ Newtons}$

$= \left(0, -9.84 * M\right) \text{ Newtons}$

☺ **COMMENTS:** Let us keep this comment short and simple—to calculate an object's weight, you must use the complete, explicit form of the Universal Law Gravitation. However, when calculating an object's weight near the surface of the Earth, you can use the approximation: $W = m \times 9.8 = m \times g$ (as we have been doing since volume 1, chapter 4). Because the radius of the Earth is so large in comparison to the height of any surface feature on the Earth, the ULG will essentially result in the same numeric answer every time. Even at the highest point on the surface of the Earth (i.e., Mount Everest) or the lowest point on the surface of the Earth (i.e., the 'Challenger Deep'), the weight of an object is essentially: $W = m \times 9.8 = m \times g$.

Let's summarize these two important simplifications, with the following remarks:

1. The weight of an object is technically the net gravitational force acting on an object from all other objects in the universe. However, near the surface of a planet, and object's weight is essentially the gravitational force between that object and the planet; the gravitational force from all other objects in the universe can be ignored.

2. Because the surface features of any planet are so small in comparison to the radius of the planet, the weight of an object is essentially its mass **m**, multiplied by some fixed number. For example, on the surface of the Earth, an object's weight is **m**∗9.8 ... on the surface of Mars, an object's weight **m**∗3.61 ... etc:

$$g_{planet} = \frac{G * M_{planet}}{R_{planet}^2}.$$

The chart from volume 1, chapter 4, reproduced on the next page, should now make more sense. Because the surface features of any planet are small compared to the radius of the planet, the acceleration due to gravity (i.e., the fourth column) for each planet in the Solar System can be calculated using the formula:

$$g_{planet} = \frac{G \times M_{planet}}{R_{planet}^2},$$

and since an object's weight on the surface of a planet is essentially its gravitational attraction to that planet, any object's weight can be calculated by multiplying its mass by the planet's gravitational constant (i.e., the fifth column).

Location:	Astronomical Symbol:	Photograph:	Acceleration due to Gravity (in Gals)	Weight of an object with mass, m (in Newtons)
Sun			274	m*(274)
Mercury			3.72	m*(3.72)
Venus			8.87	m*(8.87)
Earth			9.8	m*(9.8)
Moon			1.62	m*(1.62)
Mars			3.7	m*(3.7)
Jupiter			25.9	m*(25.9)
Saturn			11	m*(11)
Uranus			8.7	m*(8.7)
Neptune			11	m*(11)
Pluto			0.6	m*(0.6)

Photographs Credit: NASA.

Earth Image Credit: The 'Blue Marble' from Apollo 17,
Credit: NASA. Mars Image Credit: NASA/JPL-Caltech.
Mercury Image Credit: NASA. Moon Image Credit: NASA. Jupiter Image
Credit: NASA, ESA, A. Simon (Goddard Space Flight Center), and M.H.
Wong (University of California, Berkeley). Saturn Image Credit: NASA/JPL/
Space Science Institute. Sun Image Credit: NASA. Neptune Image Credit:
NASA. Pluto Image Credit: NASA. Uranus Image Credit: NASA/Space
Telescope Science Institute. Venus Image Credit: NASA/JPL-Caltech.

1.3.5 Instructional input

1.3.5.1 The 'field'

To conclude this chapter, we turn our attention to the question: 'How do two masses communicate with one another?' In other words, we can use the ULG to determine the gravitational attraction between two objects, but how does one object 'know' the other object exists? Do they 'see' each other? Do they somehow 'talk' to one another? Maybe one object transmits a signal that says, 'Calling all other objects … I'm here!' What mechanism does gravity utilize so that masses can interact with one another?

To answer this question, let us first brainstorm all the possible ways by which a force can be transmitted between two objects. Imagine we have two masses, m_1 and m_2, separated by some distance r. We know from the previous sections that these two masses will interact with each other according to Newton's Universal Law of Gravitation. Also, we know from Newton's third law of motion that this force will be equal in magnitude on both masses but opposite in direction—in the situation shown below, m_1 will be pulled to the right with a force of magnitude $(G \times m_1 \times m_2)/r^2$ while m_2 will be pulled to the left with a force of the same magnitude:

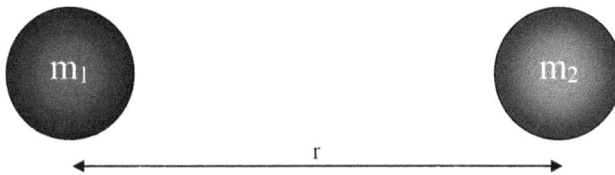

The question we now consider is how the force of gravity (as given by the ULG) is communicated between these two masses. How does m_1 'know' m_2 exists and likewise, how does m_2 'know' that m_1 exists? Let us consider several possible explanations:

1. One possible explanation is that the force of gravity is a 'contact force,' such as the normal force or the frictional force. A contact force is any force that exists as a result of two objects being in contact with one another. If gravity were a contact force, then two masses interacting, by necessity, would have to be touching in order for there to be a gravitational force between them. Look back at some of the example problems we did. Were the masses touching in order for the force of gravity to exist? Must you be in direct contact with the Earth in order for Earth's gravity to affect you? If you jump in the air and suddenly lose contact with the Earth, does gravity cease to exist? Of course not! Imagine jumping off of a cliff and the instant you lost contact with the Earth, you no longer felt the force of gravity upon you. You could just fly around without ever falling. Unfortunately, we all know that gravity acts upon us even though we may not be directly touching the Earth. We also saw in our example problems that two masses need not be touching

in order for gravity to exist between them. Therefore, we conclude the force of gravity is not a contact force. Let us try something else.

2. Another common idea is that gravity is a force that exists because of some action through a 'medium' (and no, I don't mean a spiritual medium). In other words a medium, such as air or water, must exist between the two particles in order for the force of gravity to exist. For example, a sound wave requires the presence of some kind of medium between the transmitter and the receiver; otherwise the sound wave cannot propagate. In this instance, we can say that sound waves exert a force only through the presence of a medium. If we look back at our examples, must a medium exist between the two objects in order for gravity to exist? Must there be air or water between you and the Earth in order for gravity to exist? If we remove the atmosphere, would gravity still act to attract us to the surface of the Earth? Perhaps you know that no atmosphere envelopes the Earth's moon. The surface of the Moon is surrounded entirely by a vacuum yet objects still fall to its surface without the presence of some intermediate medium—gravity still acts to attract objects to the surface of the Moon even without an intermediate medium! Therefore we conclude the force of gravity does not need to act through the medium. Let us try something else.

3. To summarize, we have found that masses can exert the force of gravity upon one another 'instantly' and 'over any distance.' Furthermore, we have found that masses do not need a medium, nor do they need to be in contact with each other, in order to exert the force of gravity upon one another. Overall, this is a difficult concept to wrap our heads around because this is unlike any of the other everyday forces we've dealt with (i.e., tension, friction, the normal force, etc). One way to summarize our findings is to state that masses interact via 'action at a distance'—it is the simple concept that an object's motion can be instantly affected by another object without being in physical contact. If you struggle with this concept, you are not alone. Newton himself had difficulty wrapping *his* head around this idea! While he was able to formulate a cohesive theory of gravity, he was deeply uncomfortable with the concept of 'action at a distance' and struggled to explain the mechanism by which gravity acted. In 1692, he wrote: 'That one body may act upon another at a distance through a vacuum without the mediation of anything else, by and through which their action and force may be conveyed from one another, is to me so great an absurdity that, I believe, no man who has in philosophic matter a competent faculty of thinking could ever fall into it.' Therefore, scientists needed to create one final idea or physical model in order to explain 'action at a distance.' Physicists now use the idea of a 'field' to explain forces, such as gravity, which act as a distance with no contact between two objects. The idea was first proposed by British physicist Michael Faraday and his model is still used in today's classical interpretation of gravity. The idea suggests that a mass exists with a 'field' everywhere and at all times. This field fills space and interacts with other masses, causing them to experience

forces even when they are not in direct contact with one another. Let us look at the model a bit closer because it is also used to explain such forces as electricity and magnetism (perhaps like the 'gravitational field,' you've heard of the 'electric field' and 'magnetic field').

Biographical Information:
MICHAEL FARADAY

BORN: Sept. 22, 1791 (Surrey, England)
DIED: Aug. 25, 1867 (Hampton, England)

Michael Faraday's father was a chronically ill, extremely poor blacksmith whose income could barely provide the family with basic necessities. Once, young Michael was given a loaf of bread and told to live off of it for a week. His parents were members of the Sandemanian Church and Faraday was raised under strict discipline. Faraday had no formal education as a child. He was an apprentice to a bookbinder and thus learned by reading the books that he was binding. In fact, Faraday's friends called him "a useless reservoir of statistics and ideas." Faraday became interested in science by accident: every week, he attended free public seminars by **Humphrey Davy** during which Faraday took meticulous notes. During one lecture, Davy was temporarily blinded by a chemical explosion and Faraday was able to complete the lecture using his own notes. Davy was so impressed that Faraday was appointed laboratory assistant at the Royal Institution. Faraday, not aware of the great opportunity given to him, only reluctantly gave-up his bookbinding apprenticeship

Faraday was a meticulous note-taker throughout his life. He discovered benzene in 1825 and served as "an expert chemical witness" at a number of trials. Faraday's other scientific contributions include: The Law of Induction (1831) and the Law of Electrolysis (1834). He introduced the terms: "Electrode," "Anode," "Cathode," "Anion," "Cation," "Ion," and "Ionization." Unlike his contemporaries, Michael Faraday did not believe that electricity, magnetism, and light were separate entities. Instead, he believed them to be different manifestations of the attractive and repulsive forces between charges. In Faraday's mind, these forces could be transmitted by fields. His religious belief in the fallibility of man caused Faraday to question contemporary thinking and develop the idea of the "field"—just because humans cannot see fields doesn't mean that they don't exist. On August 29, 1831, Faraday wrote in his notebook: "Should not the electric field be detectable as the magnetic field?"

Faraday is described as "a kind, gentle, proud and simple man in both manner and attitude," although he considered himself "ugly and unattractive." He married **Sarah Bernard** in 1821 and though he adored children, he had none of his own. Faraday suffered a nervous breakdown from 1839-1844. His mind deteriorated after 1850 at which time he retreated from the world and died 17 years later in almost complete isolation. The unit of capacitance—the **FARAD**—is named in his honor.

Michael Faraday Image Credit: Pixabay by GDJ.

The idea of a 'field' explains the gravitational attraction between two masses by invoking a two stage process: (i) Each mass (every mass in the Universe—for our purposes, let us call this mass m_1) carries with it an invisible *'gravitational field'* that exists everywhere and at all times. The field did not 'grow' from m_1. The field exists as part of m_1. In other words, masses carry fields with them. If m_1 moves, its field moves instantly everywhere with it. (ii) Another mass m_2 located at point (P.) a distance r from m_1, interacts with the gravitational field from m_1 and therefore experiences the gravitational force given by the ULG:

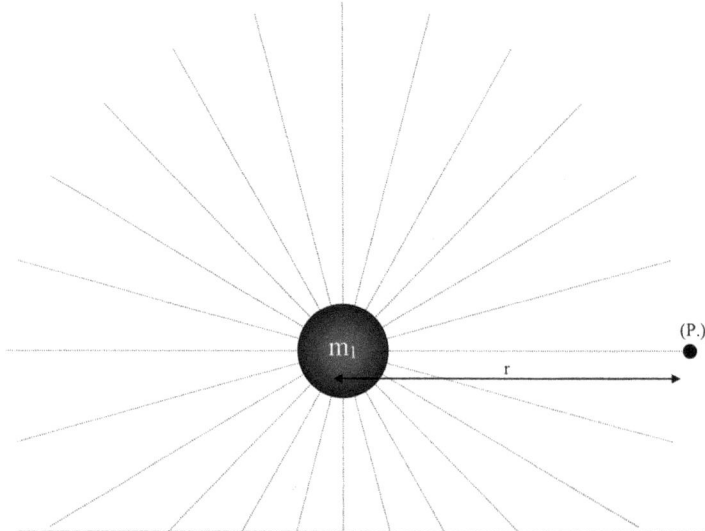

STEP #1:

A mass m_1 (m_1 could be any mass in the universe—you, me, the Earth, the Sun, this book, etc.) carries an invisible "gravitational field," which, even though it may actually be invisible, we depict with the blue dashed lines. This field exists "everywhere," meaning the blue dashed lines should extend well off the paper, above the paper, below the paper, etc. We can't possibly draw all of the field lines so we only draw a representative few. Also, the field exists "at all times," meaning the blue dashed lines were always present and always will be present they are part of m_1.

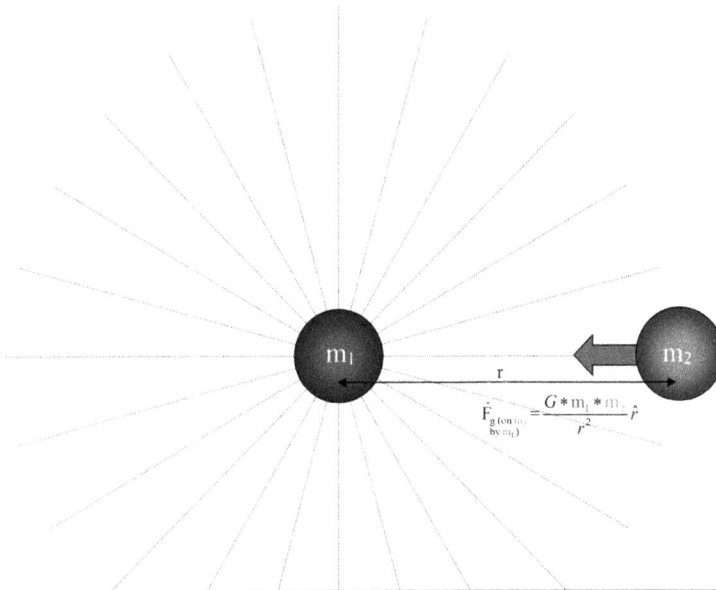

$$\vec{F}_{g\,(\text{on }m_2\,\text{by }m_1)} = \frac{G*m_1*m_2}{r^2}\hat{r}$$

STEP #2:
Another mass m_2 (m_2 could also be any mass in the universe-you, me, the Earth, the Sun, this book, etc.) now is located at point (P.) which we said is a distance r from m_1. This second mass $m2$, is "caught" by m_1's gravitational field and therefore experiences the gravitational force given by Newton's ULG.

Conceptually, an analogy that might help you visualize the concept of a 'field' is that of a fly being captured by spider. Imagine a fly and a spider are separated by some distance r—in this analogy, the spider represents m_1 and the fly represents m_2. The question we wish to answer is 'How is the spider able to capture the fly and exert a force on it? ' Very similar to the concept of a gravitational field, the spider invokes a two stage process: (i) The spider carries with it an invisible 'web' that exists everywhere and at all times, and (ii) The fly interacts or 'gets caught' by this web and is pulled by the spider:

A spider invokes a two stage process in order to exert a force on a fly: (i.) The spider carries with it an invisible web that we can imagine exists everywhere and at all times, and (ii.) The fly gets caught in the web allowing the spider to pull on the fly. The important concept to recognize is that even though the spider and fly are not in direct contact, they are able to exert a force on one another through the web—the web is the mechanism by which a mutual force is exerted.

Spider Image Credit: Jarp2/Shutterstock.com.

Similarly, every mass regardless of its size and location, carries with it a 'gravitational field' (think of the spider's web) that exists everywhere and at all times. This field exists everywhere, meaning that the field may be weaker at greater distances but it still exists at all locations. Students struggle to believe that at the very instant they are reading this text, they are indeed exerting a gravitational pull on Pluto—Yes, it's true! The gravitational force may be small, but their gravitation field exists everywhere. Also the field exists at all times, meaning it doesn't somehow 'grow' or 'develop.' The field already exists everywhere and has existed at all times. Now the second mass enters the first mass's gravitational field and experiences a force as it is 'caught' in the first mass's 'web.' Likewise the second mass, by virtue of its being a mass, has its own field that captures the first mass so that it too exerts a force of equal magnitude but opposite direction, on the first mass. In other words, you can view the analogy as running in the opposite direction—as the spider spins its web and captures the fly, so too does the fly simultaneously spin its own web and captures the spider (although the spider doesn't seem too bothered by this prospect). This way, the force is mutual as required by Newton's third law.

Perhaps the easiest conceptual example is that of your own weight, your gravitational interaction with the Earth. Notice how we explain your weight through the two-step process outlined above: The Earth, by virtue of its being a mass, carries with it the gravitational field that exists everywhere and at all times. Think of the Earth as just a giant spider that carries with it a web that exists everywhere and at all times. You, assuming the role of the fly, interact with the Earth's gravitational field and are therefore pulled towards the Earth. Likewise you, by virtue of your being a mass, carry your own gravitational field. You are now playing the role of the spider. The Earth, now playing the role of the fly, interacts with your gravitational field and feels an equal yet oppositely directed force of gravity.

1.3.5.2 Calculating the gravitational field

The recipe that we use to calculate the gravitational field is quite easy. As we develop this two-step recipe, keep in mind that the gravitational field is vector quantity so we have to be careful and use vector notation as well as the rules of vector addition and subtraction. The general recipe is outlined below:

Problem: Consider a mass M located at the origin of some coordinate system. A point (P.) is located a distance r, from the mass. Calculate the gravitational field of the mass M at the point (P.)

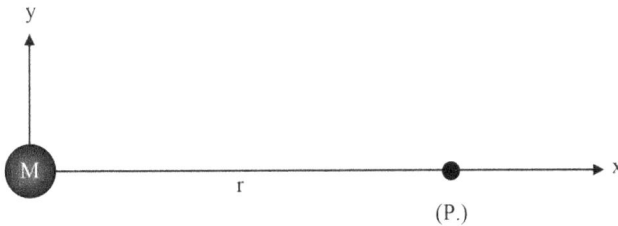

Recipe:

1. Drop-off a small 'test-mass' at the point (P.) and calculate the gravitational force exerted on that test-mass, by M, using Newton's ULG. The test-mass can be any value but we'd like to think of this step as placing a small grain of salt at point (P.) then calculating the gravitational force on that small particle.

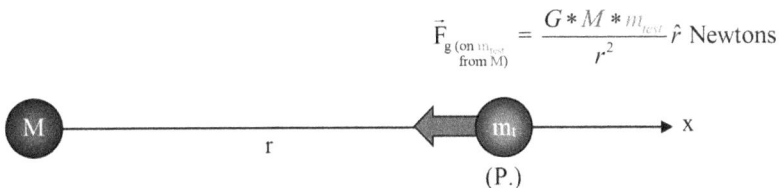

$$\vec{F}_{g\,(\text{on } m_{test},\text{ from } M)} = \frac{G*M*m_{test}}{r^2}\hat{r} \text{ Newtons}$$

2. Now remove the test-mass by dividing your answer by the value of the test-mass. As you perform this division, your answer will change from a 'force' (symbolized by $\vec{F}$) to a 'field' (symbolized by $\vec{g}$) and the units will change from 'Newtons,' to 'Newtons/kg.'

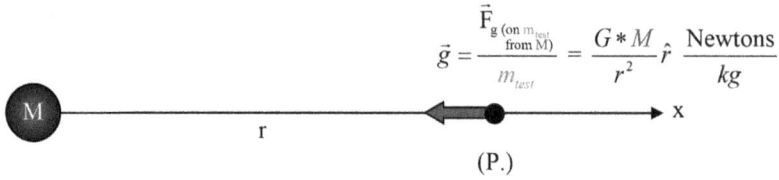

$$\vec{g} = \frac{\overset{\vec{F}_{g\,(on\,m_{test}\,from\,M)}}{}}{m_{test}} = \frac{G*M}{r^2}\hat{r} \quad \frac{Newtons}{kg}$$

M •————————————r————————————⬅•————————▶ x

(P.)

1.3.6 Modeling

❶ **EX 1:** A point-sized mass $m_1 = 3$ kg is placed at the origin. Another point-sized mass $m_2 = 4$ kg is placed at the location (8,0) m. No other masses are near these two masses, not even the Earth. To visualize this problem, imagine that it takes place in some remote location in outer space where we can assume all other masses in the Universe are infinitely far away. Calculate the gravitational field at the point long the x-axis that is midway between the two masses.

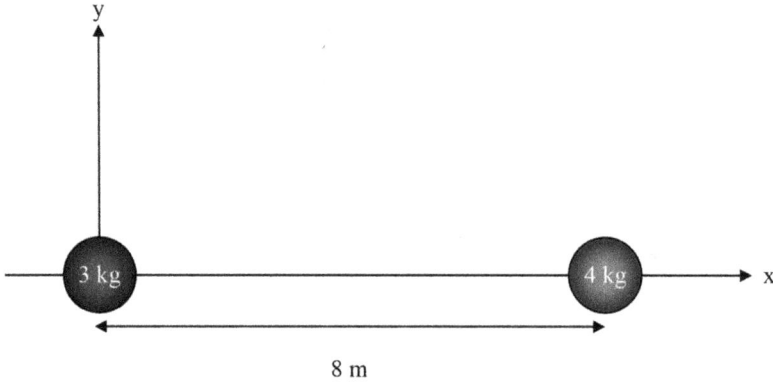

8 m

☑ **ANSWER:** Let us put a small grain of salt, with a mass denoted by m_{test} at the point (4,0) m. Now we need to compute the net gravitational force on m_{test} from both m_1 and m_2—notice that m_1 would pull a test-mass to the left while m_2 would pull is to the right:

$$\vec{F}_{g\,(on\,m_{test}\,from\,3\,kg)} = \frac{G*m_1*m_{test}}{r^2}\hat{r}$$

$$= \frac{\left(6.673 \times 10^{-11}\right)*3*m_{test}}{4^2}\hat{r}$$

$$= 1.25 \times 10^{-11}*m_{test}\,\hat{r}\ Newtons$$

$$= \left(-1.25 \times 10^{-11}*m_{test},0\right)\ Newtons$$

$$\vec{F}_{g\,(on\,m_{test}\,from\,4\,kg)} = \frac{G*m_2*m_{test}}{r^2}\hat{r}$$

$$= \frac{\left(6.673 \times 10^{-11}\right)*4*m_{test}}{4^2}\hat{r}$$

$$= 1.67 \times 10^{-11}*m_{test}\,\hat{r}\ Newtons$$

$$= \left(+1.66 \times 10^{-11}*m_{test},0\right)\ Newtons$$

To get the net gravitational force, just add these two vectors according to our rules for vector addition.

$$\vec{F}_{g\,(net\,on\,m_{test})} = \left(+4.18 \times 10^{-12}*m_{test},0\right)\ Newtons$$

To complete the gravitational field calculation, we simply divide-out the test-mass and convert our units into Newtons kg^{-1}:

$$\vec{g} = \frac{\vec{F}_{g\ (net\ on\ m_{test})}}{m_{test}} = (+4.18 \times 10^{-12}, 0) \text{ Newtons } kg^{-1}$$

☺ **COMMENTS:** Notice that the value of the test-mass never enters into the calculation; you can make it any value that you like, as long as you divide your final answer by that same number. Also, notice that the units of the gravitational field are ' Newtons kg^{-1}' and suggest interpreting the gravitational field as a 'force per unit of mass.' In the previous example, we determined that the gravitational field at the point (4,0) m was $\vec{g} = (+4.18 \times 10^{-12}, 0)$ Newtons kg^{-1}. This value suggests that every kilogram of mass located at the point (4,0) m, experiences a pull of 4.18×10^{-12} Newtons to the right. We can now ask: 'What would the force be on 2 kg of mass located at that point?' Obviously, 2 kg of mass would experience $2 \times 4.18 \times 10^{-12}$ Newtons or 8.36×10^{-12} Newtons to the right. Likewise, 3 kg of mass would experience a force of $3 \times 4.18 \times 10^{-12}$ Newtons or 1.25×10^{-11} Newtons to the right. Once the gravitational field is determined at a point (P.), we can easily multiply our answer by a mass to determine the force acting on that mass. Think of the gravitational field as a nifty vector quantity that tells us the **force per unit of mass** at a point (P.)—we can then simply multiply the gravitational field by a mass to determine the actual force acting on that mass. Let us try another example but this time, we'll actually choose a numeric value for our test-mass and divide it out to get our final answer.

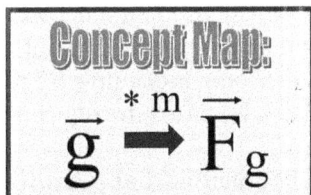

Concept Map:

$$\vec{g} \xrightarrow{\ *\ m\ } \vec{F}_g$$

❷ **EX 2:** A point-sized mass $m_1 = 3$ kg is placed at the origin. Another point-sized mass $m_2 = 4$ kg is placed at the location (8,0) m. Finally, a point (P.) is located at (0,6) m. At this time, no mass is located at point (P.) No other masses are near these two masses, not even the Earth. To visualize this problem, imagine that it takes place in some remote location in outer space where we can assume all other masses in the Universe are infinitely far away.
  a. Calculate the net gravitational field at point (P.).
  b. Calculate the net gravitational force on a 17 kg mass—if it were placed at point (P.).
  c. Calculate the net gravitational force on a 23 kg mass—if it were placed at point (P.).

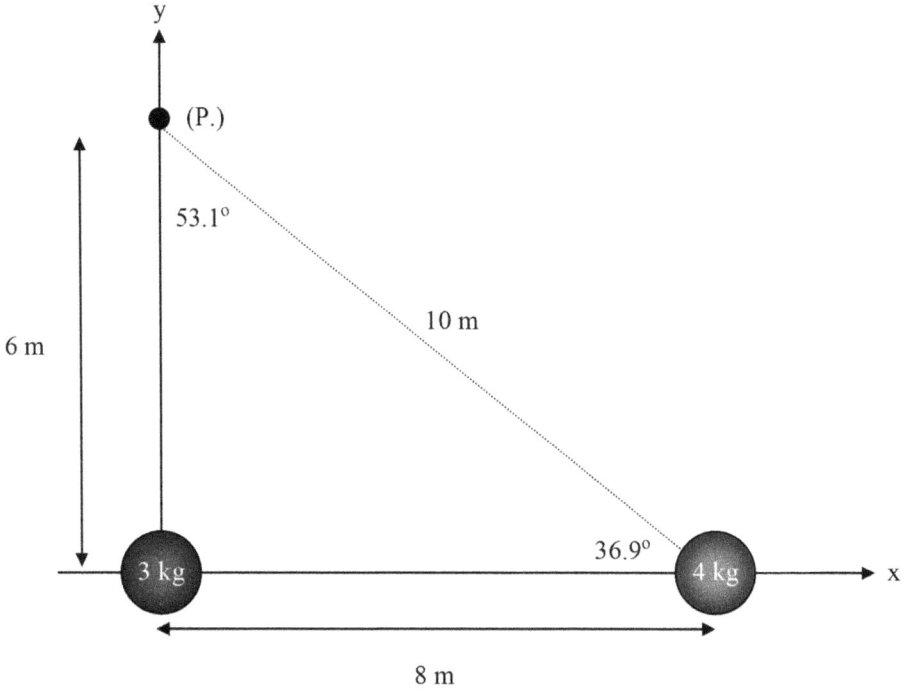

☑ **ANSWER:** Let us place a test-mass at point (P.) then determine the net gravitational force on that test-mass based on our ULG formula. We're free to place **any** test-mass at point (P.) as long as we divide it out as the last step in our recipe. Since we've already done this problem as example #2 (from the Modeling portion of the ULG section of this chapter), let us arbitrarily choose to put a 5 kg test-mass at point (P.):

   a. Calculate the net gravitational field at point (P.).

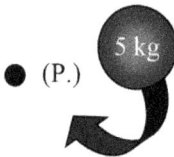

Imagine placing a 5 kg test-mass (*i.e.,* a bowling ball or stone) at the point (P.). The test-mass will "sample" the gravitational field at point (P.) for us—then we'll remove the test-mass by dividing it out of the final answer.

$$\vec{F}_{g\,(on\,5\,kg\atop from\,3\,kg)} = \frac{\left(6.673 \times 10^{-11}\right) * 3 * 5}{6^2}\,\hat{r}$$

$$= 2.78 \times 10^{-11}\,\hat{r}\ \text{Newtons}$$

$$= \left(0, -2.78 \times 10^{-11}\right)\ \text{Newtons}$$

The 3 kg mass attracts the 5 kg mass along the negative y-axis.

$$\vec{F}_{g\,(on\,5\,kg\atop from\,4\,kg)} = \frac{\left(6.673 \times 10^{-11}\right) * 4 * 5}{10^2}\,\hat{r}$$

$$= 1.33 \times 10^{-11}\,\hat{r}\ \text{Newtons}$$

The 4 kg mass attracts the 5 kg mass along direction of 36.9° below the positive x-axis.

$$= \left(1.33 \times 10^{-11} * \cos\left(36.9°\right), -1.33 \times 10^{-11} * \sin\left(36.9°\right)\right)\ \text{Newtons}$$

$$= \left(1.06 \times 10^{-11}, -7.99 \times 10^{-12}\right)\ \text{Newtons}$$

Therefore, the net force on the 5 kg test-mass is the vector sum: $\vec{F}_{g\,(total\,on\,5\,kg)} = (1.06 \times 10^{-11}, -3.58 \times 10^{-11})$ Newtons.

Now dividing the net force by 5 kg, the value of our test-mass, we get a value for the net gravitational field at point (P.):

$$\vec{g} = \frac{\vec{F}_{g\,(net\atop on\,5\,kg)}}{5\,kg} = \frac{(1.06 \times 10^{-11}, -3.58 \times 10^{-11})\ \text{Newtons}}{5\,kg}$$

$$= (2.13 \times 10^{-12}, -7.16 \times 10^{-12})\ \text{Newtons}\ kg^{-1}$$

b. Calculate the net gravitational force on a 17 kg mass—if it were placed at point (P.).

Simply multiply our answer in part (a) by 17 kg:

$$\vec{F}_{g\,(net\atop on\,17\,kg)} = (17\ \text{kg}) \times \vec{g}$$

$$= (17\text{kg}) \times (2.13 \times 10^{-12}, -7.16 \times 10^{-12})\ \text{Newtons}\ kg^{-1}$$

$$= (3.62 \times 10^{-11}, -1.22 \times 10^{-10})\ \text{Newtons}$$

c. Calculate the net gravitational force on a 23 kg mass—if it were placed at point (P.).

Simply multiply our answer in part (a.) by 23 kg:

$$\vec{F}_{g\,(net\atop on\,23\,kg)} = (23\ \text{kg}) \times \vec{g}$$

$$= (23\text{kg}) \times (2.13 \times 10^{-12}, -7.16 \times 10^{-12})\ \text{Newtons}\ kg^{-1}$$

$$= (4.90 \times 10^{-11}, -1.64 \times 10^{-10})\ \text{Newtons}$$

☺ **COMMENTS:** We now see the full usefulness of the gravitational field: For any system of masses, the gravitational field $\vec{g}$, at a point (P.) is the **force per unit mass** acting on an **imaginary** test-mass located at that point. If an actual mass m, is then placed at point (P.), the actual force on that mass is simply: $\vec{F} = m \times \vec{g}$.

❸ **EX 3:** A point-sized mass $m_1 = 3$ kg is placed at the origin. A point-sized mass $m_2 = 4$ kg is placed at the location (8,0) m. Another, a third point-sized mass $m_3 = 5$ kg is placed at the location (0,6) m. Finally, a fourth point-sized mass $m_4 = 6$ kg is placed at the location (5,7) m. No other masses are near these two masses, not even the Earth. To visualize this problem, imagine that it takes place in some remote location in outer space where we can assume all other masses in the Universe are infinitely far away.
  a. Calculate the net gravitational force on the 3 kg mass.
  b. Calculate the net gravitational field at the origin if the 3 kg mass were removed.
  c. Calculate the net gravitational force on a 44 kg mass if it were placed at the origin instead of the 3 kg mass.

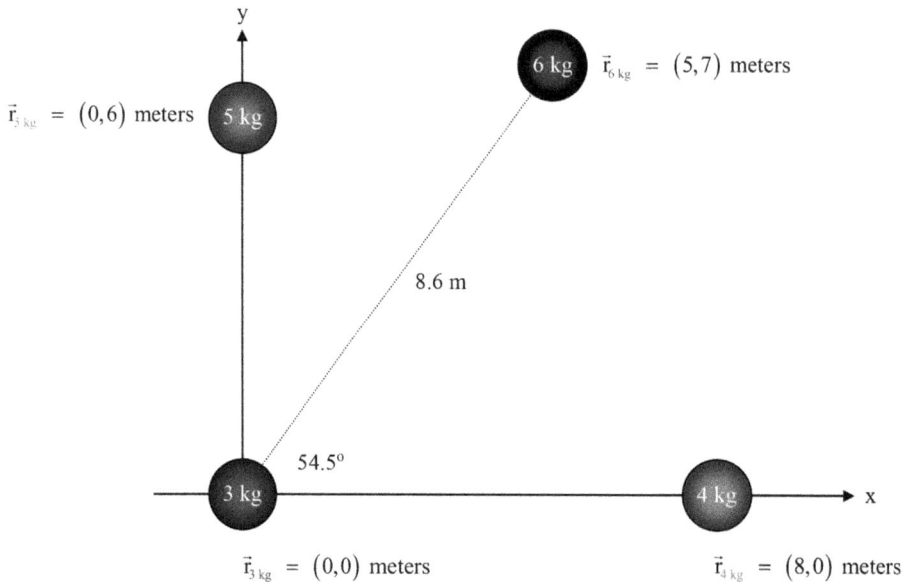

$\vec{r}_{5\,kg} = (0,6)$ meters 5 kg

6 kg $\vec{r}_{6\,kg} = (5,7)$ meters

8.6 m

54.5°

3 kg

4 kg

$\vec{r}_{3\,kg} = (0,0)$ meters

$\vec{r}_{4\,kg} = (8,0)$ meters

☑ **ANSWER:**
  a. Calculate the net gravitational force on the 3 kg mass.
      We'll use the ULG as we did in previous example. The 3 kg mass is attracted separately to the other three masses:

The 3 kg mass is attracted up towards the 5 kg mass.

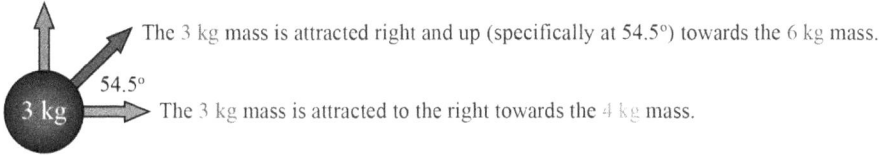

The 3 kg mass is attracted right and up (specifically at 54.5°) towards the 6 kg mass.

54.5°

3 kg

The 3 kg mass is attracted to the right towards the 4 kg mass.

$$\vec{F}_{g\,(on\,3\,kg\,from\,4\,kg)} = \frac{\left(6.673 \times 10^{-11}\right) * 3 * 4}{8^2}\,\hat{r}$$

$$= 1.25 \times 10^{-11}\,\hat{r}\text{ Newtons}$$

$$= \left(1.25 \times 10^{-11}, 0\right)\text{ Newtons}$$

The 3 kg mass is attracted to the 4 kg mass along the positive x-axis.

$$\vec{F}_{g\,(on\,3\,kg\,from\,5\,kg)} = \frac{\left(6.673 \times 10^{-11}\right) * 3 * 5}{6^2}\,\hat{r}$$

$$= 2.78 \times 10^{-11}\,\hat{r}\text{ Newtons}$$

$$= \left(0, 2.78 \times 10^{-11}\right)\text{ Newtons}$$

The 3 kg mass is attracted to the 5 kg mass along the positive y-axis.

$$\vec{F}_{g\,(on\,3\,kg\,from\,6\,kg)} = \frac{\left(6.673 \times 10^{-11}\right) * 3 * 6}{\left(8.6\right)^2}\,\hat{r}$$

The 3 kg mass is attracted to the 6 kg mass at 54.5° up from the positive x-axis.

$$= 1.62 \times 10^{-11}\,\hat{r}\text{ Newtons}$$

$$= \left(1.62 \times 10^{-11} * \cos\left(54.5°\right), 1.62 \times 10^{-11} * \sin\left(54.5°\right)\right)\text{ Newtons}$$

$$= \left(9.43 \times 10^{-12}, 1.32 \times 10^{-11}\right)\text{ Newtons}$$

Therefore, the net force is the vector sum:

$$\vec{F}_{g\,(total\,on\,3\,kg)} = (2.19 \times 10^{-11}, 4.10 \times 10^{-11})\text{ Newtons.}$$

b. Calculate the net gravitational field at the origin if the 3 kg mass were removed.

The gravitational field is simply the force per unit mass at a point (P.). Therefore, to determine the gravitational field at the origin, simply divide the force at the origin by the mass at that location:

$$\vec{g} = \frac{\vec{F}_{g\,(total\,on\,3\,kg)}}{3\text{ kg}} = \frac{(2.19 \times 10^{-11}, 4.10 \times 10^{-11})\text{ Newtons}}{3\text{ kg}}$$

$$= (7.3 \times 10^{-12}, 1.37 \times 10^{-11})\text{ Newtons kg}^{-1}$$

c. Calculate the net gravitational force on a 44 kg mass if it were placed at the origin instead of the 3 kg mass.

Now convert the gravitational field into a force by multiplying by mass:

$$\vec{F}_{g \text{ (total on 44 kg)}} = (44 \text{ kg}) \times \vec{g} = (44 \text{ kg}) \times (7.3 \times 10^{-12}, 1.37 \times 10^{-11}) \text{ Newtons kg}^{-1}$$
$$= (3.21 \times 10^{-12}, 6.03 \times 10^{-10}) \text{ Newtons}$$

1.3.7 Checking for understanding

QUESTION: A planet (it need not be the Earth) of mass M_p, revolves about the Sun which has a mass given by $M_\odot$. Let us assume the orbit is perfectly circular with a radius of R (pretend the planet is Venus with an elliptical eccentricity of $\varepsilon = 0.0068$, so the orbit is almost a perfect circle). By applying Newton's second law of motion to the orbiting planet, prove Kepler's third law of planetary motion; namely, the time for a planet to revolve once around the Sun is proportional to its mean distance from the Sun raised to the 3/2 power or: $T_{\text{one orbit}} \propto R^{3/2}$. Also, Kepler's third law of planetary motion is valid for any planet revolving about the Sun. Therefore, your proof should be independent of the orbiting planet's mass, M_p.

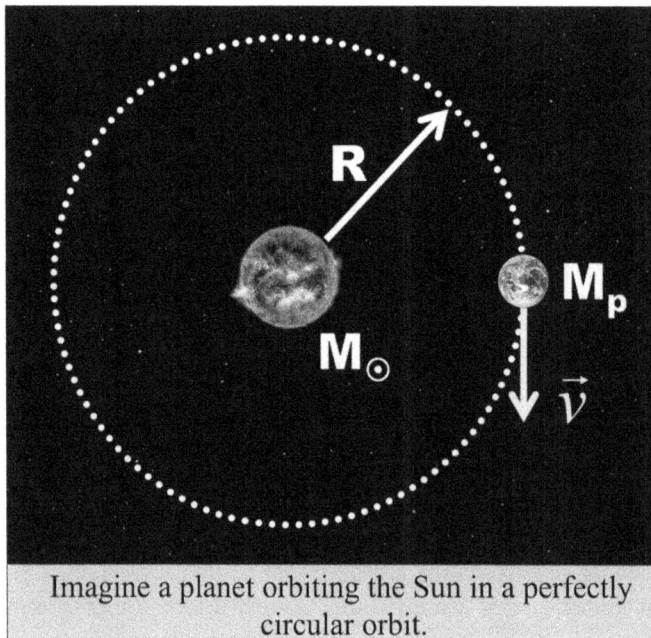

Imagine a planet orbiting the Sun in a perfectly circular orbit.

Earth Image Credit: The 'Blue Marble' from Apollo 17, Credit: NASA. Sun Image Credit: NASA.

Hints: To solve this problem, you will need information from many of the previous chapters:

- **Volume 1, chapter 1:** You will need to know how to write vectors.
- **Volume 1, chapter 2:** You will need to know the definition of speed.
- **Volume 1, chapter 4:** You will need to know Newton's second law of motion and the six common forces that are used on the left-hand-side of this law.
- **Volume 1, chapter 6:** You will need to know the formula for centripetal acceleration.
- **Volume 2, chapter 1:** You will need to know how to use Newton's ULG.

☑ **ANSWER:**

Start by applying Newton's second law to the orbiting planet:

$$\sum \vec{F} = M_p * \vec{a}$$

We know from volume 1, chapter 4 that the left-hand-side of Newton's second law of motion involves terms from 6 common forces—3 applied forces, 2 contact forces, and 1 field force.

$$\overbrace{\vec{P} + \vec{T} + \vec{F}_{sp} + \vec{f} + \vec{N} + \vec{W}}^{\sum \vec{F}} = M_p * \vec{a}$$

The only force acting on the planet is the gravitational force.

$$\vec{W} = M_p * \vec{a}$$

We now know the explicit form of the gravitational force as given by Newton's ULG.

$$\frac{G * M_p * M_\oplus}{R^2}\hat{r} = M_p * \vec{a}$$

As drawn, gravity pulls the planet to the left so the ULG formula becomes a vector directed to the left.

$$\left(-\frac{G * M_p * M_\oplus}{R^2}, 0 \right) = M_p * \vec{a}$$

From volume 1, chapter 6, we know that for circular motion, we may replace the acceleration with v^2/R directed toward the center of the circle.

$$\left(-\frac{G * M_p * M_\oplus}{R^2}, 0 \right) = M_p * \left(-\frac{v^2}{R}, 0 \right)$$

Equate the x-components and cancel M_p and one factor of R on both sides.

$$\frac{G * M_\oplus}{R} = v^2$$

The last step in the proof is to make a nifty substitution for the value of the speed. The speed, by definition from volume 1, chapter 2, is just the total distance traveled by the planet in an interval of time. In the time the planet takes to travel once around the Sun T, the planet travels the circumference of the circle $2\pi R$. Thus, we may make the substitution: $v = 2\pi R / T_{\text{one orbit}}$.

$$\frac{G * M_{\oplus}}{R} = \left(\frac{2\pi R}{T_{\text{one orbit}}} \right)^2$$

Solve for $T_{\text{one orbit}}$.

$$T_{\text{one orbit}} = \sqrt{\frac{4\pi^2 R^3}{G * M_{\oplus}}}$$

$$T_{\text{one orbit}} = \sqrt{\frac{4\pi^2}{G * M_{\oplus}}} * R^{3/2} \quad \checkmark$$

Kepler's Third Law of Planetary Motion—independent of M_p!

☺ **COMMENTS:** Notice that the final result is independent of the mass of the planet, as we required. Also, one can't help but marvel at Kepler's tenacity in determining his Third Law of Planetary Motion. Remember, Kepler's Third Law was published in 1619, almost 70 years before Newton's *Principia*. Even without Newton's second law of motion or the Universal Law of Gravitation, Kepler was able to determine this law based entirely on observation. Today, physicists and astronomers use Kepler's Third Law to determine the mass of planets, asteroids and newly-discovered stars. For example, let us say you discover a new star and track the motion of a planet orbiting this star. If you can measure the time required for the planet to complete one of its orbits and also measure the average distance between the star and planet, you can use Kepler's Third Law to determine the star's mass:

$$M_{\text{STAR}} = \frac{4\pi^2 \times R^3}{G \times T_{\text{one orbit}}^2}.$$

1.4 Keeping information

1.4.1 Closure

The anticipatory set that kicked-off this chapter involved our observing of various patterns in the night sky ... the 'analemma of the Sun' and the 'retrograde of Mars.' Also, as part of our discussion of the history of modern astronomy, we listed Kepler's three laws of planetary motion. Let us close this chapter by examining a problem that demonstrates how Kepler's laws help explain some of the patterns we observe in the night sky.

❶ **EX 1:** Below is a series of snapshots taken of our Solar System by an alien visitor from a distant galaxy. The photograph shows the positions of the Earth and Mars at seven **equally-spaced** instants of time, as they orbit the Sun. For instance, the locations of the Earth and Mars at time #1 are labeled by the number '1,' at time #2 by the number '2,' etc. The locations of ten fixed stars are also captured on the photograph; however, since only the planets move from month-to-month (remember, the word 'planet' translates into 'wanderer'), the positions of the ten stars remain fixed. Use the photograph to demonstrate how an observer on Earth, viewing Mars against the backdrop of the fixed stars, would see Mars trace out a retrograde motion over the course of the seven moments captured by the alien's photograph.

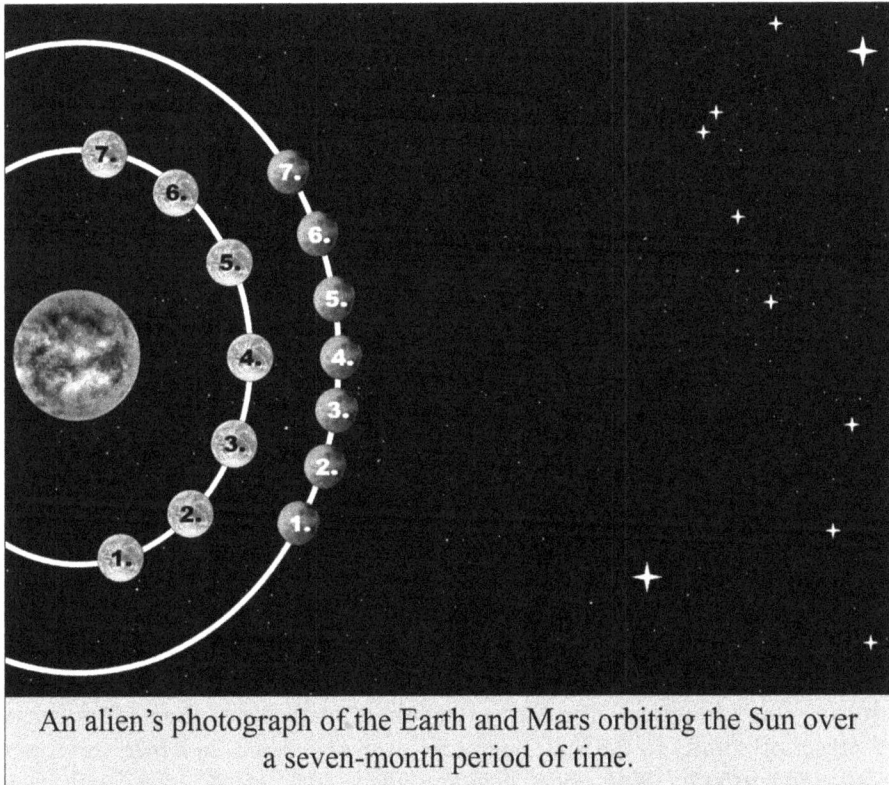

An alien's photograph of the Earth and Mars orbiting the Sun over a seven-month period of time.

Earth Image Credit: The 'Blue Marble' from Apollo 17, Credit: NASA. Mars Image Credit: NASA/JPL-Caltech.

☑ **ANSWER:** First of all, notice how all three of Kepler's laws of planetary motion are neatly encompassed in the photograph:

- **Law #1 (published in 1609):** Planets move in elliptical orbits about the Sun with the Sun located at one focus.
  Notice that both Earth and Mars orbit the Sun in elliptical orbits
- **Law #2 (published in 1609):** An imaginary line connecting a planet to the Sun sweeps out equal areas in equal time intervals.

Notice that both the Earth and Mars are orbiting the Sun with velocities that change over the course of their orbits. For instance, between times #1 and #2, Earth is far from the Sun and travels with a slow orbital velocity—it travels a certain distance between times #1 and #2. However, between times #3 and #4, Earth is much closer to the Sun and travels with a faster orbital velocity than in did between times #1 and #2—it covers more distance when it is closer to the Sun.

- **Law #3 (published in 1619):** The time for a planet to revolve once around the Sun is proportional to its mean distance from the Sun raised to the 3/2 power or: $T_{one\ orbit} \propto R^{3/2}$.

Notice that the farther a planet is from the Sun, the slower it orbits the Sun. Clearly, since Mars is farther from the Sun than the Earth, Mars is traveling much slower than the Earth—Earth has almost traced out half of its orbit over the seven months while Mars has barely completed a quarter of its orbit in the same amount of time.

To see how Kepler's three laws of planetary motion demonstrate the retrograde of Mars, imagine that an observer on the Earth looks up into the night sky and locates Mars against the backdrop of the fixed stars. At time #1, Mars is ahead of Earth in their respective orbits and if the Earth and Mars were in a track race, you'd say that Mars 'was in the lead.' The observer's line of sight would project Mars to be located next to the lower grouping of the fixed stars in our photograph. You can use a ruler and draw a line from the Earth (at time #1), through Mars (at time #1), and into the field of fixed stars. Label the endpoint of your line with the number '1' as shown on the next drawing. Repeat this process for times #2 through #7.

From times #1 through #3, an observer on the Earth would see Mars 'ahead' of the Earth and the projection of Mars against the backdrop of the fixed stars would continue to move forward from week to week. However, at time #4, an interesting thing happens—the Earth (traveling much faster than Mars because of its closer proximity to the Sun) begins *to 'overtake' Mars. In a track race, you'd say that the earth and Mars were 'neck and neck.'* At time #5, the Earth clearly pulls into the lead and an observer on the Earth would have to look back along his line of sight to see Mars. In our planetary track race, we'd say that the Earth 'has taken the lead.' Mars' projection against the backdrop of the fixed stars would begin to move backwards—or 'retrograde.' Eventually, at times #6 and #7, the projection of Mars against the backdrop of the fixed stars would again move forward.

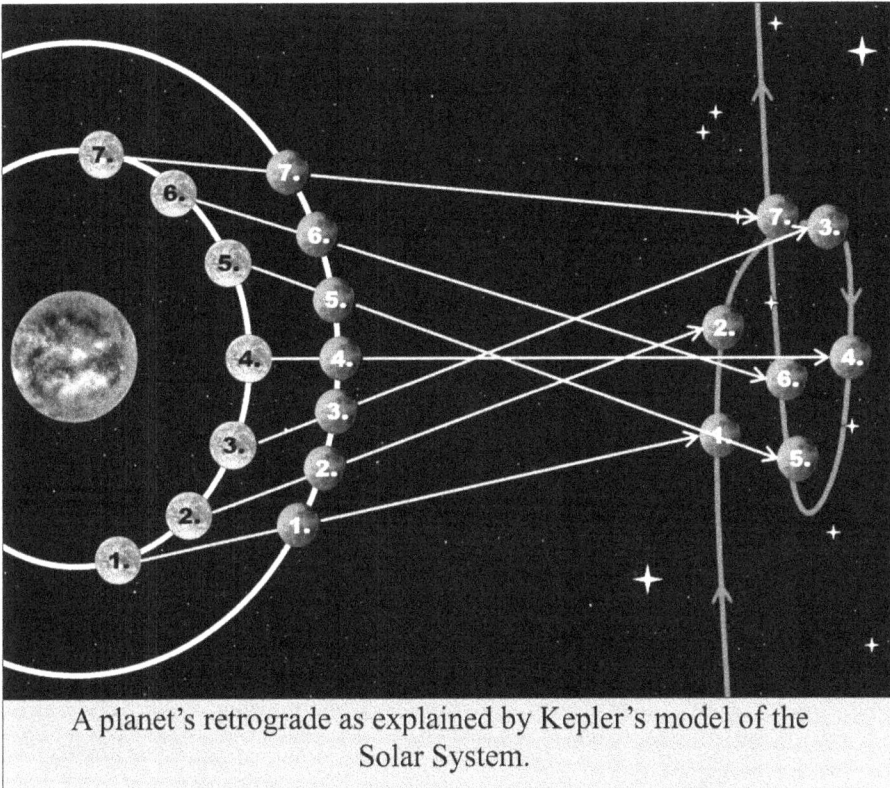

A planet's retrograde as explained by Kepler's model of the Solar System.

Earth Image Credit: The 'Blue Marble' from Apollo 17,
Credit: NASA. Mars Image Credit: NASA/JPL-Caltech.

☺ **COMMENTS:** What about the 'analemma of the Sun?' The Earth is closest to the Sun (perihelion) in early January, about two weeks after the winter solstice. On the other hand, the Earth is farthest from the Sun (aphelion) in early July, about two weeks after the summer solstice. According to Kepler's Third Law of Planetary Motion, when the Earth is near perihelion, less than 24 h occur between when the Sun is directly overhead (astronomical noon) on two consecutive days. When the Earth is far from the Sun, the time between astronomical noons on two consecutive days is slightly more than 24 h. Over the course of the year the time between consecutive astronomical noons evens out, 24 h is the average length of a solar day. The sum of these slight offsets causes the Sun to *become 'ahead' of our clocks for part of the year, and 'behind'* our clocks for part of the year. The Sun is usually not directly overhead at noon, and sunrise and sunset are usually not perfectly centered on 6 o'clock. Combining this offset with the angle north or south of the equator that the Sun is at on a given day and the length of the day due to the season yields the earth's analemma. Essentially, the analemma is a figure that shows where in the sky the Sun would be at noon throughout the course of the year for an observer at a fixed point. By taking a picture of the Sun from a camera in the same place at fixed intervals, you would find that the Sun traces out the pattern of an analemma, as shown in our Anticipatory Set.

1.4.2 Independent practice

1. Using the planetary data for Mercury, Venus, Earth, Mars, Jupiter, and Saturn shown in the section on the History of Modern Astronomy (when discussing the biography of Johannes Kepler and his Third Law of Planetary Motion), plot the period of each planet (T) along the y-axis and the mean distance of each planet (R) along the x-axis. What does this tell you about the relationship between T and R?

2. Using the planetary data for Mercury, Venus, Earth, Mars, Jupiter, and Saturn shown in the section on the History of Modern Astronomy (when discussing the biography of Johannes Kepler and his Third Law of Planetary Motion), plot the period of each planet (T) along the y-axis and the 3/2 power of the mean distance of each planet ($R^{3/2}$) along the x-axis. What does this tell you about the relationship between T and R?

3. In a remote region of space, very far from any planets or stars, a device is made which has 1 kg at the point $(0, 0)$m; 2 kg at the point $(2, 2)$m; and 3 kg at the point $(-3, 3)$m. What is the net gravitational force on the mass at the origin?

4. You have a mass m (i.e., all answers will be in terms of m) and decide to check your weight as you hop around the galaxy. You are also given the following astronomical data:

Object	Mass (kg)	Radius (m)
Earth	5.98×10^{24}	6.38×10^6
Mars	6.42×10^{23}	3.37×10^6
Saturn	5.68×10^{26}	5.85×10^7
Sun	2.00×10^{30}	7.00×10^8

Let us check your weigh at different places on Earth and in the Solar System:

a) Denver is exactly 1 mile (1610 m) above sea level, what is your weight in Denver?

b) Mount Everest is exactly 29 035 ft (8850 m) above sea level, what is your weight there?

c) The Challenger Deep is a Pacific Trench exactly 2 miles (11 033 m) below sea level, what is your weight in the Trench?

d) Is the gravitational acceleration of the Earth really constant at 9.8 m s^{-2}?

e) What is your weight on Mars?

f) What is your weight on Saturn?

g) What is your weight on the surface of the Sun?

5. A friendly elderly student named Perry (mass = 70 kg) stands on the surface of the Earth. He is exactly 1 m from Fred (mass = 60 kg) who is to his right; 0.5 m from Tom (mass = 65 kg) who is to his left; and 1.0×10^{17} from the Northern Star (mass = 1×10^{30} kg) which is directly overhead. What is Perry's weight?

6. Is the gravitational acceleration of the Earth really constant at 9.8 m s^{-2}?

7. Given: Mass of the Earth = 5.98×10^{24} kg and radius of the Earth = 6.38×10^6 m. Assume that a student named Craig has a mass of 50 kg.

    a) Calculate the force of gravity between Craig and the Earth if he were standing on the surface of the Earth.

    b) What is the acceleration due to gravity at the surface of the Earth?

    c) Calculate the force of gravity between Craig and the Earth if he were standing in downtown Mexico City which is 4 miles above sea level (keep a consistent set of units).

    d) What is the acceleration due to gravity at Mexico City?

8. Given: Mass of the Sun = 2.0×10^{30} kg and radius of the Sun = 7.0×10^8 m; Assume that a friendly dog named Snoopy has a mass of 40 kg.

    a) Calculate the force of gravity between the Snoopy and the Sun if he were standing on the surface of the Sun.

    b) What is the acceleration due to gravity at the surface of Sun?

9. A research satellite, having a mass of 2500 kg, is in a circular orbit 3.20×10^5 m above the Earth's surface.

    a) Find the force of gravity on the satellite.

    b) What is the acceleration of gravity at this altitude?

1.4.3 Peer teaching

Below is a sampling of '**Ah-ha! Moments**' from a recent introductory physics course taught at the college level. Following each comment is my remark in red.

1.4.3.1 Universal law of gravitation and weight

1. *The ULG formula only determines the magnitude of the gravitational force, the direction of the gravitational force must be determined from the geometry of the problem.*

    Correct—this is again why we spent so much time in volume 1. chapter 1 perfecting the conventions for vector-notation. The ULG formula indeed-only numerically gives us the magnitude of the gravitational force between two masses. However, our ULG formula also contains the r (spoken as 'r hat') symbol which is a **reminder** to us that the direction of the gravitational force must be part of our answer and must be determined from the geometry of the problem. For any given problem, you simply determine the distance between two objects as well as the angle of the line joining those two objects (relative to your coordinate system). Once these two quantities are known, you can then write the gravitational force by employing the conventions we established in volume 1, chapter 1 for vector-notation.

2. *Big-G, appearing in the ULG formula, is indeed 'universal'—it remains the same constant for every problem and at every location in the Universe. Little-g, the acceleration due to gravity on the surface of the Earth, is not 'universal'—it changes depending on your location on the surface of the Earth or some other place in the Universe.*

   Correct—the reason why Newton's ULG is called his 'Universal' Law of Gravitation is that the formula remains the same for any two masses, at any location in the universe. **G ALWAYS** has the value of 6.673×10^{-11}N m^2 kg^{-2}. The acceleration due to gravity, what we called g=9.8 m s^{-2} in volume 1, chapter 4, is not 'universal.' The value of g = 9.8 m s^{-2} is only valid near the surface of the Earth—and we saw in one of the modeling sections of this chapter that even g varies slightly as we move from Cleveland to Denver, to Mount Everest, to the Challenger Deep. When we visit another planet, the value of g changes completely. The ULG is valid everywhere! The value of g (determined bythe ULG) changes with location!

1.4.3.2 The gravitational field

3. *Does the gravitational field—this notion of an invisible field that exists at all places and all times that allows two masses to exert a force via 'action at a distance'—really exist? It seems a bit far-fetched.*

   Good question. The best way to answer this question is to simply state that the concept of the gravitation 'field' is just that—it's a concept! It's a model that physicists use to explain a physical phenomenon that definitely exists but that is difficult to grasp. If you wonder how two objects can pull on one another without even being in contact or connected by a surrounding medium, you'll start to realize that such a phenomenon is unlike anything we experience in everyday life. Thus, explaining 'action at a distance' is completely un-intuitive to us. Therefore, we construct a model to help us explain a physical phenomenon we know **does happen** since we experience the Earth pulling on us even though we may not be in contact with it. We imagine that every mass—as part of the definition of being a mass—carries a field with it that exists everywhere and always has. When another mass interacts with this field, the two masses pull on one another. The model explains what we indeed observe and it assists usin visualizing how the phenomenon can occur. Does it really exist? Who knows?

Chapter 2

The conservation laws

The concept of 'conservation laws' is first addressed in a typical course on *Classical Mechanics*. The first two of these conservation laws, *'The Conservation of Energy'* and *'The Conservation of Linear Momentum,'* are addressed in this chapter as well as the motivation for studying conservation laws. Two new physical quantities, *'work'* (or *'energy'*) and *'linear momentum,'* are defined and discussed. Because these quantities are conserved under certain circumstances, the motion of objects in a new set of everyday physical situations can be analyzed. Emphasis is placed on the vector *'Dot-Product,'* a method of multiplying the parallel components of two vectors. Two problem-solving strategies are described and modelled through a number of illustrative problems.

Anyone who can walk to the welfare office can walk to work.

—Al Capp

Work is the refuge of people who have nothing better to do.

—Oscar Wilde

Work is the curse of the drinking class.

—Oscar Wilde

Pleasure in the job puts perfection in the work.

—Aristotle

Love and work are the cornerstones of our humanness.

—Sigmund Freud

doi:10.1088/978-0-7503-6402-7ch2

If A equals success, then the formula is: A = X + Y + Z, X is work. Y is play. Z is keep your mouth shut.

—Albert Einstein

I find that the harder I work, the more luck I seem to have.

—Thomas Jefferson

There is no substitute for hard work.

—Thomas Edison

The reason why worry kills more people than work is that more people worry than work.

—Robert Frost

It is your work in life that is the ultimate seduction.

—Pablo Picasso

Work saves us from three great evils: boredom, vice and need.

—Voltaire

Nobody realizes that some people expend tremendous energy merely to be normal.

—Albert Camus

Nature abhors a hero. For one thing, he violates the law of conservation of energy. For another, how can it be the survival of the fittest when the fittest keeps putting himself in situations where he is most likely to be creamed?

—Solomon Short

Make it a rule of life never to regret and never to look back. Regret is an appalling waste of energy; you can't build on it; it's only for wallowing in.

—Katherine Mansfield

I merely took the energy it takes to pout and wrote some blues.

—Duke Ellington

Success in almost any field depends more on energy and drive than it does on intelligence. This explains why we have so many stupid leaders.

—Sloan Wilson

A positive attitude may not solve all your problems, but it will annoy enough people to make it worth the effort.

—Herm Albright

Life is an effort that deserves a better cause.

—Karl Kraus

Theology is the effort to explain the unknowable in terms of the not worth knowing.

—H L Mencken

When something can be read without effort, great effort has gone into its writing.

—Enrique Jardiel Poncela

Effort only fully releases its reward after a person refuses to quit.

—Napoleon Hill

There is no expedient to which a man will not go to avoid the labor of thinking.

—Thomas A Edison

A man is not idle because he is absorbed in thought. There is a visible labor and there is an invisible labor.

—Victor Hugo

He who labors diligently need never despair; for all things are accomplished by diligence and labor.

—Menander

2.1 Motivation

In volume 1, chapter 1, we developed some mathematical tools. The focus of that chapter was to develop our conventions for scalar notation and vector notation.

In volume 1, chapter 2, we defined four kinematic quantities (speed, velocity, acceleration, and jerk) that will be used throughout this series to describe how objects move.

In volume 1, chapter 3, we remained focused on describing **how** objects move and developed the UAM technique for solving problems. We saw that this technique involved 3 vector equations and is most applicable to 'ideal situations,' short-term accelerations, and free-falling objects (i.e., projectile motion). This was our first technique allowing for numerical analysis of motion.

In volume 1, chapters 4 and 5, we tackled the question of **why** objects move. Thanks to Isaac Newton and his three laws of motion, we saw that objects move because a net external force exists on them. This was our second technique allowing for numerical analysis of motion.

In volume 1, chapter 6, we adapted Newton's laws of motion to the special situation of 'Uniform Circular Motion' or '**UCM**.' In a sense, UCM is a sub-strategy of Newton's laws of motion. The focus of volume 1, chapters 4 and 5 was the general motion of objects when they are subjected to various common forces while volume 1, chapter 6 was a specific case of objects moving in a circle.

In volume 2, chapter 1, we digressed a little bit from our problem-solving strategies to examine more closely the origin of the gravitational field force that was introduced earlier in the series. We found that the formula for 'weight' that we have been using from volume 1 is really a simplification of a much more elaborate law; namely, Newton's 'Universal Law of Gravitation' (abbreviated as the 'ULG'). The ULG determines the gravitational interaction of any two bodies—whether they are on the surface of the Earth or not.

In volume 2, chapter 2, we dive right back into our goal of developing problem-solving strategies. We will define two new physical quantities called 'work' (also known as 'energy') and 'momentum' and see how their conservation (i.e., their knack for staying fixed) will enable us to analyze a new set of everyday, but interesting, physical situations. Before we define these new quantities, however, we need to develop a new mathematical operation—a method for multiplying two vectors.

2.2 Getting ready

2.2.1 Anticipatory set

If someone said to you, 'This pile of food must be conserved,' or 'The total amount of money in the bank must be conserved,' would you know what that person meant? The purpose of this anticipatory set is to build your intuition associated with the words 'conserved' and 'conservation' so that the words convey some type of physical meaning to you. In the context of a physics course, the words 'conserved' and 'conservation' must convey a very specific physical insight to you—one that we wish to develop with this anticipatory set.

Physicists usually do not discuss the concept of conservation with a rigid definition. Don't expect to see a definition such as, 'Conservation is the state of an object in which ...' Instead, when physicists say something is 'conserved,' they simply mean that something 'doesn't change with time.' In other words, a conserved quantity is something that *stays constant in time!* The take-home message here is that the words 'conserved' and 'conservation' always refer to *time* (i.e., 'What it is now ... is what it will be later.'). A conserved quantity can change from place to place, but not from 9:00 am to 10:15 am.

Now that you know what the word 'conserved' means in the context of a physics problem, you probably already have a good sense of how such quantities might be able to help you. Let's take a look at a list of 'conserved' and 'non-conserved' quantities that were compiled by a class of undergraduate sophomores and juniors. The students were asked to make a list of things which 'change in time' and things that 'remain fixed in time.' The items listed by the students could come from any area of their lives—the items did not have to somehow connect to physics, mathematics, or a science class. After each item, the students were also asked to briefly explain why they thought the item 'changed' or 'remained fixed' in time. After looking at this list, we'll then briefly discuss how these notions translate to a physics context.

Non-Conserved Quantities—i.e., things that change in time:
- **Some of my friends**—'It seems like every week, someone is fighting with me or angry about something I did. My friendships seem to change week by week.'
- **My life**—'Every day, something happens to me that I didn't expect. Nothing in my life stays constant.'
- **The weather**—'The weather is so unpredictable. One day it's sunny, the next it's pouring rain. The weathermen can't even accurately predict the weather.'
- **My college classes**—'No matter how hard I study, I keep struggling with certain classes. I never have had an easy semester and every week is so unpredictable.'
- **Time itself**—'Time itself is not conserved. By definition, time changes over time. One minute, the time is 3:00 pm; the next minute, it's 3:01 pm; the next minute, it's 3:02 pm; etc.'

Conserved quantities—i.e., things that remain fixed in time:
- **The Sun**—'The Sun always rises in the east and sets in the west.'
- **My favorite sports team**—'They always end up losing.'
- **My Mom**—'I can always count on my Mom to come through for me.'
- **Gravity**—'What goes up, always comes down.'
- **The days of the week and the months of the year**—'Tuesday always follows Monday. September always follows August.'

Take a moment to add your own ideas to the lists above—think of any aspect of your life or the world around you in which you believe *things 'remain fixed over time' or 'change over time.'* Like the list above, provide a brief, one- or two-sentence, rationale for why you believe these quantities behave as they do.

As you can see, the previous lists cover a diverse range of areas and interactions. For instance, 'some of my friends,' 'my life' and 'my Mom' deal with interpersonal relationships. The 'Sun,' 'gravity' and 'time' are physical quantities that would seem to pop up in a science course. 'My favorite sports team' and 'my college classes' deal with social settings. Some of these examples may seem humorous, but they all nicely illustrate the point that we want to make. **Namely, conserved quantities are predictable ... non-conserved quantities are not predictable!** When something is conserved, it remains fixed in time and we can therefore predict its value at a later time. Because the Sun is conserved in certain ways (i.e., the rotation of the Earth guarantees the Sun will always rise and set in certain predictable locations), we know that tomorrow, it will rise in the east and set in the west. We also know that 20 years from today, it will again rise in the east and set in the west. The Sun's predictability means that we can count on its behavior over time and therefore possibly use this predictability to our advantage —we can plan the direction our new house will face, the orientation of a new baseball stadium, or what the sun-glare during the morning rush hour will be.

Some of our friendships, on the other hand, are non-conserved. We can't always count on some of our friends to even talk to us tomorrow, let alone 20 years from now. The unpredictability of some of our friends means that we simply cannot rely on their behavior over time. Think about how useful the conserved quantities are in your life—ultimately, you can make plans based on the predictability of the Sun, but not on your friends.

Now let's take a look at how this notion of 'conservation' will translate into a physics setting. As you would probably guess, most physical quantities change over time, especially when we deal with objects in motion. In volume 1, chapter 2, we saw that an object's position changes in time (giving us a velocity); an object's velocity changes in time (giving us an acceleration); and an object's acceleration changes in time (giving us a jerk). Thus, position, velocity and acceleration are all quantities that are typically not conserved—we simply cannot assume they will remain unchanged over time. Wouldn't that have been nice? We could determine an object's velocity at one instant and simply use that value at any other time during a problem. In volume 1, chapter 3, we saw that we can assume, in certain situations, that the acceleration of an object is constant (i.e., the UAM assumption). Since the acceleration is constant, we could use its fixed value at any time during the problem. The UAM assumption makes the acceleration a *predictable quantity* that we can use to our advantage when calculating an object's motion. In volume 1, chapter 4, we saw that forces (i.e., $\vec{P}$, $\vec{T}$, $\vec{N}$, $\vec{f}$, $\vec{F}_{sp}$, and $\vec{W}$) cause objects to accelerate. Thus, forces cause an object's position and velocity to change. Again, we cannot simply calculate an object's position or velocity and use them at any time in a problem. In volume 1, chapter 6, every object we considered was moving in a circular path. These objects were changing position and velocity at every instant during the problems so we know these quantities were changing over time.

Unfortunately, we see that in most of the previous problem-solving techniques that we developed, the physical quantities that we are interested in, like position, velocity and acceleration, change over time. In a sense, we can easily be discouraged when calculating these quantities—to invest all of our computational time and effort, and then realize that our answers are only valid for a split instant in time. However, as we are about to see, physicists have determined several physical quantities that actually remain fixed over time. Physicists say these quantities are 'conserved' and we say that these quantities obey 'Conservation Laws.' We may need a long, tedious problem-solving strategy to calculate these quantities; but once we have them, they remain fixed over time— ready for us to exploit any time we wish. In this chapter, you will see that the physical quantity of 'energy' (also called 'work') and 'momentum' are two of these magical, conserved quantities. Other conserved quantities exist but we will not explore them in this series. The conserved quantities of 'energy' and 'momentum' are powerful tools to add to our arsenal of problem-solving strategies. These two quantities will indeed allow us to describe a number of physical situations.

2.2.2 Objective

By the end of this chapter, you will be able to:

- Multiply two vectors using the vector 'Dot'-product.
- Interpret the physical meaning of the vector 'Dot'-product.
- Define the 'Work' (or 'Energy') done by a constant force.
- State the 'Law of Conservation of Energy' and use this conservation law as a problem-solving technique.
- Define the 'Linear Momentum' of a moving object.
- State the 'Law of Conservation of Momentum' and use this conservation law as a problem-solving technique.

2.2.3 Purpose

This information is needed:

- Because the vector 'Dot'-product will give us additional practice in using our vector notation (which will help us gain speed and accuracy in manipulating vectors), but more importantly, the vector 'Dot'-product will now enable us to calculate how much of one vector is *parallel* to another vector (i.e., what fraction of one vector is parallel to the direction of another vector).
- Because conservation laws are some of the most important and widely-used problem-solving techniques used by physicists. Perhaps more than any other problem-solving technique we will discuss, physicists use (or at least consider using) the 'Law of Conservation of Energy' or the 'Law of Conservation of Momentum' to tackle problems.

2.3 Giving information

2.3.1 Instructional input

2.3.1.1 Multiplying vectors by scalars

In volume 1, chapter 1, we spent a great deal of time discussing vector notation and the rules for vector addition and subtraction. We also spent a great deal of time practicing vector manipulation so we could acclimate ourselves to seeing how vectors may arise in certain physical situations—essentially, we used volume 1, chapter 1 to get *comfortable* with adding and subtracting vectors. We then saw in subsequent chapters that this investment in time quickly paid off! Vectors pop up in a bunch of physical situations (i.e., projectile motion, Newton's laws of motion, circular motion, etc) and we'd better get used to not only writing them, but adding and subtracting them as well. Once we were comfortable adding and subtracting vectors, we are able to develop numerous problem-solving techniques which describe various physical situations.

However, maybe you wondered: 'What about vector multiplication and division? '

We left volume 1, chapter 1 without exploring the possibility of vector multiplication or division. The reason is simple: because we did not need these types of vector operations until now!

In the simplest case, we already know how to multiply and divide vectors by scalars. We have been doing so throughout all the chapters in this series:

►**Multiplying a vector by a scalar:** If $\vec{A} = (A_x, A_y)$ and k is a scalar, then $k \cdot \vec{A} = (k \cdot A_x, k \cdot A_y)$. This is nothing new—we simply multiply the components of $\vec{A}$ by the scalar k, making sure we keep x and y components separate. We have already done this for problems involving UAM, Newton's laws of motion, Uniform Circular Motion, and the Universal Law of Gravitation.

❶ **EX 1:** Let's look at a problem involving Newton's laws of motion. Pretend we have solved for the acceleration of a moving ball and now we want to multiply our answer by its mass of 5 kg—in other words, we are performing the right-hand-side calculation of Newton's second law of motion.

If $\vec{a} = (3, 4)$ m s^{-2}, what is 5 kg$\cdot\vec{a}$?

☑ **ANSWER:** Simply multiply each component of the acceleration by the scalar 5 kg:

$$5 \text{ kg}\cdot\vec{a} = 5 \text{ kg}\cdot(3, 4) \text{ ms}^{-2} = (15, 20) \text{ Newtons.}$$

► **Dividing a vector by a scalar:** If $\vec{A} = (A_x, A_y)$ and k is a scalar, then $\vec{A} \div k = (A_x/k, A_y/k)$. Even though this situation never actually appeared in our previous problem-solving techniques, the methodology is nothing new—again, we simply divide the components of $\vec{A}$ by the scalar k, making sure we keep x and y components separate.

❷ **EX 2:** Let's look at a problem involving Newton's motion laws, run in reverse. Pretend we have solved for the force acting on a moving ball and now we want to divide our answer by its mass of 5 kg so we can state its acceleration—in other words, we are performing an $\vec{a} = \vec{F}/m$ calculation of Newton's second law of motion.

If $\vec{F} = (3, 4)$ Newtons, what is $\vec{a} = \vec{F}/m$?

☑ **ANSWER:** Simply divide each component of the force by the scalar 5 kg:

$$\vec{a} = \frac{(3, 4) \text{ Newtons}}{5 \text{ kg}} = \frac{1}{5 \text{ kg}} \times (3, 4) \text{ Newtons} = \left(\frac{3}{5}, \frac{4}{5}\right) \text{ m s}^{-2} = (0.6, 0.8) \text{ m s}^{-2}.$$

2.3.1.2 Multiplying vectors by vectors: the vector 'dot'-product
In this chapter, we are now going to turn our attention to multiplying or dividing two vectors. To start off on an easy note, you'll be happy to know that you cannot divide two vectors. Period. End of story. You cannot divide one vector by another vector! We need not explore vector division any further.

Rule for vector division:
You cannot divide a vector by another vector!

What about multiplication? First of all, notice that normally, when you multiply two numbers (i.e., two scalars), you can use the symbol of the 'Dot' interchangeably with the symbol of the 'Cross' (or asterisk). Aren't the following two lines completely equivalent ways of representing the multiplication two numbers?

The Interchangeability of the symbols of the 'Dot' and 'Cross' when Multiplying Two Scalars:

'Dotting' the numbers 4 and 5 to get 20: $4 \cdot 5 = 20$, or $4 \cdot 5 = 20$, or $4 \bullet 5 = 20$

'Crossing' the numbers 4 and 5 to get 20: $4 \times 5 = 20$, or $4 \times 5 = 20$, or $4 \otimes 5 = 20$

To make up for the lack of vector division, we physicists have developed two ways to multiply vectors. The first method is called the vector 'Dot'-product and the second is called the vector 'Cross'-product. These are two completely different ways of multiplying two vectors and you must be careful from the outset to know the difference. So how do we tell the difference between the 'Dot'-product and the 'Cross'-product? How do I know when to follow the procedure for 'Dotting' two vectors versus following the procedure for 'Crossing' two vectors? The answer is simple—the *symbol* between the two vectors is the key:

Warning about vector multiplication:
When multiplying two vectors, the symbol of the 'Dot' or the 'Cross' signals which procedures must be used!

In other words, $a \bullet b = a \times b$ (two scalars), but $\vec{A} \bullet \vec{B} \neq \vec{A} \times \vec{B}$ (two vectors).

▶ **Multiplying two scalars:** $a \bullet b = a \times b$

▶ **Multiplying two vectors:** $\vec{A} \bullet \vec{B} \neq \vec{A} \times \vec{B}$

So the first concept you have to understand is that the **SYMBOL BETWEEN THE TWO VECTORS DETERMINES WHICH PROCEDURE TO USE!** In this chapter, we will focus *only* on the vector 'Dot'-product. We will start off slowly and master this one method of multiplying a vector by another vector. The vector 'Cross'-product will be introduced in the next chapter, when discussing rotational motion and is especially helpful during a course in electricity and magnetism when describing the force exerted by a magnetic field on a moving, charged particle. For now, you only need to focus on the vector 'Dot'-product for multiplying two vectors.

▶ **RULES FOR THE VECTOR 'DOT'-PRODUCT:** Two vectors are written in vector notation, $\vec{A} = (A_x, A_y)$ and $\vec{B} = (B_x, B_y)$:

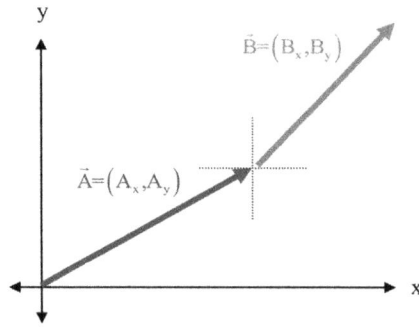

The vector 'Dot'-product is defined as:

$$\vec{A} \bullet \vec{B} = \left(A_x, A_y\right) \bullet \left(B_x, B_y\right) = A_x \cdot B_x + A_y \cdot B_y$$

or

$$\vec{A} \bullet \vec{B} = \begin{pmatrix} A_x \\ A_y \end{pmatrix} \bullet \begin{pmatrix} B_x \\ B_y \end{pmatrix} = A_x \cdot B_x + A_y \cdot B_y$$

Notice the result is not a vector—the answer is not in parentheses.

Notice the following four features of the vector 'Dot'-product:

1. The vector 'Dot'-product takes two vectors, multiplies them, and results in a scalar quantity! When completed, the vector 'Dot'-product is a scalar quantity.

2. The units of the 'Dot'-product are the units of vector $\vec{A}$, times the units of vector $\vec{B}$. For instance, if $\vec{A}$ has units of 'Cats' and $\vec{B}$ has units of 'Dogs,' then $\vec{A} \bullet \vec{B}$ has units of 'Cats·Dogs.'

3. The 'Dot'-product is 'commutative.' In other words, the order of vectors is not important: $\vec{A} \bullet \vec{B}$ is the same as $\vec{B} \bullet \vec{A}$.

4. Physically, *the 'Dot'-product multiplies **parallel** components of two vectors.* In other words, the x-component of $\vec{A}$ multiplies the x-component of $\vec{B}$, while the y-component of $\vec{A}$ multiplies the y-component of $\vec{B}$. *We will need to explore this in more detail in the upcoming examples but we will now interpret the 'Dot'*-product as a method for multiplying the parallel components of two vectors.

Let's try some examples and check that each of these four features is true.

2.3.2 Modeling

❶ **EX 1:** Two vectors are given: $\vec{A} = (3, 4)$Newtons and $\vec{B} = (5, 6)$m. What is the vector 'Dot'-product $\vec{A} \cdot \vec{B}$?

☑ **ANSWER:** Let's follow the multiplication rules stated above. Also, for reasons that will become clear in a moment, let's write our vectors as column-vectors rather than row-vectors:

$$\vec{A} \cdot \vec{B} = \begin{pmatrix} 3 \\ 4 \end{pmatrix} \cdot \begin{pmatrix} 5 \\ 6 \end{pmatrix} = 3 \cdot 5 + 4 \cdot 6 = 15 + 24 = 39 \text{ Newton} \cdot \text{m}$$

☺ **COMMENTS:** Let's check each of the four features:

1. ✓ The result is a scalar quantity—yes, our answer is 39 Newton·m and not written as a vector. The answer is a simple scalar quantity containing only information on the magnitude. No directional information is included.

2. ✓ The result has units of vector $\vec{A}$ (which are 'Newtons'), times the units of vector $\vec{B}$ (which are 'meters').

3. ✓ The procedure is commutative—we could have done $\vec{B} \cdot \vec{A}$ and ended up with the same answer:

$$\vec{A} \cdot \vec{B} = \begin{pmatrix} 3 \\ 4 \end{pmatrix} \cdot \begin{pmatrix} 5 \\ 6 \end{pmatrix} = 3 \cdot 5 + 4 \cdot 6 = 15 + 24 = 39 \text{ Newton} \cdot \text{meters}$$

$$\vec{B} \cdot \vec{A} = \begin{pmatrix} 5 \\ 6 \end{pmatrix} \cdot \begin{pmatrix} 3 \\ 4 \end{pmatrix} = 5 \cdot 3 + 6 \cdot 4 = 15 + 24 = 39 \text{ Newton} \cdot \text{meters}$$

Same result!

4. ✓ The result multiplies parallel components of the two vectors. Notice that vector notation keeps x and y components separate. Then performing the 'Dot'-product, we multiply x components separately—then we multiply y components separately—then we add the two products.

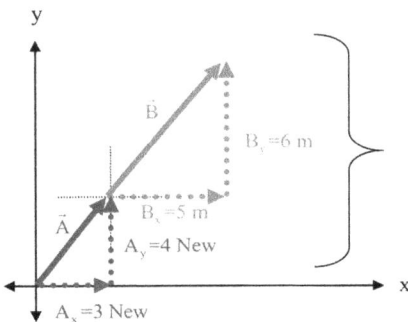

Multiply the x-components: 3 New · 5 m = 15 N·m
Multiply the y-components: 4 New · 6 m = 24 N·m
Add the two products: 15 N·m + 24 N·m = 39 N·m

If you look over this first example of a vector 'Dot'-product, you'll notice that we preferred to write the vectors as column-vectors rather than row-vectors (although writing them as row-vectors is perfectly fine). The reason for this preference is because a neat and useful mnemonic device results. When performing the 'Dot'-product and using column-vectors, you'll notice that the x and y components of the two vectors are aligned side-by-side. When multiplying the side-by-side components, you will naturally form the 'parallel' symbol of two parallel lines:

$$\vec{A} \bullet \vec{B} = \begin{pmatrix} A_x \\ A_y \end{pmatrix} \bullet \begin{pmatrix} B_x \\ B_y \end{pmatrix} = \begin{matrix} A_x \cdot B_x \\ + \\ A_y \cdot B_y \end{matrix} = A_x \cdot B_x + A_y \cdot B_y$$

or

$$\vec{A} \bullet \vec{B} = \underline{\hspace{3cm}}$$

This is a very useful mnemonic device because it constantly reminds us that the 'Dot'-product physically multiplies the parallel components of two vectors. This mnemonic reminds us that if we ever encounter a physical situation in which we must ensure that two vectors are parallel or if we need to check whether or not two vectors are parallel, then the vector 'Dot'-product is the way to go!

❷ **EX 2:** Four vectors are given: $\vec{A} = (3, 4)$Newtons, $\vec{B} = (5, 6)$m, $\vec{C} = (-7, 8)$Cats, and $\vec{D} = (-4, 3)$Dogs. Calculate the following vector 'Dot'-products:

    a) $\vec{A} \bullet \vec{B}$
    b) $\vec{A} \bullet \vec{C}$
    c) $\vec{A} \bullet \vec{D}$
    d) $\vec{C} \bullet \vec{D}$
    e) $\vec{C} \bullet \vec{A}$

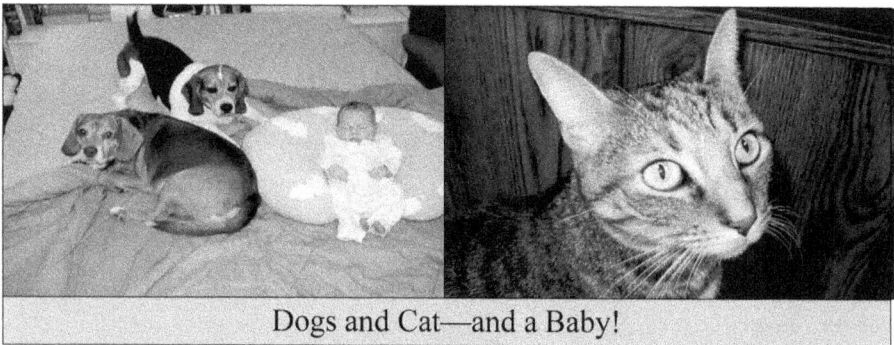

Dogs and Cat—and a Baby!

Image Credit: Author.

☑ **ANSWER:** Let's follow the multiplication rules stated above and we'll write our vectors as column-vectors rather than row-vectors:

a) $\vec{A} \cdot \vec{B} = \begin{pmatrix} 3 \\ 4 \end{pmatrix} \cdot \begin{pmatrix} 5 \\ 6 \end{pmatrix} = 3 \cdot 5 + 4 \cdot 6 = 15 + 24 = 39$ Newton $\cdot$ m

b) $\vec{A} \cdot \vec{C} = \begin{pmatrix} 3 \\ 4 \end{pmatrix} \cdot \begin{pmatrix} -7 \\ 8 \end{pmatrix} = 3 \cdot (-7) + 4 \cdot 8 = -21 + 32 = 11$ Newton $\cdot$ Cats

c) $\vec{A} \cdot \vec{D} = \begin{pmatrix} 3 \\ 4 \end{pmatrix} \cdot \begin{pmatrix} -4 \\ 3 \end{pmatrix} = 3 \cdot (-4) + 4 \cdot (3) = -12 + 12 = 0$ Newton $\cdot$ Dogs

d) $\vec{C} \cdot \vec{D} = \begin{pmatrix} -7 \\ 8 \end{pmatrix} \cdot \begin{pmatrix} -4 \\ 3 \end{pmatrix} = (-7) \cdot (-4) + 8 \cdot (3) = 28 + 24 = 52$ Cats $\cdot$ Dogs

e) $\vec{C} \cdot \vec{A} = \begin{pmatrix} -7 \\ 8 \end{pmatrix} \cdot \begin{pmatrix} 3 \\ 4 \end{pmatrix} = (-7) \cdot 3 + 8 \cdot 4 = -21 + 32 = 11$ Newton $\cdot$ Cats

☺ **COMMENTS:** Let's check each of the four features:

1. ✓ The results are all scalar quantities!
2. ✓ The results have units of vector $\vec{A}$ times the units of vector $\vec{B}$.
3. ✓ The procedure is commutative—notice that answers (b) and (e) are the same. We could have skipped a calculation in part (e) because we know immediately that it must be the same answer as in part (a). As we know, order is not important when 'Dotting' two vectors.
4. ✓ The results multiply parallel components of two vectors. Notice that part (c) has an answer of zero, or no 'Dot'-product. We would say that, 'No Dot-product exists.' This should not surprise us because vectors $\vec{A}$ and $\vec{D}$ **are perpendicular**—and the 'Dot'-product only multiplies parallel components. Even if we didn't realize that $\vec{A}$ and $\vec{D}$ *are perpendicular*, the 'Dot'-product would tell us that, 'No part of vector $\vec{A}$ is parallel to vector $\vec{D}$' or 'No part of vector $\vec{D}$ is parallel to vector $\vec{A}$.'

PART (C) ILLUSTRATES HOW THE 'DOT'–PRODUCT SHOULD BE PHYSICALLY INTERPRETED—IT TELLS HOW MUCH OF ONE VECTOR IS PARALLEL TO ANOTHER VECTOR!

❸ **EX 3:** Two vectors are given: $\vec{A} = (3, 4)$Newtons and $\vec{B} = (5, 6)$m. What is the vector 'Dot'-product $-5\vec{A} \cdot 2\vec{B}$?

☑ **ANSWER:** Let's just follow the conventions we've used in the previous two examples.

$$-5\vec{A} \cdot 2\vec{B} = -5\begin{pmatrix} 3 \\ 4 \end{pmatrix} \cdot 2\begin{pmatrix} 5 \\ 6 \end{pmatrix} = \begin{pmatrix} -15 \\ -20 \end{pmatrix} \cdot \begin{pmatrix} 10 \\ 12 \end{pmatrix}$$

$$= -15 \cdot 10 + -20 \cdot 12 = -150 - 240 = -390 \text{ Newton} \cdot \text{m}$$

☺ **COMMENTS:** Let's check each of the four features:

1. ✓ The result is a scalar quantities ... and yes, scalars can be negative!

2. ✓ The result has units of vector $\vec{A}$ (which are 'Newtons'), times the units of vector $\vec{B}$ (which are 'meters').

3. ✓ The procedure is commutative—we could have done $\vec{B} \cdot \vec{A}$ and ended up with the same answer:

4. ✓ The result multiplies parallel components of the two vectors.

Also notice in this example, we can still multiply vectors by scalars (i.e., $-5\vec{A}$ and $2\vec{B}$), before using them in a 'Dot'-product.

2.3.3 Instructional input

2.3.3.1 Work or energy

Now that we've established the rules and conventions for calculating a vector 'Dot'-product, we are in a position to calculate a new physical quantity called 'Work' or 'Energy.' For the moment, we really don't care why we choose the definition that we do—the reason for our choice will become apparent later. Instead, we want to focus on three things: (i) Knowing (possibly even 'memorizing') the definition of work; (ii) mastering how to calculate work; and (iii) gaining an intuition or physical interpretation of work.

The 'Work' (also called 'Energy') done on an object, by a constant force $\vec{F}$, as the object is displaced by $\vec{d}$, is given by:

▶ **Definition of Work:** $W = \vec{F} \cdot \vec{d}$

▶ **Definition of Energy:** $U = \vec{F} \cdot \vec{d}$

Etymology:

The word 'energy' comes from the Greek word 'energia' that translates into 'a moving force.'

The word 'work' is Germanic in origin and comes from the work 'werk' which means 'labor.'

This simple definition incorporates a number of features that are worth itemizing and clarifying. As usual, we'll check these features when we actually start calculating work in example problems:

1. Work is symbolized by the scalar W while energy is symbolized by the scalar U. You might wonder why energy isn't more conveniently symbolized by the scalar E. Energy gets its funny symbol so as not to be confused with the

electric field $\vec{E}$. Think of it this way—the electric field got first bids on the letter 'E.'

2. The words 'Work' and 'Energy' are completely interchangeable—sometimes, a problem asks for 'work,' sometimes it asks for 'energy,' and sometimes the words 'work' and 'energy' flip-flop in the same problem. Don't let this interchangeability confuse you—the two words are synonymous.

3. Work is the result of a vector 'Dot'-product; thus work is a scalar! In other words, two vectors ($\vec{F}$ and $\vec{d}$) are multiplied to give us a scalar (W or U).

4. Recall that the vector 'Dot'-product is a method by which the parallel components of two vectors are multiplied. Since we compute work through a vector 'Dot'-product, we are physically only multiplying the parallel components of $\vec{F}$ and $\vec{d}$. If $\vec{F}$ and $\vec{d}$ are perpendicular to one another, no 'Dot'-product will exist. If $\vec{F}$ and $\vec{d}$ are parallel to one another, the 'Dot'-product will be maximized. If $\vec{F}$ and $\vec{d}$ are at an angle to one another, the 'Dot'-product will be some value between zero and the maximum.

5. Since we have defined work as a new physical quantity, we will need new units to measure work:

System	Unit of force	Unit of displacement:	Unit of work or energy
m.k.s.	Newton	m	Newton · m = Joule
c.g.s.	dyne	cm	dyne · cm = Erg
English	Pound	foot	Pound · ft = Foot−Pound

Of course, we will always use the **m.k.s.** system for our calculations so work will have units of joules, named in honor of James Prescott Joule. Interestingly, the **English** unit of work is not renamed to honor someone—it is simply the 'Foot-Pound.'

Etymology
The word 'erg' comes from the Greek word 'ergon' which simply translates into 'work.'

Biographical Information:
JAMES PRESCOTT JOULE

BORN: December 24, 1818 (Salford, England)
DIED: October 11, 1889 (Sale, England)

James Prescott Joule came from a wealthy family of English brewers. His grandfather was the founder of a brewery and Joule's father, **Benjamin**, eventually became owner. Joule himself became an active manager of the brewery until the business was sold in 1854. Joule had some formal education and always enjoyed pursuing science as a hobby. However, his interest in science became more serious when he began experimenting with replacing the brewery's steam engines with the newly-invented electric motor. In fact, many of Joule's experiments were carried out in his grandfather's brewery using its various instruments. Because of his practical knowledge as a brewer, Joule was able to make very precise measurements of temperature. In the 1840s, motivated by theological beliefs, he attempted to demonstrate the *unity of forces in nature*. He determined the mechanical equivalent of heat by measuring change in temperature produced by the friction of a paddlewheel attached to a falling weight (see below). Joule's experiments primarily focused on the conversion of work into heat, and his experiments were fundamental to the ultimate understanding to the **Law of Conservation of Energy** and the **First Law of Thermodynamics**. Unfortunately, Joule's lack of a formal academic or engineering position, as well as his reliance on precise measurements (which many physicists at the time doubted were accurate), made his findings unacceptable to his contemporaries. Much of the initial resistance to Joule's work stemmed from its dependence upon extremely precise measurements. He claimed to be able to measure temperatures to within $1/200$ of a degree Fahrenheit. Such precision was certainly uncommon at that time but he had experience in the art of brewing and access to state-of-the-art technologies. His various publications were not met with much enthusiasm. In 1847, he married **Amelia Grimes**, but she died in 1854 as she over-worked to support his scientific pursuits. Joule's health was always described as weak and delicate. The unit of work or energy—**the JOULE**—is named in his honor.

Joule's heat apparatus

As an interesting side note, on the day of Joule's birth, **Joseph Moore**, a priest, gave a poem that he had written to his friend **Franz Grubber** who was the organist and choir-master in Moore's parish. Grubber used the poem to write the words and melody for a new song to be played that night—the song was written for guitar accompaniment because the parish organ was broken and the two men desperately wanted music for their Christmas Eve service. The song has since become one of the most beloved of Christmas songs, "*Silent Night*."

2.3.4 Modeling

❶ **EX 1:** A force of 300 Newtons (this could be a rope, a push, a pull, a spring, etc), directed 30° up from the positive x-axis, is used to pull a block along the ground for 4 m. What is the work done on the block, by the force?

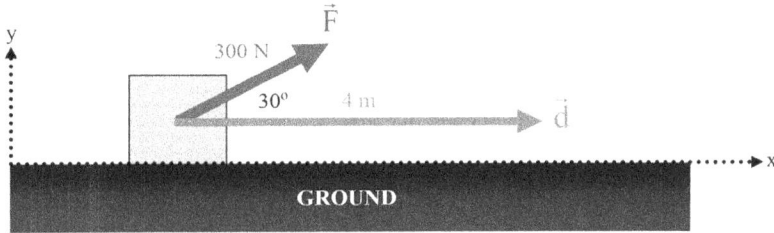

☑ **ANSWER:** We simply convert the force and displacement depicted in the drawing into vector notation and plug into our formula for work:

$$\vec{F} = \begin{pmatrix} 300 \times \cos(30°) \\ 300 \times \sin(30°) \end{pmatrix} = \begin{pmatrix} 260 \\ 150 \end{pmatrix} \text{ Newtons and } \vec{d} = \begin{pmatrix} 4 \\ 0 \end{pmatrix} \text{ m.}$$

$$W = \vec{F} \cdot \vec{d} = \begin{pmatrix} 260 \\ 150 \end{pmatrix} \cdot \begin{pmatrix} 4 \\ 0 \end{pmatrix} = 260 \cdot 4 + 150 \cdot 0 = 1040 + 0 = 1040 \text{Joules}$$

☺ **COMMENTS:** Let's check each of the five features we discussed when we first defined work:

1. ✓Work is symbolized by the scalar quantity, W.
2. ✓The question could have asked for the 'Energy' and our answer would have been the same.
3. ✓Our answer is a scalar quantity—1040 joules is not a vector. We only have stated a number and have not specified any directional information.
4. ✓The procedure for multiplying a vector 'Dot'-product has guaranteed that only parallel components of $\vec{F}$ and $\vec{d}$ have been multiplied. Notice that since $\vec{d}$ is entirely along the x-axis, only the x-component of $\vec{F}$ entered into the calculation.
5. ✓Our answer uses the new **m.k.s.** units of 'joules.'

❷ **EX 2:** A 50 kg crate has the following four forces acting on it: friction, normal, pull, and weight. The forces are depicted in the drawing below:

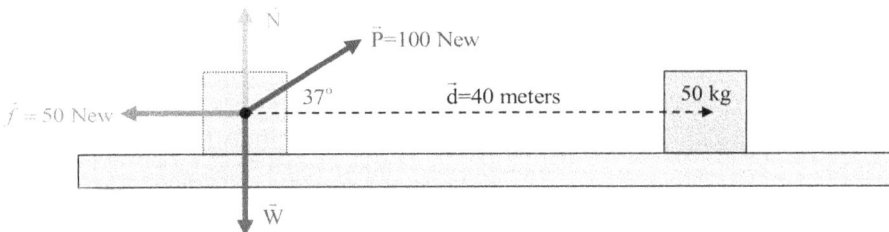

The result of these four forces acting on the crate is to displace the crate 40 m to the right. Calculate the work done by:
   a) gravity;
   b) the normal force;
   c) friction;
   d) the pull;
   e) calculate the total work done by all four forces acting on the block.

☑ **ANSWER:** Again, we simply convert the forces and displacement depicted in the drawing into vector notation and plug into our formula for work:

a) $\vec{W} = \begin{pmatrix} 0 \\ -50 \times 9.8 \end{pmatrix} = \begin{pmatrix} 0 \\ -490 \end{pmatrix}$ Newtons and $\vec{d} = \begin{pmatrix} 40 \\ 0 \end{pmatrix}$ m.

Therefore, $W_{\text{gravity}} = \vec{W} \cdot \vec{d} = \begin{pmatrix} 0 \\ -490 \end{pmatrix} \cdot \begin{pmatrix} 40 \\ 0 \end{pmatrix} = 0$ Joules.

b) $\vec{N} = \begin{pmatrix} 0 \\ +N \end{pmatrix}$ and $\vec{d} = \begin{pmatrix} 40 \\ 0 \end{pmatrix}$ m.

Therefore, $W_{\text{normal}} = \vec{N} \cdot \vec{d} = \begin{pmatrix} 0 \\ +N \end{pmatrix} \cdot \begin{pmatrix} 40 \\ 0 \end{pmatrix} = 0$ Joules.

c) $\vec{f} = \begin{pmatrix} -50 \\ 0 \end{pmatrix}$ Newtons and $\vec{d} = \begin{pmatrix} 40 \\ 0 \end{pmatrix}$ m.

Therefore, $W_{\text{friction}} = \vec{f} \cdot \vec{d} = \begin{pmatrix} -50 \\ 0 \end{pmatrix} \cdot \begin{pmatrix} 40 \\ 0 \end{pmatrix} = -2000$ Joules.

d) $\vec{P} = \begin{pmatrix} 100 \times \cos(37^o) \\ 100 \times \sin(37^o) \end{pmatrix} = \begin{pmatrix} 80 \\ 60 \end{pmatrix}$ Newtons and $\vec{d} = \begin{pmatrix} 40 \\ 0 \end{pmatrix}$ m.

Therefore, $W_{\text{pull}} = \vec{P} \cdot \vec{d} = \begin{pmatrix} 80 \\ 60 \end{pmatrix} \cdot \begin{pmatrix} 40 \\ 0 \end{pmatrix} = +3200$ Joules.

e) Since work is a scalar, we can get the total by just doing a summation of scalar quantities:
$$W_{\text{total}} = W_{\text{gravity}} + W_{\text{normal}} + W_{\text{friction}} + W_{\text{pull}} = +1200 \text{ Joules.}$$

☺ **COMMENTS:**
Notice in part (b) that even though the normal force was not given, we could still compute the work it does. Since we know something about the normal force's direction from the drawing, we are able to write it in vector notation and still complete the 'Dot'-product. This type of situation arises often in work/energy problems. Namely, not all the information is given to fully write a force in vector notation— perhaps just the magnitude is known or only the direction is known. However, enough information exists in the problem to still determine the work done by that force.

Also, notice that the total work was computed from simple scalar addition. That's the beauty of using work/energy—it's a scalar quantity! Once work/energy is computed, we simply use the regular rules or addition and subtraction to manipulate our answers.

2.3.5 Instructional input

2.3.5.1 *Work done by a non-constant force*

Before we start developing a problem-solving strategy that somehow uses work/energy, we need to address one lingering question. Namely, when we defined work, we were careful to insist that the definition only held for a constant force. How do we define work for a non-constant force? Certainly, the work done by a constant force is more specialized than the work done by a non-constant force—so shouldn't we define work for the most general situation? For instance, as you drive along the highway, the force exerted by your car on the surface of the highway varies considerably over the length of your journey. As you speed-up and slow-down, the force of your car on the surface of the highway is anything but constant. In such a situation, how can we compute the work done by the car if its force is not constant?

Fortunately, for such a situation, we don't need a new definition of work. Instead, we can use our current definition of work, but apply it in a more careful way.

Let's consider the situation where some non-constant force $\vec{F}$ (perhaps you prefer the notation $\vec{F}(x)$ or $\vec{F}(\vec{r})$, to symbolize that varies as a function of position), is exerted on an object and causes the object to move from point (A) to point (B). The displacement from (A) to (B) is represented by $\vec{d}$. We can imagine that this displacement is actually comprised of a series of smaller displacements that add up to the total displacement (i.e., $\vec{d}$ is comprised of bunch of smaller displacements, $\Delta \vec{r}$):

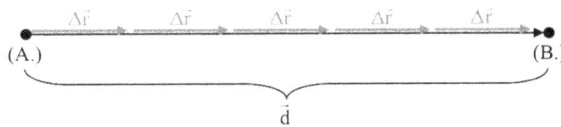

We can imagine further that these smaller displacements, $\Delta \vec{r}$ are comprised of even smaller displacements, $d\vec{r}$:

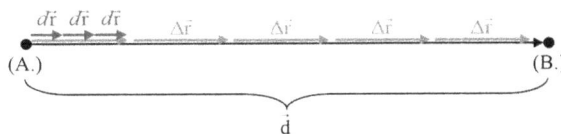

We can continue this 'cutting' procedure until we convince ourselves that $d\vec{r}$ **is so small**, the force would not vary over such a small distance. In other words, the force does indeed vary as the object moves from point (A) to point (B), but we make $d\vec{r}$ to be so infinitesimally small that the force couldn't possibly vary over such a small distance. To use the car analogy again, the force exerted by your car on the surface of the highway varies considerably as you drive the 300 miles from your home to college. However, if we break-up the 300 mile trip into a series of 1-inch displacements (you would have to make over 19 million cuts), you can believe that 'over the infinitesimally small 1-inch displacement, the force of the car on the surface of the highway does not vary.'

Now, we simply apply our definition of work/energy to these infinitesimally small displacements (because the force is constant) and calculate the work done by the force over that small displacement. To get the total work done by the non-constant force over the entire displacement $\overrightarrow{d}$, we just add up all of the small works done over the small displacements:

$$W_{\text{total}} = W_{\text{step 1}} + W_{\text{step 2}} + W_{\text{step 3}} + W_{\text{step 4}} + \cdots$$

$$W_{\text{total}} = \overrightarrow{F} \cdot d\overrightarrow{r_1} + \overrightarrow{F} \cdot d\overrightarrow{r_2} + \overrightarrow{F} \cdot d\overrightarrow{r_3} + \overrightarrow{F} \cdot d\overrightarrow{r_4} + \cdots$$

This 'cutting and adding' procedure is the basis of integral calculus and can be formalized with the following notation:

▶ **Definition of Work done by a Constant Force:** $W = \overrightarrow{F} \cdot \overrightarrow{d}$

▶ **Definition of Work done by a Non-Constant Force:** $W = \int_{(A.)}^{(B.)} \overrightarrow{F} \cdot d\overrightarrow{r}$

The discussion of the work done on an object by a non-constant force will not be further explored in this chapter—it was included here to answer a potentially lingering question as well as for your educational enjoyment. We will be able to do a lot of sophisticated physics with just the simpler form of work done by a constant force.

2.3.5.2 The 'work–energy theorem' or the 'law of conservation of energy'

We are about to embark on a rather lengthy derivation. However, the investment in time is well worth the headache. Ultimately, we will derive an equation that will be one of the easier and more widely-used of our problem-solving techniques. Remember to always keep the bigger picture in mind—and the bigger picture is this: we will start with Newton's second law of motion and manipulate it to show that total energy is conserved in time. Recalling our anticipatory set, when we say a physical quantity is 'conserved,' we mean that it is 'unchanged' over time. Thus, starting with Newton's second law of motion, we will show that total energy does not change over time.. This result is one of the most widely-used and powerful problem-solving techniques used by physicists.

▶ **Derivation of the Law of Conservation of Energy (abbreviated as the 'LCE'):**
Start with Newton's second law of motion for a single moving mass, m:

$$\sum \vec{F}_{\text{external}} = m \cdot \vec{a}$$

From volume 1, chapter 4, we know that the external forces come in three forms: applied, contact and field forces.

$$\underbrace{\vec{P} + \vec{T} + \vec{F}_{\text{sp}}}_{\text{applied forces}} + \underbrace{\vec{N} + \vec{f}}_{\text{contact forces}} + \underbrace{\vec{W}}_{\text{field force}} = m \cdot \vec{a}$$

Let's assume that the external forces cause the mass to be displaced by some value $\vec{d}$. For fun, let's vector multiply (using the "Dot"-product) $\vec{d}$ into both sides.

$$\left(\vec{P} + \vec{T} + \vec{F}_{\text{sp}} + \vec{N} + \vec{f} + \vec{W} \right) \bullet \vec{d} = m \cdot \vec{a} \bullet \vec{d}$$

Let's break the left-hand-side into a bunch of separate terms that can be analyzed individually.

$$\underbrace{\left(\vec{P} \bullet \vec{d} \right)}_{\text{term 1}} + \underbrace{\left(\vec{T} \bullet \vec{d} \right)}_{\text{term 2}} + \underbrace{\left(\vec{F}_{\text{sp}} \bullet \vec{d} \right)}_{\text{term 3}} + \underbrace{\left(\vec{N} \bullet \vec{d} \right)}_{\text{term 4}} + \underbrace{\left(\vec{f} \bullet \vec{d} \right)}_{\text{term 5}} + \underbrace{\left(\vec{W} \bullet \vec{d} \right)}_{\text{term 6}} = \underbrace{m \cdot \vec{a} \bullet \vec{d}}_{\text{term 7}}$$

So now we have seven terms that are all forces 'Dotted' into the displacement $\overrightarrow{d}$. Hence, each of these seven terms is a work/energy term. The seven work terms must now be analyzed individually and hopefully simplified. Some of the terms will, in fact, simplify while others will have to remain as they are. Let's take a quick digression to simplify the following three terms—term #3, #6 and #7. Once we make the simplifications, we'll pick up the derivation from the above equation.

▶ **Term 7: The Moving Object:** Term 7 contains the net acceleration of the moving object and therefore represents the net work done on the object. In other words, the external forces acting on the object have combined to produce the net acceleration buried in term 7. However, term 7 can be simplified thanks to a previous result from our discussions on acceleration in volume 1, chapter 3 of this series:

UAM Equation #3 already contains a simplified expression for $\vec{a} \bullet d$:

$$v^2 = v_o^2 + 2\vec{a} \bullet (\vec{r} - \vec{r}_o)$$
$$= v_o^2 + 2\vec{a} \bullet \vec{d}$$

or

$$\vec{a} \bullet \vec{d} = \frac{v^2 - v_o^2}{2}$$

$$W_{term\,7} = m \cdot \vec{a} \bullet \vec{d}$$

Since we are dealing with a constant force, we can use any of the UAM results from volume 1, chapter 3; namely, UAM equation #3.

$$W_{term\,7} = m \cdot \left(\frac{v^2 - v_o^2}{2} \right)$$

Rearranging

$$W_{term\,7} = \frac{1}{2}mv^2 - \frac{1}{2}mv_o^2$$

Since this term contains the net acceleration (or motion) of the object, this term is called the 'Kinetic Energy' term. Thus, term 7 can **always** be simplified as follows:

Term 7:

Kinetic energy:

$$\Delta KE = \left(\tfrac{1}{2}m \cdot v_{final}^2 \right) - \left(\tfrac{1}{2}m \cdot v_{initial}^2 \right)$$

1. The term is called 'the change in Kinetic Energy.'
2. The term is a work/energy and is therefore a scalar quantity.
3. The term is a work/energy and therefore has units of joules.
4. Term 7 can always be replaced by the expression above.

Etymology:
The word 'kinetic' comes from the Greek word 'kinetikos' that translates into 'moving.'

▶ **Term 6: The Work Done by Gravity:** Term 6 is simply the work done by the gravitational field of the Earth and can be simplified using our expression for the weight of an object, $\vec{W} = m \cdot \vec{g}$.

$$W_{\text{term 6}} = \vec{W} \bullet \vec{d}$$

We know that any mass **m**, has a weight **mg**.

$$W_{\text{term 6}} = m \cdot \vec{g} \bullet \vec{d}$$

We also know that weight is always down and that any displacement can be represented as separate displacements along the x- and y-axes.

$$W_{\text{term 6}} = \begin{pmatrix} 0 \\ -mg \end{pmatrix} \bullet \begin{pmatrix} \Delta x \\ \Delta y \end{pmatrix}$$

Rearranging

$$W_{\text{term 6}} = -mg \cdot \Delta y$$

Since this term contains the force of gravity acting on the object, this term is called the 'Gravitational Potential Energy' (i.e., gravity has the 'potential' to do work on the object if it experiences a displacement). Thus, term 6 can **always** be simplified as follows:

Term 6:

Gravitational Potential Energy:
$$\Delta U_g = (-mgy_{\text{final}}) - (-mgy_{\text{initial}})$$

1. The term is called 'the change in Gravitational Potential Energy.'
2. The term is a work/energy and is therefore a scalar quantity.
3. The term is a work/energy and therefore has units of joules.
4. Term 6 can always be replaced by the expression above.

Etymology:
The word 'potential' comes from the Latin word 'potentialus' that translates into 'powerful.'

▶ **Term 3: The Work Done by a Spring:** Term 3 is the work done on the object by a spring. When discussion Newton's laws of motion, we saw that we can use Hooke's Law to model the force of a spring by: $\vec{F}_{sp} = -K \cdot x$, where K is called the 'spring constant' and x is the amount by which the spring has been 'stretched' or 'squished.'

$$W_{\text{term 3}} = \vec{F}_{sp} \bullet \vec{d}$$

Since the force of a spring varies as the object moves (*i.e.*, as x changes, the force changes), we have to use the formula for the work done on an object by a non-constant force (Don't worry—this won't happen again).

$$W_{\text{term 3}} = \int_{x_o}^{x} \vec{F}_{sp} \bullet d\vec{r}$$

Use the formula for the force exerted by a spring and write the infinitesimal displacement as separate infinitesimal displacements along the x- and y-axes.

$$W_{\text{term 3}} = \int_{x_{\text{initial}}}^{x_{\text{final}}} \begin{pmatrix} -K \cdot x \\ 0 \end{pmatrix} \bullet \begin{pmatrix} dx \\ dy \end{pmatrix}$$

Use the result from calculus for the integral of a simple polynomial.

$$W_{\text{term 3}} = \int_{x_{\text{initial}}}^{x_{\text{final}}} \left(-K \cdot x \right) dx = \frac{-K \cdot x^2}{2} \bigg|_{x_{\text{initial}}}^{x_{\text{final}}} = \frac{-K}{2} \left(x_{\text{final}}^2 - x_{\text{initial}}^2 \right)$$

Since this term contains the force of a spring acting on the object, this term is called the 'Elastic Potential Energy' (i.e., the spring has the 'potential' to do work on the object if it is stretched or squished). Thus, term 3 can **always** be simplified as follows:

Term 3:

Elastic Potential Energy:

$$\Delta U_e = \left(-\frac{K \times x_{\text{final}}^2}{2} \right) - \left(-\frac{K \times x_{\text{initial}}^2}{2} \right)$$

1. The term is called 'the change in Elastic Potential Energy.'
2. The term is a work/energy and is therefore a scalar quantity.
3. The term is a work/energy and therefore has units of joules.
4. Term 3 can always be replaced by the expression above.

We left our derivation at the following line, but were able to simplify three of the seven terms:

$$\underbrace{\left(\vec{P}\bullet\vec{d}\right)}_{\text{term 1}}+\underbrace{\left(\vec{T}\bullet\vec{d}\right)}_{\text{term 2}}+\underbrace{\left(\vec{F}_{sp}\bullet\vec{d}\right)}_{\text{term 3}}+\underbrace{\left(\vec{N}\bullet\vec{d}\right)}_{\text{term 4}}+\underbrace{\left(\vec{f}\bullet\vec{d}\right)}_{\text{term 5}}+\underbrace{\left(\vec{W}\bullet\vec{d}\right)}_{\text{term 6}}=\underbrace{m\cdot\vec{a}\bullet\vec{d}}_{\text{term 7}}$$

simplifies   simplifies simplifies

To simplify the equation even further, let's rename the four work terms that do not simplify. Let's group these four terms together and collectively call them 'the work done by other forces.' This will allow us to rearrange our equation as:

$$\underbrace{\underbrace{\left(\vec{P}\bullet\vec{d}\right)}_{\text{term 1}}+\underbrace{\left(\vec{T}\bullet\vec{d}\right)}_{\text{term 2}}+\underbrace{\left(\vec{N}\bullet\vec{d}\right)}_{\text{term 4}}+\underbrace{\left(\vec{f}\bullet\vec{d}\right)}_{\text{term 5}}}_{W_{\text{other}}}+\underbrace{\left(\vec{F}_{sp}\bullet\vec{d}\right)}_{\text{term 3}}+\underbrace{\left(\vec{W}\bullet\vec{d}\right)}_{\text{term 6}}=\underbrace{m\cdot\vec{a}\bullet\vec{d}}_{\text{term 7}}$$

$$W_{\text{other}}+\underbrace{\left(\vec{F}_{sp}\bullet\vec{d}\right)}_{\text{term 3}}+\underbrace{\left(\vec{W}\bullet\vec{d}\right)}_{\text{term 6}}=\underbrace{m\cdot\vec{a}\bullet\vec{d}}_{\text{term 7}}$$

Now use the three simplifications that we just derived.

$$W_{\text{other}}+\underbrace{\left(\left(-\frac{K\cdot x_{\text{final}}^2}{2}\right)-\left(-\frac{K\cdot x_{\text{initial}}^2}{2}\right)\right)}_{\text{term 3}}+\underbrace{\left(\left(-mgy_{\text{final}}\right)-\left(-mgy_{\text{initial}}\right)\right)}_{\text{term 6}}=\underbrace{\frac{1}{2}mv_{\text{final}}^2-\frac{1}{2}mv_{\text{initial}}^2}_{\text{term 7}}$$

Finally, let's move every term with a 'final' subscript to the right-hand-side of the equation while moving every term with an 'initial' subscript to the left-hand-side of the equation:

$$W_{\text{other}}+\frac{1}{2}mv_{\text{initial}}^2+mgy_{\text{initial}}+\frac{K\cdot x_{\text{initial}}^2}{2}=\frac{1}{2}mv_{\text{final}}^2+mgy_{\text{final}}+\frac{K\cdot x_{\text{final}}^2}{2}$$

Use our short-hand notation.

The Law of Conservation of Energy:

$$W_{\text{other}}+KE_{\text{initial}}+U_{g_{\text{initial}}}+U_{e_{\text{initial}}}=KE_{\text{final}}+U_{g_{\text{final}}}+U_{e_{\text{final}}}$$

This final equation represents the 'Law of Conservation of Energy' (abbreviated as '**LCE**') and will serve as one of the most powerful weapons in our arsenal of problem-solving techniques. The derivation of the LCE did require several pages of mathematical manipulation, but as we'll soon see, the derivation will be well worth the effort.

Last, before tackling some example problems, let's enumerate some of the features of the LCE as well as some of the items to monitor whenever we attempt to solve problems:

1. The LCE is a scalar equation—terms in the LCE may be positive or negative, but direction is unimportant. Scalars are easier to use than vectors so the LCE.

2. At any time, we can substitute for 'Kinetic Energy' with the following expression: $KE = 1/2 \cdot mv^2$.
   - This term is **always** positive since mass is always positive and v is squared.
   - We derived the LCE for forces exerted on a single mass; however, this term exists for each mass appearing in a problem. If a problem involves two masses, two KE terms must appear in the LCE; if a problem involves three masses, three KE terms must appear in the LCE; etc.

3. At any time, we can substitute for 'Gravitational Potential Energy' with the following expression: $U_g = mgy$.
   - This term can be positive or negative depending on the sign of y. For example, if the object is at ground-level, y is zero; if the object moves above ground-level, y is positive; and if the object moves below ground-level, y is negative.
   - We derived the LCE for forces exerted on a single mass; however, this term exists for each mass appearing in a problem. If a problem involves two masses, two U_g terms must appear in the LCE; if a problem involves three masses, three U_g terms must appear in the LCE; etc.

4. At any time, we can substitute for 'Elastic Potential Energy' with the following expression: $U_e = 1/2 \cdot Kx^2$.
   - This term is always positive since the spring constant is always positive and x is squared.
   - This term was derived for a single spring; however, this term exists for each spring appearing in a problem. If a problem involves two springs, two U_e terms must appear in the LCE; if a problem involves three springs, three U_e terms must appear in the LCE; etc.

5. The W_{other} term can often cause trouble in our problem-solving attempts. Remember that W_{other} collectively accounts for the work done by any forces that are not due to gravity or springs. Thus, if we see a push, tension, friction, or   in a problem, we must consider if these forces produce a W_{other} term. Fortunately for us, a vast number of problems can be considered in which W_{other} terms can be ignored. Such problems will become clearer as we model our problem-solving strategy.

2.3.6 Modeling

❶ **EX 1:** A 1 kg block, initially at rest, is released 50 m above a spring that is initially neither stretched nor squished. The block compresses the spring by 1.75 m and comes to rest 0.25 m above the ground. The situation is drawn below in 'before' and 'after' photographs, but not to scale. What is the spring constant?

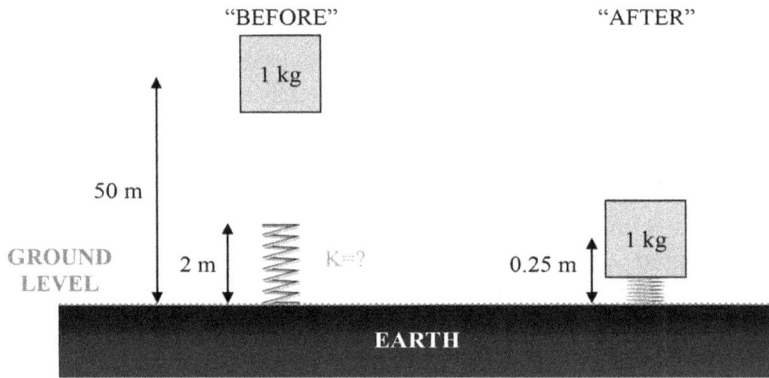

Spring Image Credit: yusufdemirci/Shutterstock.com.

☑ **ANSWER:** Let's plug into the LCE and explain each of the terms after we solve for the spring constant:

$$\cancel{W}_{other} + \cancel{KE}_{initial} + U_{g_{initial}} + \cancel{U}_{e_{initial}} = \cancel{KE}_{final} + U_{g_{final}} + U_{e_{final}}$$

$$1*9.8*(+50) = 1*9.8*(+0.25) + \frac{1}{2}K*(1.75)^2$$

$$490 = 2.45 + 1.53*K$$

$$487.55 = 1.53*K$$

$$K = 318.7 \;{}^{New}\!/_{meter}$$

☺ **COMMENTS:** Let's comment on each of the terms appearing in the LCE:
LEFT-HAND-SIDE OF LCE:
- The only forces appearing in the problem come from springs and gravity. Nowhere in the problem do we see ropes, pushes, pulls, friction, etc. Therefore, $W_{other} = 0$.
- Initially, the mass is at rest. Therefore, $KE_{initial} = 0$.
- Initially, the mass is 50 m off of the ground. Therefore, its y-value is $+$ 50 m, making $U_{g_{initial}} = 1 \times 9.8 \times (+50) = 490$ joules.
- Initially, the spring is neither stretched nor squished. Therefore, $U_{e_{initial}} = 0$.

RIGHT-HAND-SIDE OF LCE:

- In the final state, the mass is at rest. Therefore, $KE_{final} = 0$.
- In the final state, the mass is 0.25 m off the ground. Therefore, its y-value is $+ 0.25$ m, making $U_{g_{initial}} = 1 \times 9.8 \times (+0.25) = 2.45$ joules.
- In the final state, the spring is squished by 1.75 m. Therefore, $x = 1.75$ m, making $U_{e_{initial}} = 1/2 \cdot K(1.75)^2$.

❷ **EX 2:** Repeat the previous example, but now assume the mass is initially thrown **downward** with a speed of 4 m s^{-1}. The situation is drawn below in 'before' and 'after' photographs, but not to scale. What is the spring constant?

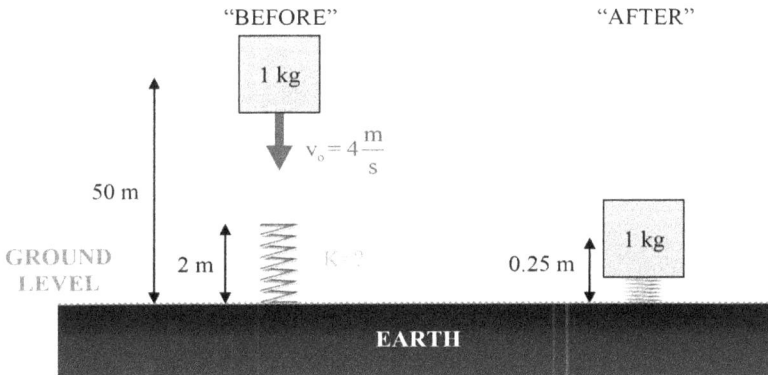

Spring Image Credit: yusufdemirci/Shutterstock.com.

☑ **ANSWER:** Everything remains the same as in the previous example except the Kinetic Energy term is no longer initially zero:

$$\cancel{W_{other}} + KE_{initial} + U_{g_{initial}} + \cancel{U_{e_{initial}}} = \cancel{KE_{final}} + U_{g_{final}} + U_{e_{final}}$$

$$\frac{1}{2}*1*(4)^2 + 1*9.8*(+50) = 1*9.8*(+0.25) + \frac{1}{2}K*(1.75)^2$$

$$8 + 490 = 2.45 + 1.53*K$$

$$495.55 = 1.53*K$$

$$K = 324 \text{ New}\Big/\text{meter}$$

☺ **COMMENTS:** The initial Kinetic Energy adds an extra 8 joules of energy to the left-hand-side. The net effect is that the spring constant has to be bigger (i.e., the spring has to be stiffer) to stop the mass that is carrying this extra 8 joules of energy.

❸ **EX 3:** Repeat the previous example, but now assume the mass is initially thrown **upward** with a speed of 4 m s^{-1}. The situation is drawn below in 'before' and 'after' photographs, but not to scale. What is the spring constant?

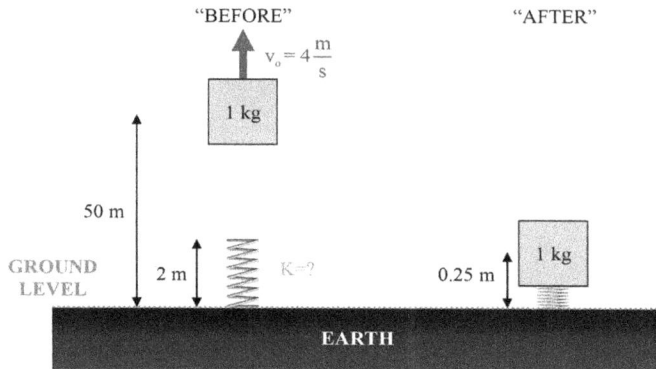

Spring Image Credit: yusufdemirci/Shutterstock.com.

☑ **ANSWER:** The LCE is a scalar equation; therefore, direction is unimportant. The mass has the same initial Kinetic Energy whether it is thrown upward or downward. The answer to this problem is identical to the answer in the previous example.

$$W_{other} + KE_{initial} + U_{g\,initial} + U_{e\,initial} = KE_{final} + U_{g\,final} + U_{e\,final}$$

$$\frac{1}{2}*1*(4)^2 + 1*9.8*(+50) = 1*9.8*(+0.25) + \frac{1}{2}K*(1.75)^2$$

$$8 + 490 = 2.45 + 1.53*K$$

$$495.55 = 1.53*K$$

$$K = 324 \; \text{New}/_{\text{meter}}$$

☺ **COMMENTS:** Notice that despite the lengthy derivation, the LCE is a somewhat easy problem-solving technique because energy is a scalar quantity.

❹ **EX 4:** A 1 kg block, initially at rest, is released 50 m above the ground. A spring is initially neither stretched nor squished. The block falls into a ditch that is 10 m deep. The mass lands on the spring and compresses it by 1.75 m. The mass comes to rest 0.25 m above the bottom of the ditch. The situation is drawn below in 'before' and 'after' photographs, but not to scale. What is the spring constant?

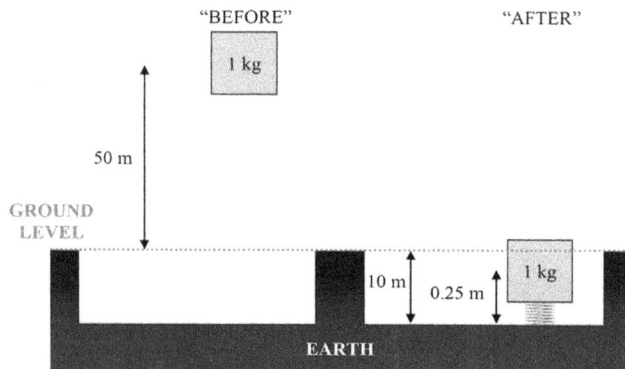

Spring Image Credit: yusufdemirci/Shutterstock.com.

☑ **ANSWER:** Again, we'll plug-n-chug and explain our answer with some subsequent comments:

$$\cancel{W_{other}} + \cancel{KE_{initial}} + U_{g_{initial}} + \cancel{U_{e_{initial}}} = \cancel{KE_{final}} + U_{g_{final}} + U_{e_{final}}$$

$$1*9.8*(+50) = 1*9.8*(-9.75) + \frac{1}{2}K*(1.75)^2$$

$$490 = -95.55 + 1.53*K$$

$$585.55 = 1.53*K$$

$$K = 382.7 \; \text{New}/\text{meter}$$

☺ **COMMENTS:** The only terms that need to be explained are the Gravitational Potential Energy terms.

LEFT-HAND-SIDE OF LCE:
- Initially, the mass was 50 m off of the ground. Therefore, its y-value is $+50$ m, making $U_{g_{initial}} = 1 \times 9.8 \times (+50) = 490$ joules.

RIGHT-HAND-SIDE OF LCE:
- In the end, notice that the mass is 9.75 m below the ground. Therefore, its y-value is -9.75 m, making $U_{g_{initial}} = 1 \times 9.8 \times (-9.75) = -95.55$ joules.

Notice that U_g can be negative if the location of the mass is below ground–level!

---

❺ **EX 5:** Repeat the same problem except this time, assume ground-level is at the bottom of the ditch. The situation is drawn below in 'before' and 'after' photographs, but not to scale. What is the spring constant?

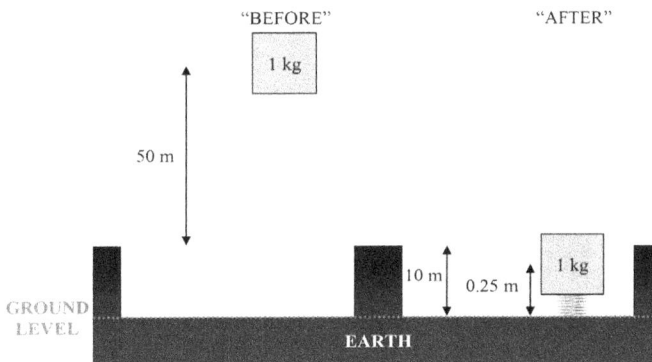

Spring Image Credit: yusufdemirci/Shutterstock.com.

☑ **ANSWER:** As we are about to see, the location of ground-level is completely arbitrary:

$$\cancel{W_{other}} + \cancel{KE_{initial}} + U_{g\,initial} + \cancel{U_{e\,initial}} = \cancel{KE_{final}} + U_{g\,final} + U_{e\,final}$$

$$1*9.8*(+60) = 1*9.8*(+0.25) + \frac{1}{2}K*(1.75)^2$$

$$588 = 2.45 + 1.53*K$$

$$585.55 = 1.53*K$$

$$K = 382.7 \; \text{New}\big/\text{meter}$$

☺ **COMMENTS:** If the bottom of the ditch is chosen to be ground-level, then the mass is initially +60 m above ground (instead of only +10 m) and likewise ends-up +0.25 m above ground (instead of −9.75 m below ground). Notice that the individual terms change values, but ultimately, the same total energy remains in the system as in the previous example. Thus, the value of the spring constant remains the same.

> When computing U_g, the choice of ground−level is completely arbitrary and will not affect the LCE.

2.3.7 Checking for understanding

The best way to gain speed and accuracy in solving LCE types of problems is through practice. Let's try several examples, and to make things interesting, we'll include some examples in which W_{other} makes a contribution to the final answer.

❶ **EX 1:** A 10 kg block is initially at rest. A monkey named Cornelius (a reference here to the 1968 movie, 'Planet of the Apes') pushes on the block with a constant force of 100 Newtons directed 35° below the x-axis as shown below. Assuming the ground is a frictionless surface, how fast is the block moving after it has been displaced 40 m to the right? The situation is drawn below in 'before' and 'after' photographs, but not to scale.

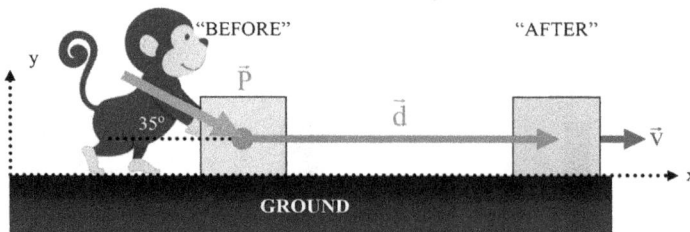

Monkey Image Credit: Studio_G/Shutterstock.com.

☑ **ANSWER:** The tell−tale sign that you should consider an LCE type of solution is that the statement of the problem describes a 'before' and 'after'situation—or an 'initial' and 'final' situation—or a 'point (A.)' and 'point (B.)' situation. In this problem, we see a monkey pushing on a block in the 'before' photograph, and the block moving at some speed in the 'after' photograph. Our job is to find the speed in the 'after' photograph.

$$W_{other} + KE_{initial} + U_{g_{initial}} + U_{c_{initial}} = KE_{final} + U_{g_{final}} + U_{c_{final}}$$

$$W_{push} = \frac{1}{2} \cdot 10 \cdot v^2$$

$$\vec{P} \bullet \vec{d} = 5v^2$$

$$\begin{pmatrix} 100 * \cos(35°) \\ -100 * \sin(35°) \end{pmatrix} \bullet \begin{pmatrix} 40 \\ 0 \end{pmatrix} = 5v^2$$

$$\begin{pmatrix} 81.9 \\ -57.4 \end{pmatrix} \bullet \begin{pmatrix} 40 \\ 0 \end{pmatrix} = 5v^2$$

$$3,276.6 = 5v^2$$

$$v = 25.6 \; \text{m}\!/\!\text{s}$$

☺ **COMMENTS:** Notice that the only term putting energy into the system is the work done by the applied push of the monkey. All of the 3276.6 joules of energy put into the system by the monkey is converted into the final Kinetic Energy.

❷ **EX 2:** Repeat the previous problem but assume the ground is not a frictionless surface. Instead, the coefficient of friction between the block and ground is 0.4. How fast is the block moving after it has been displaced 40 m to the right? The situation is drawn below in 'before' and 'after' photographs, but not to scale.

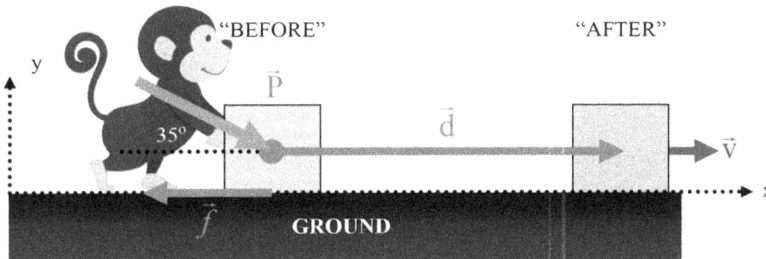

Monkey Image Credit: Studio_G/Shutterstock.com.

☑ **ANSWER:** Again, we know to use LCE as our problem-solving strategy because the statement describes a 'before' and 'after' situation. This problem is also complicated by the presence of two W_{other} terms—namely, W_{push} and $W_{friction}$.

$$W_{other} + \cancel{KE_{initial}} + \cancel{U_{g_{initial}}} + \cancel{U_{c_{initial}}} = KE_{final} + \cancel{U_{g_{final}}} + \cancel{U_{c_{final}}}$$

$$W_{friction} + W_{push} = \frac{1}{2} \cdot 10 \cdot v^2$$

$$\vec{f} \bullet \vec{d} + \vec{P} \bullet \vec{d} = 5v^2$$

$$\begin{pmatrix} -0.4*155.4 \\ 0 \end{pmatrix} \bullet \begin{pmatrix} 40 \\ 0 \end{pmatrix} + \begin{pmatrix} 100*\cos(35°) \\ -100*\sin(35°) \end{pmatrix} \bullet \begin{pmatrix} 40 \\ 0 \end{pmatrix} = 5v^2$$

$$\begin{pmatrix} -62.1 \\ 0 \end{pmatrix} \bullet \begin{pmatrix} 40 \\ 0 \end{pmatrix} + \begin{pmatrix} 81.9 \\ -57.4 \end{pmatrix} \bullet \begin{pmatrix} 40 \\ 0 \end{pmatrix} = 5v^2$$

$$-2,485.7 + 3,276.6 = 5v^2$$

$$v = 12.6 \; \frac{m}{s}$$

SIDE CALCULATION:

$$\vec{P} + \vec{f} + \vec{N} + \vec{W} = m \cdot \vec{a}$$

$$\begin{pmatrix} 100*\cos(35°) \\ -100*\sin(35°) \end{pmatrix} + \begin{pmatrix} -0.4*N \\ 0 \end{pmatrix} + \begin{pmatrix} 0 \\ +N \end{pmatrix} + \begin{pmatrix} 0 \\ -10*9.8 \end{pmatrix} = 10 \cdot \begin{pmatrix} a \\ 0 \end{pmatrix}$$

across y : $N = 155.4$ Newtons

across x : $a = 2 \; \frac{m}{s^2}$

☺ **COMMENTS:** Notice that an entire side calculation, using Newton's laws of motion, was needed just to determine the normal force, which in turn was needed to compute the frictional force. Often, when W_{other} terms are present, we must resort to some other problem-solving technique (usually Newton's laws of motion) to determine the force acting in these W_{other} terms. The lesson to learn here is that LCE types of problems are easy if $W_{other} = 0$, but can quickly get complicated if W_{other} terms are present.

❸ **EX 3:** Here's a problem that involves only symbols. A spring of spring constant K (you know the value of K) is in equilibrium and attached between a brick wall and a block of mass m (you know the value of m). The block is attached to another block of mass 2m by a rope. The mass m rests on a frictionless surface while 2m hangs over a frictionless pulley. We hold everything in place so that both masses are motionless. Suddenly, we release the masses—the block of mass m moves to the right while the block of mass 2m drops. What is the speed of both masses after they have been displaced by d (you know the value of d)? You should be able to solve for the speed of the blocks v, in terms of K, m and d. The 'initial' set-up is drawn below:

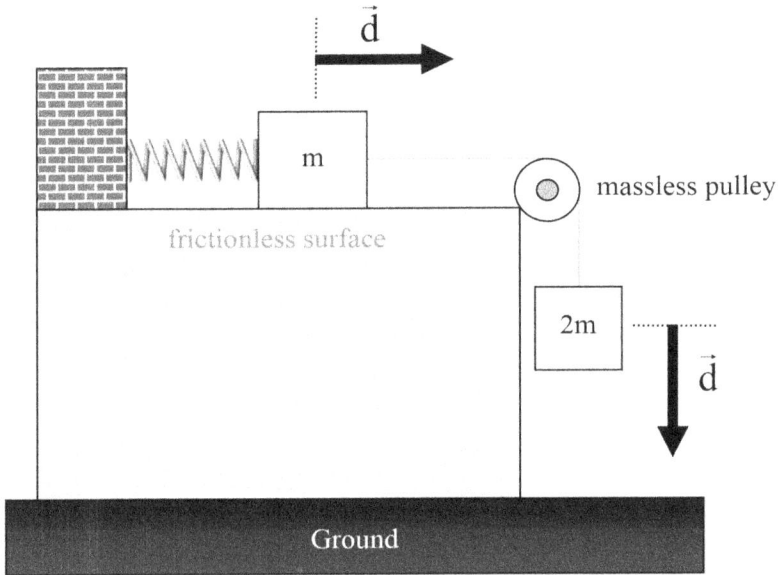

Spring Image Credit: yusufdemirci/Shutterstock.com.

☑ **ANSWER:** Again, we know to use LCE as our problem-solving strategy because the statement describes a 'before' and 'after' situation. However, this problem is more complicated than previous problems because we have two masses and multiple W_{other} terms—namely, W_{normal} and $W_{tension}$, where the normal force acts only on mass m and tension works on both mass m and mass 2m. To simplify our solution, let's make 'ground-level' the table-top for mass m, and the initial height of mass 2m as its 'ground-level.'

For every mass we see in the problem, we need to consider Kinetic Energy and Gravitational Potential Energy terms—thus, we need *two KE terms on both sides of the LCE equation* and *two U_g terms on both sides of the LCE equation.* For every spring we see in the problem, we need to consider an Elastic Potential Energy term—thus, we only need *one U_e term on both sides of the LCE equation.*

- Both masses are initially at rest so both initial Kinetic Energy terms are zero.
- Both masses are initially at their respective 'ground-levels' so both initial Gravitational Potential Energy terms are zero.
- The spring is initially at equilibrium so the initial Elastic Potential Energy term is zero.
- Finally, notice in the final state, the mass m is still on the table-top (its 'ground-level') so the final Gravitational Potential Energy is zero.

$$\underbrace{W_{\text{other}}}_{W_{\text{normal}} + W_{\text{tension}}} + \cancel{KE_{\text{initial-mass 1}}} + \cancel{U_{g\text{initial-mass 1}}} + \cancel{KE_{\text{initial-mass 2}}} + \cancel{U_{g\text{initial-mass 2}}} + \cancel{U_{e\text{initial}}} =$$

$$KE_{\text{final-mass 1}} + \cancel{U_{g\text{final-mass 1}}} + KE_{\text{final-mass 2}} + U_{g\text{final-mass 2}} + U_{e\text{final}}$$

$$\underbrace{W_{\text{normal}}}_{N \cdot d} + \underbrace{W_{\text{tension}}}_{T_1 \cdot d + T_2 \cdot d} = \underbrace{KE_{\text{final-mass 1}}}_{\frac{1}{2} \cdot m \cdot v^2} + \underbrace{KE_{\text{final-mass 2}}}_{\frac{1}{2} \cdot 2m \cdot v^2} + \underbrace{U_{g\text{final-mass 2}}}_{m \cdot g \cdot (-d)} + U_{e\text{final}}$$

$$\underbrace{\begin{pmatrix} 0 \\ +N \end{pmatrix} \cdot \begin{pmatrix} +d \\ 0 \end{pmatrix}}_{\substack{\text{Work done by normal-force} \\ \text{acting on mass m}}} + \underbrace{\begin{pmatrix} +T \\ 0 \end{pmatrix} \cdot \begin{pmatrix} +d \\ 0 \end{pmatrix}}_{\substack{\text{Work done by tension} \\ \text{acting on mass m}}} + \underbrace{\begin{pmatrix} 0 \\ +T \end{pmatrix} \cdot \begin{pmatrix} 0 \\ -d \end{pmatrix}}_{\substack{\text{Work done by tension} \\ \text{acting on mass 2m}}} = \frac{1}{2} \cdot m \cdot v^2 + \frac{1}{2} \cdot 2m \cdot v^2 + m \cdot g \cdot (-d) + \frac{1}{2} \cdot K \cdot d^2$$

$$0 + Td - Td = \frac{3}{2} \cdot m \cdot v^2 - m \cdot g \cdot d + \frac{1}{2} \cdot K \cdot d^2$$

$$0 = \frac{3}{2} \cdot m \cdot v^2 - m \cdot g \cdot d + \frac{1}{2} \cdot K \cdot d^2$$

$$\frac{3}{2} \cdot m \cdot v^2 = m \cdot g \cdot d - \frac{1}{2} \cdot K \cdot d^2$$

Now solve for v.

$$v = \sqrt{\frac{2}{3} \cdot \left(g \cdot d - \frac{Kd^2}{2m} \right)}.$$

To practice working with numbers, try plugging in these values to see that you get the correct answer: if $K = 150 \text{ N m}^{-1}$ (a pretty stiff spring), $m = 10$ kg (roughly 22 lbs) and $d = 0.25$ m (almost 10 inches), then the velocity of the mass is: $v = \sqrt{4/3}$ or 1.2 m s^{-1}.

☺ **COMMENTS:** A few items should be highlighted:

- Overall, no energy initially existed in the system. The energy gained by moving both blocks and stretching the spring is balanced by the energy lost by the gravitational field of the earth as block m moves down—thus, zero initial energy means we must have zero final energy.
- Only one rope exists in the problem so only one tension exists in the problem. However, the work done by the tension on mass m is canceled by its work done on mass 2m. The net result is that the rope does no work.
- Clearly the normal force is perpendicular to the displacement, so it too does no work.
- When computing the two final Kinetic Energy terms, students often choose to denote the speed of mass m by v_1 and the speed of mass 2m by v_2. With this notation, the two final Kinetic Energy terms have separate variables and cannot be combined. However, the masses are connected by a rope; thus any displacement of mass 2m is matched by a similar displacement of mass m (for example, imagine that if 2m moves 1 m in 1 s, then m must also move one meter in one second). The speed of mass m is equal to the speed of mass 2m— that's why we only need to denote one speed v, in the solution.
- We set the 'ground-level' of mass 2m to be its initial position. In the final state, the mass has *dropped* a distance, *d below* its initial position. Thus, the final Gravitational Potential Energy term must include a negative sign:

$m \cdot g \cdot (-d)$. Regardless of its 'ground-level,' mass 2m always ends-up a distance d *below* its starting point.

- Finally, in the final state, the spring is stretched by d. Since the formula for U_e involves the square of the displacement of the spring, whether the spring is squished or stretched by d, it would make the same contribution to the LCE equation.

2.3.8 Instructional input

2.3.8.1 Linear momentum

At the beginning of this chapter, we stated that we would define *two* new physical quantities that result in problem-solving strategies involving conservation principles. Now that we have defined 'work' (or 'energy') and solved problems using the Law of Conservation of Energy (which we abbreviated for convenience as the 'LCE'), we can define our second conserved physical quantity—namely, 'Linear Momentum.' As we'll see in the next chapter, in certain situations involving rotational motion, we will need to also define a physical quantity labeled 'Angular Momentum,' but for now, we will restrict our motion to linear motion and therefore assume that the word 'momentum' implies 'linear momentum.' We will use the same approach we used when developing the LCE: First, we will define momentum and list its various properties. Next, we will practice computing it in a context that is disconnected from problem-solving. Next, we will state the conditions under which it is conserved. After all, only because momentum is conserved is why we bothered to define it in the first place; thus, knowing the conditions under which momentum is conserved is key to understanding its utility. Finally, we will develop a problem-solving strategy for characteristic momentum situations and practice using our problem-solving strategy on various sample problems.

Consider a single particle (or object) of mass m, moving with velocity $\vec{v}$:

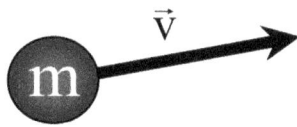

We define its 'Linear Momentum' as:

▶ **Definition of Linear Momentum:** $\vec{p} = m\vec{v}$

Etymology:
The word 'momentum' comes from the Latin word 'movimentum' or 'movere' that translates into 'to move,' 'to excite,' or 'to set into motion.'

Similar to our definition of work, the simple definition of linear momentum incorporates a number of sophisticated features that are worth itemizing and clarifying. As usual, we will check these features, one by one, when we actually calculate momentum in example problems:

1. To start, the adjective 'linear' is used in our definition because the particle is translating; that is to say, moving along a straight line. Even if the object is following a complicated curving path, the path can be thought of as being made of a series of smaller linear paths.

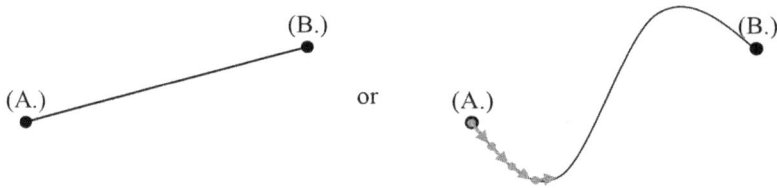

Situations do arise when instead of translating, the object is purely rotating (i.e., fixed at a single point yet spinning about that point). In these cases, we discuss 'Angular Momentum,' as opposed to 'Linear Momentum.' Again, in this chapter, we only consider situations in which an object is translating. Therefore, the object only has 'Linear Momentum.' For the sake of convenience, we will suspend the use of the word 'linear' with the understanding that the particle or object under consideration is indeed moving along a straight line or series of straight lines.

2. Momentum is symbolized by the vector, $\vec{p}$. You might wonder why it isn't more conveniently symbolized by the vector, $\vec{m}$ for 'momentum.' First, since scalar m is used for mass, be thankful that something other than vector $\vec{m}$ was chosen for momentum. If vector $\vec{m}$ were indeed used for momentum, its definition would be: $\vec{m} = m\vec{v}$, which I think we can all agree would be too confusing. Using the same letter for two distinct physical quantities, in the same definition, would be a really bad idea! However, the historical reason why vector $\vec{p}$ is used to symbolize momentum is actually quite interesting. As we are about to find out, the concept of momentum was invented by Isaac Newton. In his *Philosophiae Naturalis Principia Mathematica* (which we introduced in volume 1, chapter 4 of this series), Newton originally called 'Momentum' by a different name, 'Impetus.' The word 'Impetus' comes from the Latin word 'petere,' that translates into 'to go,' 'to seek,' or 'to pursue.' Thus, the vector $\vec{p}$ was a logical choice to symbolize 'Impetus.' As the concept of momentum evolved after Newton, the vector $\vec{p}$ was kept.

3. **Momentum is a vector!** In other words, momentum has x- and y-components that need to be separated by a comma and enclosed by a set of parentheses—that is how we agreed to write vectors in volume 1, chapter 1. Also, momentum obeys the rules for vector addition, subtraction, and multiplication.

4. The plural of 'momentum' is 'momenta.'
5. For a system of many particles (imagine a bunch of marbles ricocheting off of one another), the total momentum is just the vector sum of the individual momenta:

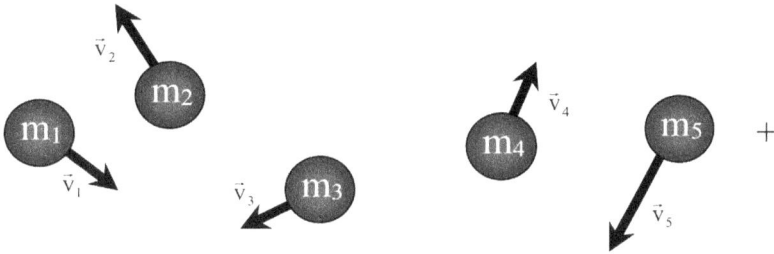

▶ **Definition of total linear momentum for a system of particles:**

$$\vec{p}_{\text{total}} = \vec{p_1} + \vec{p_2} + \vec{p_3} + \vec{p_4} + \vec{p_5} + \cdots$$
$$= m_1 \vec{v_1} + m_2 \vec{v_2} + m_3 \vec{v_3} + m_4 \vec{v_4} + m_5 \vec{v_5} + \cdots$$
$$= \sum_{i=1}^{n} m_i \vec{v_i}, \text{ where } n \text{ is the total number of particles in the system.}$$

6. Since we have defined momentum as a new physical quantity, we *might* need new units to measure momentum:

System	Unit of mass	Unit of velocity	Unit of momentum
m.k.s.	kg	m s^{-1}	$\frac{\text{kg} \cdot \text{m}}{\text{s}}$ are Newton·sec
c.g.s.	g	cm s^{-1}	$\frac{\text{g} \cdot \text{cm}}{\text{s}}$ ordyne · sec
English	slug	feet s^{-1}	$\frac{\text{slug} \cdot \text{ft}}{\text{s}}$ are Pound·sec

As we see above, we actually do not need a new set of units to measure momentum. The grouping of units for momentum does not get its own name. Of course, we will always use the **m.k.s.** system, kg · m s^{-1} (or Newton · s) for our calculations.

2.3.8.2 The 'law of conservation of momentum'
Like before, we are about to embark on a lengthy derivation—although not as long as the one involving the Law of Conservation of Energy. Also like before, the investment in time is well worth the headache. Ultimately, we will end up with

another equation involving a 'conserved' physical quantity. Like the Law of Conservation of Energy, the Law of Conservation of Momentum is one of the most powerful and widely-used weapons that a physicist can have in his or her arsenal of problem-solving techniques.

► **Derivation of the Law of Conservation of Momentum (abbreviated as the 'LCP'):**

Consider a system of many particles. You can imagine the previous example of a bunch of marbles ricocheting off of one another ... or imagine ping pong balls bouncing around inside a lottery tumbler ... or bumper cars (i.e., 'Dodgems') banging into each other at an amusement park. Considering all of the particles in the system, let's calculate the total change in momentum over time:

$$\frac{\Delta(\vec{p}_{total})}{\Delta t} = \frac{\Delta(\vec{p}_1 + \vec{p}_2 + \vec{p}_3 + \vec{p}_4 + \vec{p}_5 + \ldots)}{\Delta t}$$

We can separate each of the momenta out of the summation. In other words, the change of a sum of terms is the same as the sum of the change in each term.

$$\frac{\Delta(\vec{p}_{total})}{\Delta t} = \frac{\Delta(\vec{p}_1)}{\Delta t} + \frac{\Delta(\vec{p}_2)}{\Delta t} + \frac{\Delta(\vec{p}_3)}{\Delta t} + \frac{\Delta(\vec{p}_4)}{\Delta t} + \frac{\Delta(\vec{p}_5)}{\Delta t} + \ldots$$

Each individual momentum is just the product of a mass and velocity.

$$\frac{\Delta(\vec{p}_{total})}{\Delta t} = \frac{\Delta(m_1 \cdot \vec{v}_1)}{\Delta t} + \frac{\Delta(m_2 \cdot \vec{v}_2)}{\Delta t} + \frac{\Delta(m_3 \cdot \vec{v}_3)}{\Delta t} + \frac{\Delta(m_4 \cdot \vec{v}_4)}{\Delta t} + \frac{\Delta(m_5 \cdot \vec{v}_5)}{\Delta t} + \ldots$$

Let's assume the mass of each particle is constant. This assumption may not necessarily be true, but if so, each mass can be pulled outside of the Δ-operator. Each mass is just a constant scalar.

$$\frac{\Delta(\vec{p}_{total})}{\Delta t} = m_1 \cdot \frac{\Delta(\vec{v}_1)}{\Delta t} + m_2 \cdot \frac{\Delta(\vec{v}_2)}{\Delta t} + m_3 \cdot \frac{\Delta(\vec{v}_3)}{\Delta t} + m_4 \cdot \frac{\Delta(\vec{v}_4)}{\Delta t} + m_5 \cdot \frac{\Delta(\vec{v}_5)}{\Delta t} + \ldots$$

Each change in velocity with respect to time is just the acceleration of each particle.

$$\frac{\Delta(\vec{p}_{total})}{\Delta t} = (m_1 \cdot \vec{a_1}) + (m_2 \cdot \vec{a_2}) + (m_3 \cdot \vec{a_3}) + (m_4 \cdot \vec{a_4}) + (m_5 \cdot \vec{a_5}) + \cdots$$

At this point, each term involving the product of a mass and acceleration (i.e., each of the $m \cdot \vec{a}$ terms) represents the sum of *external* and *internal* forces on an individual particle. For example, $m_1 \cdot \vec{a_1}$ is not only the sum of external forces on m_1, but also represents the forces between m_1 and m_2 ... between m_1 and m_3 ... between m_1 and m_4 ... and so on. The net motion of each particle is due to any external forces on it as well as any interactions the particle may have with other particles in the system.

$$\frac{\Delta(\vec{P}_{\text{total}})}{\Delta t} = (\vec{F}\text{external} + \vec{F}^{m_2}_{\text{on}m_1} + \vec{F}^{m_3}_{\text{on}m_1} + \vec{F}^{m_4}_{\text{on}m_1} + \cdots) + (\vec{F}\text{external} + \vec{F}^{m_1}_{\text{on}m_2} + \vec{F}^{m_3}_{\text{on}m_2} + \vec{F}^{m_4}_{\text{on}m_2} + \cdots) + \cdots$$

However, this seemingly messy equation greatly simplifies when we realize that according to Newton's third law of motion, all of the internal forces cancel out one another! Since the interaction of each particle with every other particle is a pair-wise interaction, each interaction is an 'action-reaction' couple. Each internal force is equal, but oppositely directed, to its reaction pair. For example,

$$\vec{F}^{m_2}_{\text{on}m_1} = -\vec{F}^{m_1}_{\text{on}m_2} \text{ and } \vec{F}^{m_3}_{\text{on}m_1} = -\vec{F}^{m_1}_{\text{on}m_3} \text{ and } \vec{F}^{m_4}_{\text{on}m_1} = -\vec{F}^{m_1}_{\text{on}m_4} \text{ and} \ldots$$

Thus, we are left with a great simplified equation:

$$\frac{\Delta(\vec{p}_{\text{total}})}{\Delta t} = \vec{F}_{\substack{\text{external} \\ \text{on } m_1}} + \vec{F}_{\substack{\text{external} \\ \text{on } m_2}} + \vec{F}_{\substack{\text{external} \\ \text{on } m_3}} + \ldots$$

$$\frac{\Delta(\vec{p}_{\text{total}})}{\Delta t} = \sum \vec{F}_{\substack{\text{external} \\ \text{total}}}$$

The right hand side of the equation is just the sum of external forces on the system of particles.

We pause here in our derivation to highlight the historical significance of the equation in its current state. If the equation at this point reminds you of Newton's second law of motion, you are absolutely correct! This is the form of Newton's second law of motion as originally stated by Newton in his *Principia* (recalling that he used the term 'Impetus' instead of 'Momentum'). Newton's second law of motion was not originally formulated as: $\sum \vec{F}_{\substack{\text{external} \\ \text{total}}} = m \cdot \vec{a}$, but rather as:

$\sum \vec{F}_{\substack{\text{external} \\ \text{total}}} = \Delta \vec{p}_{\text{total}}/\Delta t$. Although the two equations are somewhat interchangeable, you will see the advantage of Newton's original formulation below.

Returning to our derivation, let's examine the system of many particles drawn below. We've only drawn five masses because of limited space, but we can imagine that hundreds, or thousands, of particles are actually present. The question we want to consider is: *'How do we define our system?'* How do we define our 'system' in such a way as to simplify our derivation and make it somehow useful for problem-solving?

Let's start by trying something small. Let's define the 'system' as only being comprised of particles 1, 2, and 3. To indicate our choice of a system, we draw an oddly-shaped, red, dotted line around the particles of interest. This line represents a huge flexible balloon that can expand to any size, and take on any shape, depending on how particles 1, 2, and 3 move and ricochet ... but the balloon is limited to contain only particles 1, 2, and 3. Unfortunately, if we define the 'system' as only particles 1, 2, and 3, then complicated external forces exist. For instance, if the 'system' is only comprised of particles 1, 2, and 3, then the interaction of particle

4 with the system is an external force ... the interaction of particle 5 with the system is an external force ... the interaction of particle 238 with the system is an external force ... and so on. If particle 4 strikes particle 3, then something *outside of the system* has impacted our system. Particle 4 must have somehow penetrated the balloon and interacted with our small system. Therefore, defining the 'system' as particles 1, 2, and 3 means that complicated external forces need to be determined. These forces may be so complicated that they may be too difficult to even calculate. How do we estimate the force of particle 4 ricocheting off of particle 3? How do we estimate the force of particle 238 ricocheting off of particle 2? Do some of the particles stick together (i.e., pieces of clay)? Do some of the particles explode on contact (i.e., two firecrackers)? Clearly, defining the 'system' as particles 1, 2, and 3 makes life complicated—it is an unwise decision.

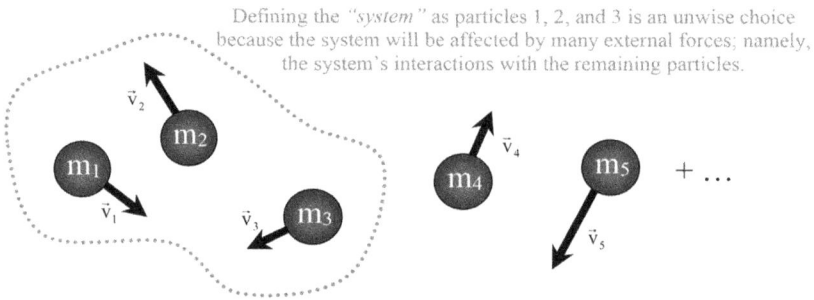

Defining the *"system"* as particles 1, 2, and 3 is an unwise choice because the system will be affected by many external forces; namely, the system's interactions with the remaining particles.

m_1 $\vec{v}_1$ $\vec{v}_2$ m_2 $\vec{v}_3$ m_3 m_4 $\vec{v}_4$ m_5 $\vec{v}_5$ + ...

However, watch what happens if the 'system' is defined as all *n* particles. Again, we construct a huge flexible balloon that can expand to any size, and take on any shape, depending on how the *n* particles move. We symbolize our choice of the 'system' by drawing an oddly-shaped, red, dotted line around all *n* particles. If we define the system as all *n* particles as shown below, then no external forces impact the system. In other words, no *outside forces* penetrate the 'system.' The particles in our system are still free to bounce around and ricochet into one another. The particles can move in any direction since the balloon can expand and take on any shape (literally imagine millions of air molecules bouncing around inside a balloon or hundreds of packing peanuts mixing inside a cardboard container). Since all of the particles are contained within the balloon, nothing remains to penetrate the balloon from the outside world. Not only are no other particles present to penetrate the balloon, but no other outside forces exist either—no springs touch the system from the outside world; no ropes or rods are connected to the system from the outside world; no frictional forces drag on the system from the outside world; etc. Thus, by making a wise choice of 'system,' we can eliminate all of the external forces acting on it. Mathematically, this translates into setting: $\sum \vec{F}^{\text{external}}_{\text{total}} = 0$.

By defining the "*system*" as all *n* particles, no external forces impact the system.

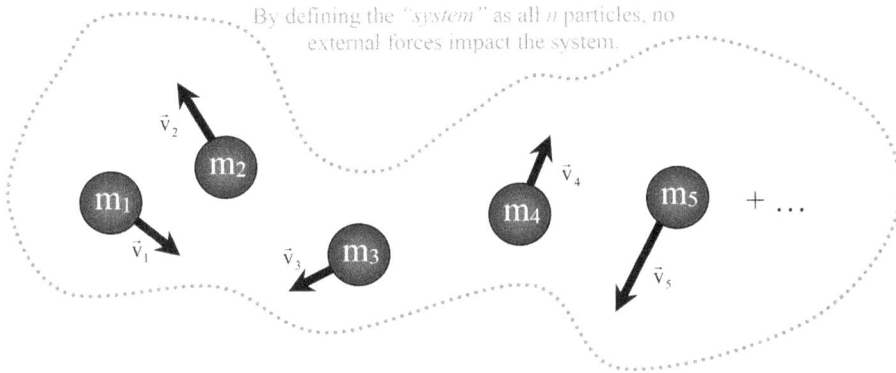

Knowing that a wise choice of 'system' will eliminate all of the external forces on the system, we can complete our derivation. Let's pick up the derivation from our last line—the original formulation of Newton's second law of motion:

$$\frac{\Delta\left(\vec{p}_{total}\right)}{\Delta t} = \sum \vec{F}_{external \atop total}$$

Multiply both sides by Δt.

$$\Delta\left(\vec{p}_{total}\right) = \left(\sum \vec{F}_{external \atop total}\right) \cdot \Delta t$$

The Δ-operator always represents subtracting the "initial" state from the "final" state.

$$\left(\vec{p}_{total}\right)_{final} - \left(\vec{p}_{total}\right)_{inital} = \left(\sum \vec{F}_{external \atop total}\right) \cdot \Delta t$$

We can choose the "system" in such a way so that no external forces exist on the system.

The Law of Conservation of Momentum:

$$\left(\vec{p}_{total}\right)_{final} - \left(\vec{p}_{total}\right)_{inital} = \left(\sum \vec{F}_{external \atop total}\right) \cdot \Delta t$$

Therefore, if $\sum \vec{F}_{external \atop total} = 0$, then: $\left(\vec{p}_{total}\right)_{initial} = \left(\vec{p}_{total}\right)_{final}$.

This final equation represents the 'Law of Conservation of Momentum' (abbreviated as '**LCP**'). The derivation of the LCP did require several pages of mathematical manipulation, but as we'll soon see, the derivation will be well worth the effort. Before tackling some example problems, let's enumerate some of the features of the LCP as well as some of the items to monitor whenever we attempt to solve problems. Most importantly, let's list the situations in which the LCP is most often used in the context of problem-solving:

1. The LCP is a vector equation. In other words, we have to keep all terms in the LCP in vector notation, with separate x- and y-components. In full-blown vector notation, the LCP is written as:

$$\begin{pmatrix} P_{x-\text{total}} \\ P_{y-\text{total}} \end{pmatrix}_{\text{final}} - \begin{pmatrix} P_{x-\text{total}} \\ P_{y-\text{total}} \end{pmatrix}_{\text{inital}} = \begin{pmatrix} \sum F_{x-\text{external} \atop \text{total}} \\ \sum F_{y-\text{external} \atop \text{total}} \end{pmatrix} \cdot \Delta t$$

2. The LCP originates from the original formulation of Newton's second law of motion and is used to solve two types of characteristic problems:

a) **Collisions and explosions:** The key to making the LCP useful is to find situations in which $\sum \vec{F}_{\text{external} \atop \text{total}} = 0$. In such cases, the right-hand side of the equation is zero and we are left with $(\vec{P}_{\text{total}})_{\text{initial}} = (\vec{P}_{\text{total}})_{\text{final}}$. Luckily, a set of situations exists that are tailor-made for just such a formulation—namely, 'collisions' and 'explosions.' We will examine these situations in upcoming examples, but the take-home message here is that collisions and explosions are problems best tackled by the LCP. Our general approach will be to label the 'initial' situation (before the collision or explosion) and the 'final' situation (after the collision or explosion), then equate the total momenta of the two situations. Even if the external forces along one axis add up to zero, the LCP may turn out to be extremely useful. Certain situations do occur in which the sum of the external forces is zero *only along one axis*. In these situations, momentum is conserved along that single axis. For example, if the external forces only along the x-axis sum to zero, the LCP simplifies as:

$$\text{If} \sum F_{x-\text{external} \atop \text{total}} = 0, \text{ then} \begin{pmatrix} P_{x-\text{total}} \\ P_{y-\text{total}} \end{pmatrix}_{\text{final}} - \begin{pmatrix} P_{x-\text{total}} \\ P_{y-\text{total}} \end{pmatrix}_{\text{inital}} = \begin{pmatrix} 0 \\ \sum F_{y-\text{external} \atop \text{total}} \end{pmatrix} \cdot \Delta t.$$

Therefore, $(P_{x-\text{total}})_{\text{final}} = (P_{x-\text{total}})_{\text{initial}}.$

b) **Impulse or quick strikes:** The other scenario in which the LCP is typically employed is called an 'Impulse' or 'Quick Strike' situation. In these instances, specific information is known about the external forces acting on the system as well as the time interval over which those forces are exerted. For example, if we know that a baseball bat exerts a certain force on a moving baseball for exactly 0.1 s, then we know everything on the right-hand side of the LCP equation and can therefore compute the change in momentum of the ball. In order for impulse or quick strike problems to be useful, detailed information has to be known about the external forces and the time interval over which they are exerted.

3. Typically, when dealing with collisions and explosions, quantities occurring after the collision or explosion are denoted with a 'prime' (i.e., the 'tick mark'). For example, the velocity of a baseball after being hit might be denoted as $v'_{baseball}$ while the momentum of a set of marbles after a collision might be symbolized by $p'_{marbles}$. Notice that a 'prime' or 'tick mark' has been added to each symbol.

2.3.9 Modeling

❶ **EX 1:** Imagine you are in a pool hall hanging from a ceiling fan directly above a pool table. You look down at the pool table and witness the following collision between a small cue ball of mass 1 kg and a much heavier 8-ball of mass 2 kg. (In reality, the cue ball is the heaviest of a standard set of pool balls at 6 ounces, or 0.170 kg, while the remaining 15 balls each weigh 5.5 ounces, or 0.156 kg.) The cue ball was initially traveling 5 m s^{-1} to the right while the 8-ball was at rest. After the collision, the 8-ball is observed to be traveling 3.33 m s^{-1} to the right. Calculate the final velocity of the cue ball, denoted with a 'prime' by v'_1. The situation is drawn below in 'before' and 'after' photographs, but not to scale.

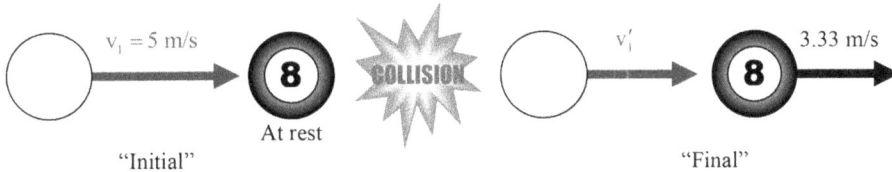

✓ ANSWER: Since this is a collision, let's attack the problem with the LCP. If we define the 'system' as both the cue ball *and* the 8-ball, then no external forces penetrate the system. In other words, if we draw an oddly-shaped, red, dotted line around both the cue ball *and* 8-ball (to represent a balloon that can expand to any size and shape), then the collision takes place totally *inside* the balloon.

Since $\sum \vec{F}_{external \atop total} = 0$, then: $\left(\vec{p}_{total}\right)_{initial} = \left(\vec{p}_{total}\right)_{final}$.

$$\left(\vec{p}_{total}\right)_{initial} = \left(\vec{p}_{total}\right)_{final}$$

The momentum of each ball is the product of its mass and velocity, written in vector notation.

$$1*\begin{pmatrix}5\\0\end{pmatrix}+2*\begin{pmatrix}0\\0\end{pmatrix}=1*\begin{pmatrix}v'_1\\0\end{pmatrix}+2*\begin{pmatrix}3.33\\0\end{pmatrix}$$

The x-axis contains the information to solve for the final velocity of the cue ball.

$$v'_1 = -1.66 \text{ m/s}$$

☺ **COMMENTS:** Let's comment on a few aspects of our analysis:

1. When you see a collision or explosion, you should immediately think of LCP as the first approach to solving the problem.

2. To start an LCP problem, you have to define your 'system.' Consider constructing a mental balloon around the particles you are examining. Ask yourself: *'Does any force penetrate this balloon? Does anything from the outside world enter this balloon?'* If you can eliminate all external forces, the LCP equation simply becomes: $(\vec{p}_{total})_{initial} = (\vec{p}_{total})_{final}$.

3. Notice that vector notation was necessary since momentum is indeed a vector. Each object's velocity was written as a vector. Each of these velocities was then multiplied by the corresponding object's mass, which is a scalar quantity.

4. The negative sign on your final answer of v'_1 indicates that the cue ball moves to the left after the collision. When we wrote the velocity of the cue ball after the collision, we wrote it as a vector with the variable v'_1 nested inside the parentheses. The variable v'_1 is to be viewed as a place-holder for the unknown magnitude of the cue ball's velocity immediately following the collision. After plugging into the LCP, v'_1 is returned with the negative value of $-1.66\ \text{m s}^{-1}$. The negative sign therefore indicates that the place-holder was storing a negative value; thus, the cue ball moved to the left after the collision.

❷ **EX 2:** Imagine you are in a pool hall hanging from a ceiling fan directly above a pool table. You look down at the pool table and witness the following collision between a small cue ball of mass 1 kg and a much heavier 8-ball of mass 2 kg. The cue ball was initially traveling 5 m s^{-1} to the right while the 8-ball was at rest. After the collision, the cue ball moves off at 30° while the 8-ball moves off at 45° as shown below. Calculate the final velocity of the cue ball (denoted with a 'prime' by v'_1) and 8-ball (denoted with a 'prime' by v'_2).

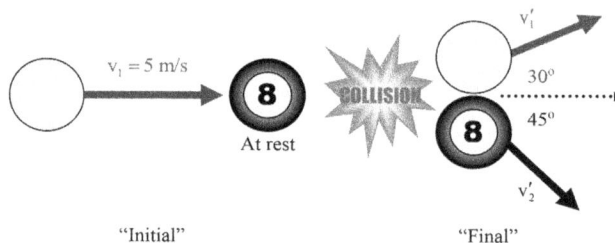

"Initial"   "Final"

☑ **ANSWER:** Again, when you see a collision, think LCP. If we define the 'system' as both the cue ball *and* the 8-ball, then no external forces penetrate the system and LCP reduces to: $(\vec{p}_{total})_{initial} = (\vec{p}_{total})_{final}$.

Since $\sum \vec{F}_{external \atop total} = 0$, then: $\left(\vec{p}_{total}\right)_{initial} = \left(\vec{p}_{total}\right)_{final}$.

$$\left(\vec{p}_{total}\right)_{initial} = \left(\vec{p}_{total}\right)_{final}$$

The momentum of each ball is the product of its mass and velocity, written in vector notation.

$$1 * \binom{5}{0} + 2 * \binom{0}{0} = 1 * \binom{v_1' * \cos\left(30^o\right)}{v_1' * \sin\left(30^o\right)} + 2 * \binom{v_2' * \cos\left(45^o\right)}{-v_2' * \sin\left(45^o\right)}$$

The expression can be simplified. Note that:

- $\cos\left(30^o\right) = \dfrac{\sqrt{3}}{2}$,
- $\sin\left(30^o\right) = \dfrac{1}{2}$, and
- $\cos\left(45^o\right) = \sin\left(45^o\right) = \dfrac{\sqrt{2}}{2}$.

$$\binom{5}{0} = \binom{\frac{\sqrt{3}}{2} \cdot v_1'}{\frac{1}{2} \cdot v_1'} + \binom{\sqrt{2} \cdot v_2'}{-\sqrt{2} \cdot v_2'}$$

Vector notation keeps information along the x-axis separate from information along the y-axis.

across x: $5 = \dfrac{\sqrt{3}}{2} \cdot v_1' + \sqrt{2} \cdot v_2'$

across y: $0 = \dfrac{1}{2} \cdot v_1' - \sqrt{2} \cdot v_2'$

Two equations with two unknowns can be solved.

$v_1' = 3.66$ m/s and $v_2' = 1.29$ m/s

☺ **COMMENTS:** Let's comment on a few aspects of our analysis:
1. Again, when you see a collision or explosion, think LCP!
2. You should start seeing a pattern in our LCP problem-solving approach: In a collision, if you define the 'system' as *all* of the particles involved in the collision, you have eliminated all of the external forces. Thus, the LCP equation reduces to: $(\overrightarrow{p}_{total})_{initial} = (\overrightarrow{p}_{total})_{final}$.
3. Finally, the momentum of both particles (the cue ball and 8-ball) before the collision and after the collision must both be written as a vector. The directions and angles are important and have to be incorporated into the notation.

❸ **EX 3:** A 10 kg cannonball is launched out of a cannon at a velocity of 100 m s^{-1}, directed 30° above the horizon. The cannonball explodes into three fragments—namely: a 2 kg fragment moves 75° above the horizon at an unknown velocity (denoted with a 'prime' by v′₁); a 4 kg fragment moves completely horizontally at 25 m s^{-1}; and a 4 kg fragment moves 60° below the horizon at an unknown velocity (denoted with a 'prime' by v'_2). Solve for the two unknown velocities.

"Initial"   "Final"

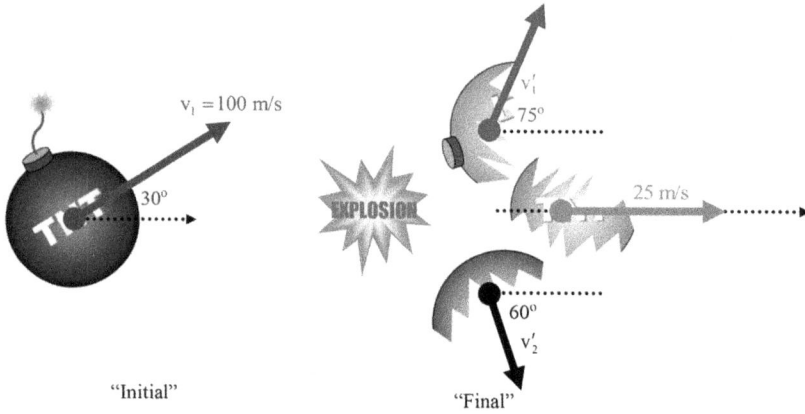

☑ **ANSWER:** First, notice that an 'explosion' is essentially a 'collision' run in reverse. Again, when you see a collision or explosion, think LCP. If we define the 'system' as the cannonball, which after the explosion becomes three fragments, then no external forces penetrate the system and LCP reduces to: $(\vec{p}_{total})$initial $= (\vec{p}_{total})$final. In other words, you can think of the LCP reducing to: 'The momentum of the cannonball before the explosion equals the total momentum of the three fragments after the explosion.'

Since $\sum \vec{F}_{\substack{external \\ total}} = 0$, then: $\left(\vec{p}_{total}\right)_{initial} = \left(\vec{p}_{total}\right)_{final}$.

$$\left(\vec{p}_{total}\right)_{initial} = \left(\vec{p}_{total}\right)_{final}$$

The momentum of each fragment is the product of its mass and velocity, written in vector notation.

$$10 * \begin{pmatrix} 100*\cos(30°) \\ 100*\sin(30°) \end{pmatrix} = 2 * \begin{pmatrix} v_1' * \cos(75°) \\ v_1' * \sin(75°) \end{pmatrix} + 4 * \begin{pmatrix} 25 \\ 0 \end{pmatrix} + 4 * \begin{pmatrix} v_2' * \cos(60°) \\ -v_2' * \sin(60°) \end{pmatrix}$$

The expression can be simplified by using the numerical values of the trigonometric quantities.

$$\begin{pmatrix} 866 \\ 500 \end{pmatrix} = \begin{pmatrix} 0.518 \cdot v_1' \\ 1.93 \cdot v_1' \end{pmatrix} + \begin{pmatrix} 100 \\ 0 \end{pmatrix} + \begin{pmatrix} 2 \cdot v_2' \\ -3.46 \cdot v_2' \end{pmatrix}$$

Vector notation keeps information along the x-axis separate from information along the y-axis.

across x: $866 = 0.518 \cdot v_1' + 100 + 2 \cdot v_2'$

across y: $500 = 1.93 \cdot v_1' - 3.46 \cdot v_2'$

Two equations with two unknowns can be solved.

$v_1' = 645.8$ m/s and $v_2' = 215.7$ m/s.

☺ **COMMENTS:** Let's comment on a few aspects of our analysis:

1. Remember, 'collisions' are 'explosions' run in reverse and 'explosions' are 'collisions' run in reverse. In other words, 'collisions' and 'explosions' are the same phenomenon as far as the LCP is concerned. For the last time, when you see a collision or explosion, think LCP!

2. The characteristic approach to solving LCP problems should be cemented in your brain by now. In a collision or explosion, if you define the 'system' as all of the particles involved in the collision, you have eliminated all of the external forces. Thus, the LCP equation reduces to: $(\overrightarrow{p_{total}})_{initial} = (\overrightarrow{p_{total}})_{final}$.

3. Finally, the momentum of all particles before the collision/explosion and after the collision/explosion must both be written as a vector. The directions and angles are important and have to be incorporated into the notation.

2.3.10 Checking for understanding

2.3.10.1 What if an object's mass changes?

Before wrapping up this chapter on conservation laws, one final remark about the LCP is worth noting. In volume 1, chapter 4, we stated that Newton's second law of motion had the form:

$$\sum \overrightarrow{F}_{external \atop total} = m \cdot \overrightarrow{a},$$

while in this chapter, we stated that the original form of his second law of motion (as published in Newton's *Philosophiae Naturalis Principia Mathematica*) was:

$$\sum \overrightarrow{F}_{external \atop total} = \frac{\Delta(\overrightarrow{P_{total}})}{\Delta t}.$$

You might be wondering if the two forms are identical? If you look at our derivation of the LCP, you will see a small, seemingly insignificant line where we assumed that the mass of each particle in our 'system' was constant. If the mass of each particle in a system is indeed constant, the two forms of Newton's second law of motion are identical. This can be derived below in only a few lines—consider a single particle (i.e., a marble) of constant mass, m:

$$\sum \vec{F}_{external \atop total} = \frac{\Delta(\vec{p}_{total})}{\Delta t}$$

The object's momentum is just the product of its mass and velocity.

$$\sum \vec{F}_{external \atop total} = \frac{\Delta(m \cdot \vec{v}_{total})}{\Delta t}$$

If the object's mass is constant, the mass can be pulled out of the Δ-operator and treated as a constant scalar.

$$\sum \vec{F}_{external \atop total} = m \cdot \frac{\Delta(\vec{v}_{total})}{\Delta t}$$

Last, use our kinematic definition of acceleration.

$$\sum \vec{F}_{external \atop total} = m \cdot \vec{a}_{total}.$$

In the case of an object with changing mass, the actual form of Newton's second law of motion is the one presented in this chapter; namely:

$$\sum \overrightarrow{F}_{external \atop total} = \frac{\Delta(\overrightarrow{P_{total}})}{\Delta t}.$$

This is the correct form of Newton's second law of motion since it takes into account the possibility of an object not having a constant mass.

$$\sum \vec{F}_{\text{external} \atop \text{total}} = \frac{\Delta(\vec{p}_{\text{total}})}{\Delta t}$$

$$= \frac{\Delta(m \cdot \vec{v})}{\Delta t}$$

$$= m \cdot \frac{\Delta(\vec{v})}{\Delta t} + \vec{v} \cdot \frac{\Delta(m)}{\Delta t}$$

For example, as a rocket flies into space, it consumes more and more fuel. Thus, the mass of a rocker *decreases* as it move along its flight. As another example, consider a snowball falling through a snowstorm. The snowball scoops-up all the snow in its path. Any snowflake that hits the snowball sticks to the snowball. Thus, the mass of the snowball *increases* as it falls during a snowstorm.

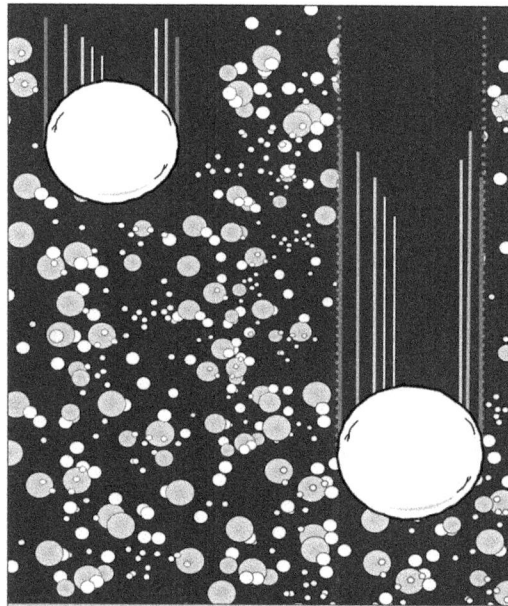

As a snowball falls through a snowstorm, it accumulates more and more snowflakes. Thus, the mass of the snowball increases as the snowball moves. The original formulation of Newton's second law of motion is needed to tackle the motion of such a snowball.

Image Credit: Author.

2.4 Keeping information

2.4.1 Closure

To close our discussion of LCE and LCP, let's focus on conceptual questions regarding the various terms appearing in the LCE and LCP equations. Some of the problems below can be solved with LCE, some with LCP, and some require both. Recognizing the characteristic patterns of LCE- and LCP-types of problems are part of the physicist's job.

2.4.1.1 LCE and LCP

We'll start with a great example that will really test your understanding of the vector 'Dot'-product and how it is used to calculate energy terms.

❶ EX 1: A mass is connected to a length of rope to form a pendulum. The photograph below shows the pendulum at two locations: points (A) and (B). The mass is released from rest from position (A). We want to find the speed of the mass at point (B).

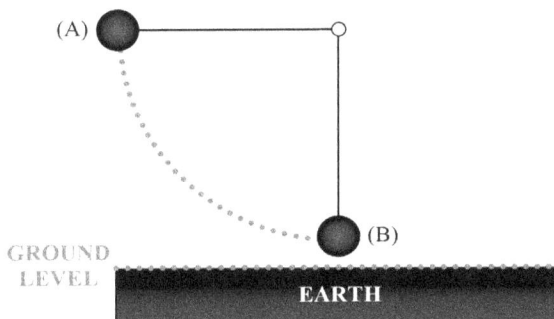

We set up the LCE equation to include: $U_{g\text{-initial}}$ at point (A), and KE_{final} at point (B). So far, so good! The only remaining question is whether or not to include W_{other} because of the presence of the rope.

Does the rope do any work in this problem—and therefore do we need to include W_{other}?

☑ **ANSWER:** Recall from our discussions on Newton's laws of motion that the tension in a rope is always directed away from the object. Also recall from our discussions on Uniform Circular Motion, that the velocity of an object in circular motion is tangent to the circle—thus, the mass is making small displacements along paths that are tangent to the circle. The tension and displacement is shown at points (A) and (B), and at two points in between, in the figure below:

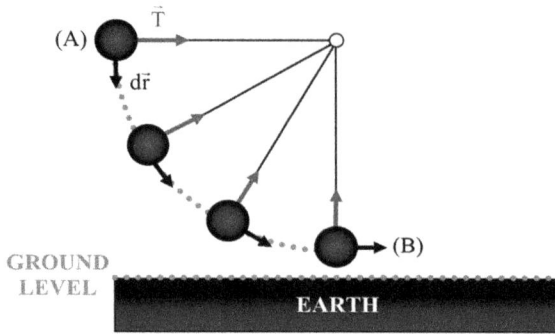

Since the displacement is always perpendicular to the tension, the tension does not do any work. Work is only done by the component of a force that is parallel to a displacement—this is the physical interpretation of the vector 'Dot'-product that was stressed when vector multiplication was first introduced earlier in this chapter.

☺ **COMMENTS:** The above problem is actually trickier than it may look at first glance. Notice that the tension in the rope varies between points (A) and (B). Use Newton's laws of motion at any point and you'll see that both its magnitude and direction change at every point between points (A) and (B). Thus, to correctly calculate the work done by the rope (i.e., W_{other}), we need to use the formula for work done by a non-constant, or varying, force: $W = \int_{(A.)}^{(B.)} \vec{F} \cdot d\vec{r}$. Fortunately, this formula doesn't change the fact that the tension and infinitesimal displacements of the mass $d\vec{r}$, are always perpendicular to one another; thus all of the infinitesimal work contributions (i.e., $\vec{T} \cdot d\vec{r}$) are zero.

❷ **EX 2:** This problem illustrates how multiple problem-solving strategies can be used to describe one situation—it will nicely demonstrate how each problem-solving strategy 'specializes' in solving certain types of problems. A 200 kg mass is lifted 30 m by a single cable with an acceleration of 1.47 m s^{-2} upward.

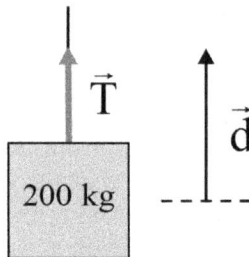

   a. Determine the tension in the cable.
   b. Determine the net work done on the mass.
   c. Determine the work done by the cable on the mass.
   d. Determine the work done by gravity on the mass.
   e. Determine the final speed of the mass assuming it started from rest.

☑ ANSWER:

a) To determine a force, our primary problem-solving strategy is Newton's laws of motion:

$$\vec{T} + \vec{W} = m \cdot \vec{a}$$

$$\begin{pmatrix} 0 \\ +T \end{pmatrix} + \begin{pmatrix} 0 \\ -200 \cdot 9.8 \end{pmatrix} = 200 \cdot \begin{pmatrix} 0 \\ 1.47 \end{pmatrix}$$

$$T - 1960 = 294$$

$$T = 2254 \text{ Newtons.}$$

b) To determine the net work, we must change our problem-solving strategy to the definition of work/energy:

$$W_{total} = \vec{F}_{total} \bullet \vec{d}$$

$$= m \cdot \vec{a} \bullet \vec{d}$$

$$= 200 \cdot \begin{pmatrix} 0 \\ +1.47 \end{pmatrix} \bullet \begin{pmatrix} 0 \\ +30 \end{pmatrix}$$

$$= 8820 \text{ joules.}$$

c) To determine the work done by the cable, we need to determine the work done by tension. Again, we need to use the definition of work/energy:

$$W_{tension} = \vec{T} \bullet \vec{d}$$

$$= \begin{pmatrix} 0 \\ +2254 \end{pmatrix} \bullet \begin{pmatrix} 0 \\ +30 \end{pmatrix}$$

$$= 67\ 620 \text{ joules.}$$

d) To determine the work done by gravity, we again need to use the definition of work/energy:

$$W_{gravity} = \vec{W} \bullet \vec{d}$$

$$= \begin{pmatrix} 0 \\ -200 \cdot 9.8 \end{pmatrix} \bullet \begin{pmatrix} 0 \\ +30 \end{pmatrix}$$

$$= -\ 58\ 800 \text{Joules.}$$

e) To determine the final speed, we must change our problem-solving strategy to the LCE equation. Assume that the mass starts at 'ground-level'— remember that when using the LCE equation, we can arbitrarily choose the 'ground-level' when computing U_g terms:

$$W_{other} + \underbrace{\cancel{KE_{initial}}}_{\substack{\text{starts from rest}}} + \underbrace{\cancel{U_{g\,initial}}}_{\substack{\text{starting point} \\ \text{is the ground}}} = KE_{final} + U_{g\,final}$$

$$W_{tension} = \frac{1}{2} \cdot 200 \cdot v^2 + 200 \cdot 9.8 \cdot (+30)$$

$$67,620 = 100 \cdot v^2 + 58,800$$

$$8,820 = 100 \cdot v^2$$

$$v = 9.4 \ ^{m}\!/_{s}$$

☺ **COMMENTS:** Notice how the individual work terms calculated in parts (b), (c) and (d), all appeared in the LCE equation.

❸ **EX 3:** Let's quickly review some quick facts about the various terms appearing in the LCE equation. 'Kinetic Energy' is the energy associated with motion while 'Potential Energy' is the energy associated with position.
  a) Can Kinetic Energy be negative?
  b) Can Gravitational Potential Energy be negative?
  c) Can Elastic Potential Energy be negative?
  d) Is the work done by friction always positive or negative?

☑ **ANSWER:**
  a) **No**—the term $KE = 1/2\, m \cdot v^2$ can never be negative because mass is always positive and the v term is squared.
  b) **Yes**—the term $U_g = m \cdot 9.8 \cdot y$ can be negative or positive depending on whether the y-coordinate is above (i.e., U_g will be positive) or below (i.e., U_g will be negative) the arbitrarily chosen 'ground-level' in the statement of the problem.
  c) **No**—the term $U_e = 1/2\, K \cdot x^2$ can never be negative because the spring constant K, is always positive and the x term is squared.
  d) **Negative**—by definition, friction is always directed opposite to the direction of motion or displacement. Therefore the frictional force and the displacement will always be oriented $180°$ to one another (technically, physicists say they are 'anti-parallel'). Thus, when we 'Dot' the frictional force and displacement (i.e., $W_{friction} = \overrightarrow{f} \cdot \overrightarrow{d}$), we will always get a negative work/energy.

❹ **EX 4:** Imagine you are in a helicopter hovering over the ocean while a strange event unfolds. A fisherman named Santiago (a reference here to Ernest Hemingway's classic novella, *The Old Man and the Sea*), in his boat, has just simultaneously speared a large swordfish and shark. Both creatures remain motionless as they are reeled aboard the boat by Santiago. Together, Santiago and his boat have a mass of 12 000 kg. As he reels the two creatures aboard his boat, you witness:
  • His boat move 75 m, at an angle of $45°$ north of east,
  • The shark move 150 m, at an angle of $30°$ north of west, and
  • The swordfish move 45 m, at an angle of $70°$ south of west.

Determine the mass of the shark and swordfish. The event is shown below.

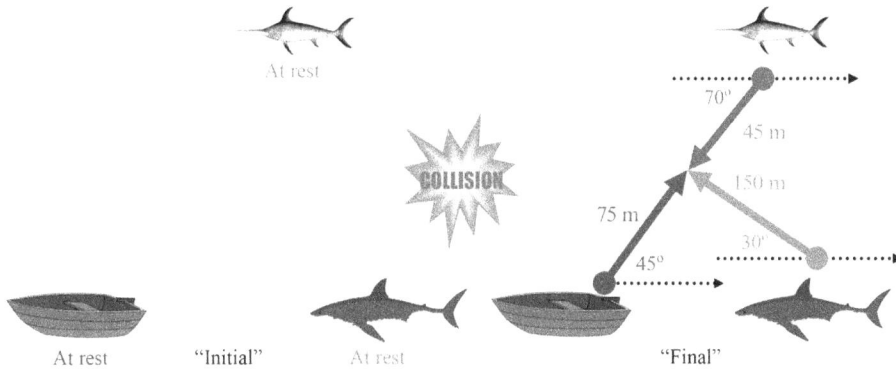

Boat Image Credit: Pixabay by Clker-Free-Vector-Images.
Marlin Image Credit: Pixabay by Clker-Free-Vector-Images. Shark Image
Credit: Pixabay by mostafaelturkey97.

☑ **ANSWER:** Looking at the drawing, you can easily convince yourself that this is nothing more than a 'collision.' We have seen that the first approach to solving any problem involving a 'collision' or 'explosion' is LCP. If we define the 'system' as the boat/fisherman, shark, and swordfish, then no external forces penetrate the system and LCP reduces to: $(\vec{p}_{total})_{initial} = (\vec{p}_{total})_{final}$. In words, you can think of the LCP reducing to: 'Since no momentum exits to start (i.e., everything is motionless), there should be no momentum in the end.'

$$\text{Since } \sum \vec{F}_{external \atop total} = 0, \text{ then: } \left(\vec{p}_{total}\right)_{initial} = \left(\vec{p}_{total}\right)_{final}.$$

$$\left(\vec{p}_{total}\right)_{initial} = \left(\vec{p}_{total}\right)_{final}$$

The momentum of each fragment is the product of its mass and velocity, written in vector notation.

$$12,000 \cdot \begin{pmatrix} 0 \\ 0 \end{pmatrix} + M_{sh} \cdot \begin{pmatrix} 0 \\ 0 \end{pmatrix} + M_{sw} \cdot \begin{pmatrix} 0 \\ 0 \end{pmatrix} = 12,000 \cdot \begin{pmatrix} 75 \cdot \cos(45°) \\ 75 \cdot \sin(45°) \end{pmatrix} + M_{sh} \cdot \begin{pmatrix} -150 \cdot \cos(30°) \\ 150 \cdot \sin(30°) \end{pmatrix} + M_{sw} \cdot \begin{pmatrix} -45 \cdot \cos(70°) \\ -45 \cdot \sin(70°) \end{pmatrix}$$

Vector notation keeps information along the x-axis separate from information along the y-axis.

across x: $0 = 636,396 - 130 \cdot M_{sh} - 15.4 \cdot M_{sw}$

across y: $0 = 636,396 - 75 \cdot M_{sh} - 42.3 \cdot M_{sw}$

Two equations with two unknowns can be solved.

$M_{shark} = 2,573 \text{ kg and } M_{swordfish} = 19,600 \text{ kg}.$

2.4.2 Independent practice

2.4.2.1 The law of conservation of energy

1. A baseball player slides into home plate. Initially, he has an unknown speed. The coefficient of friction between the player and the dirt is 0.4 for the entire length of his slide. He slides for exactly 2.0 m.
    a) Calculate the work done by each of the forces involved.
    b) Calculate the speed of the player when he starts his slide.

2. A baseball is thrown from the roof of a 27.5-meter tall building with an initial velocity of magnitude 18.5 m s^{-1} and directed at an angle of 37° above the horizontal.
    a) What is the speed of the ball just before it hits the ground?
    b) What is the answer to (a) if the angle is 37° below the horizontal?

3. A 6.0 kg package slides 4.0 m down a ramp that is inclined 53.1° above the horizontal. The coefficient of friction between the ramp and package is 0.40.
    a) Calculate the work done by friction on the package.
    b) Calculate the work done by gravity on the package.
    c) Calculate the work done by the normal force on the package.
    d) Calculate the total work done on the package.

4. A snowman whose mass is 20 kg sits on top of a giant snowball of a radius of 1.0 meter. The snowball is assumed to be frictionless.

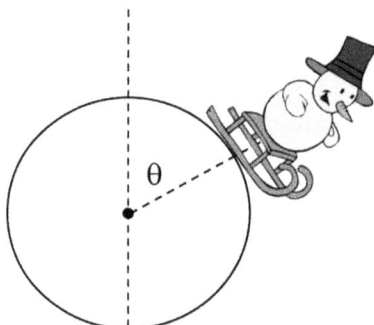

Snowman Image Credit: Pixabay by ArtRose.

    a) What is the snowman's speed when $\theta = 30°$?
    b) What is the normal force on the snowman when $\theta = 30°$?
    c) What is the snowman's speed when he hits the ground?

5. A small rock with a mass of 0.10 kg is released from rest at the top edge of a hemispherical bowl of radius 0.50 m. When it reaches the bottom of the bowl, the rock is observed to be moving at a speed of 1.8 m s^{-1}. Calculate the work done by friction on the rock when it moves from the top to bottom of the bowl. The frictional force is not constant, but you can still easily find the work done by the friction <u>without</u> having to use the formula for work done by a non-constant, or varying, force: $W = \int_{(A.)}^{(B.)} \vec{F} \cdot d\vec{r}$.

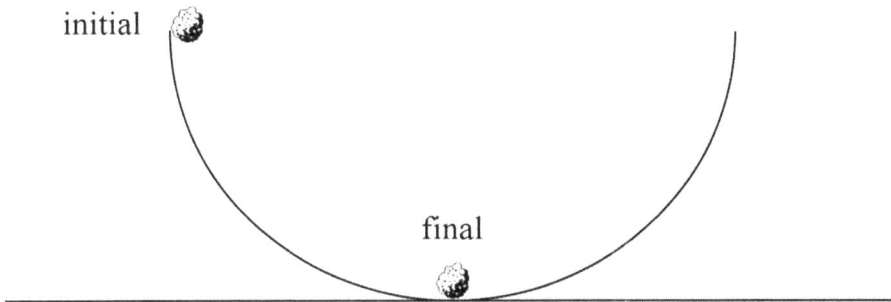

initial

final

Paper Image Credit: Pixabay by OpenClipart-Vectors.

6. A skier with a mass of 80.0 kg starts from rest at the top of a ski slope 75.0 m high.
    a) Assuming negligible friction, how fast is she going at the bottom of the slope?
    b) Now moving horizontally, the skier crosses a patch of snow where $\mu = 0.20$. If the patch is 225 m wide, how fast is she going after she crosses the patch?
    c) The skier hits a snowdrift and penetrates 2.5 m into it before coming to a stop. What is the average force exerted on her by the snowdrift as it stops her?

7. A 2 kg block is released on a 53.1° incline, 4.0 m from a long spring with a spring constant of $K = 70 \text{ N m}^{-1}$ that is attached at the bottom of the incline. The coefficient of friction is 0.20.
    a) What is the speed of the block just before touching the spring?
    b) What is the maximum compression of the spring—where the block momentarily comes to rest?
    c) How close does the block get to its original position?

8. A small glider with a mass of 0.05 kg is placed against a compressed spring that has a spring constant of $K = 150 \text{ N m}^{-1}$ at the bottom of a frictionless surface that slopes upward at an angle of 40° above the horizontal. When the spring is released, the glider leaves the spring and travels 1.80 m up the ramp.
    a) What was the original compression of the spring?
    b) What is the Kinetic Energy of the glider when it has traveled 0.8 m from its initial position?

9. A wagon of mass 2.50 kg moves in a straight line on a frictionless surface. The initial speed is 3.0 m s^{-1} and is pushed 4.0 m by a constant force of 2.5 Newtons.
    a) What is the wagon's final speed?
    b) What is the acceleration produced by the force?
    c) Use your answer in part (b) to calculate the final speed of the wagon and compare your answer to your answer in part (a).

10. A block of mass 4 kg is attached to a spring of spring constant $K = 20$ N m^{-1} and is pulled down a 30° inclined plane by a constant force of 30 Newtons that is parallel to the plane. The coefficient of friction is 0.25. If the block starts from rest, with the spring neither stretched nor compressed, how far will it move before coming to temporary rest?

11. The average human has a mass of 70 kg and can typically jump about 0.5 m into the air. What is his change in energy (or how much work does he do) when he jumps?

12. A ball is thrown from the top of a cliff. Will its velocity at the bottom be the same whether it was thrown upward or downward?

13. A pendulum is released from rest from a point that is 5 m above the ground. Can the pendulum bob ever reach a point that is higher than 5 m off the ground?

14. Jane is running at full steam at a velocity of 5.6 m s^{-1}. In order to reach Tarzan, she grabs a vine and swings upward. What is the maximum height she will be able to reach?

15. A marble is placed at the top of a 17 cm ramp that makes an angle of 35° with respect to the ground.
    a) If the marble is released from rest, what will its velocity be at the bottom of the ramp?
    b) How will your answer change if the marble is given an initial speed of 2 m s^{-1} at the top of the ramp?

16. At the 1936 Olympics, Jesse Owens reached a height of 1.1 m above the ground during the long jump. Also, his speed at the top of the jump was measured to be 6.5 m s^{-1}. how fast was he running when he started his jump?

2.4.3 Independent practice

2.4.3.1 The law of conservation of momentum

1. A block of mass 1 kg is connected by a spring, with spring constant 10 N m^{-1}, to another block of mass 3 kg. The spring is compressed and released so that the 1 kg mass moves to the left while the 3 kg mass moves to the right with a velocity of 0.5 m s^{-1}. Balance the total energy and total momentum 'before' and 'after' the spring is released to determine how much the spring was compressed?

2. A golf ball of mass 0.045 kg is moving in the $+y$-direction with a speed of 5.0 m s^{-1}, and a baseball of mass 0.145 kg is moving in the $-x$-direction (i.e., negative x) with a speed of 2.0 m s^{-1}. What are the magnitude and direction of the total momentum of the system?

3. Hockey star Wayne Gretzky is skating at 13.0 m s^{-1} toward a defender, who in turn is skating at 5.0 m s^{-1} toward Gretzky. Gretzky's weight is 756 N; that of the defender is 900 N. Immediately after the collision, Gretzky is moving at 2.5 m s^{-1} in his original direction. What is the velocity of the defender after the collision?

4. A still hockey puck B is struck by a second hockey puck A, which was originally traveling at 40 m s^{-1} to the right and deflected 30° up from its

original path. Puck B moves down at 45° from its original position. The pucks have the same mass. Calculate the speed of both pucks after the collision.

5. A wagon containing two boxes of gold and having a total mass of 300 kg has been cut loose from the horses by an outlaw when the wagon is 50 m up a 6.0° slope. The outlaw plans to have the wagon roll back down the slope and across the level ground and then crash into a canyon where confederates wait. However, in a tree 40 m from the canyon edge, wait the Lone Ranger (mass = 80 kg) and Tonto (mass = 70 kg). They drop vertically into the wagon as it passes beneath them. If they require 5.0 s to grab the gold and jump out, will they make it before the wagon goes over the edge? Assume the wagon itself has no friction in its wheels.

6. A 0.3 kg block is moving to the right on a horizontal, frictionless surface with a speed of 0.6 m s^{-1}. It makes a head-on collision with a 0.2 kg block that is moving to the left with a speed of 1.50 m s^{-1}. Find the final velocity (magnitude and direction) of each block if the collision is *'elastic.'* Since the collision is head-on, all motion is along a straight line. Note: In general, you tackle a collision−type problem because you cannot account for the forces involved during the collision process. (i.e., You cannot account for the W$_{other}$term during the collision because you do not know if the particles stick together. In other words, you do not know if there is some type of interactive force occurring the collision.) However, for the special case of an 'elastic' collision, kinetic energy is conserved because the particles bounce off of one another without any interaction. This is a special classification of collision. Therefore, for this problem, you may conserve kinetic energy.

7. A railroad handcar is moving along a straight horizontal frictionless track. In each of the following cases the car initially has a total mass (contents and car) of 200 kg and is traveling east with a velocity of magnitude 5.0 m s^{-1}. Find the final velocity of the car in each case.
   a. A 20 kg mass is thrown sideways out of the car with a velocity of magnitude 2.0 m s^{-1} relative to the initial velocity of the car.
   b. A 20 kg mass is thrown backward out of the car with a velocity of magnitude 5.0 m s^{-1} relative to the initial motion of the car.
   c. A 20 kg mass is thrown into the car with a velocity of 6.0 m s^{-1} relative to the ground and opposite in direction to the initial velocity of the car.

8. A fisherman (Santiago) in a boat catches a great white shark with a harpoon. The shark struggles for a while then dies at a distance of 1000 feet from the boat. The fisherman pulls in the shark by the rope attached to the harpoon. During this operation, the boat (initially at rest) moves 150 feet in the direction of the shark. If the weight of the boat is 12 000 pounds, what is the weight of the shark? Assume that the water is frictionless.

9. A block of mass m rests on a horizontal frictionless surface and is attached to two springs, each with spring constant K. A bullet of mass μ (μ is the bullet's mass not the coefficient of friction) moving horizontally with speed v hits the block, passes entirely through it, and emerges on the other side moving horizontally but now with a reduced speed of $v/2$. The bullet misses the

springs and we ignore gravity throughout the analysis. The springs are initially neither compressed nor stretched. What is the maximum distance the block moves after the block goes through it?

2.4.4 Peer teaching

Below is a sampling of **'Ah-ha! Moments'** from a recent introductory physics course taught at the college level. Following each comment is my remark in red.

2.4.4.1 Gravitational potential energy
1. *I finally realized that the ground-level can be moved to any convenient location when calculating the Gravitational Potential Energy.*

Yes—but let us be a bit more careful in the phrasing of this idea. When you initially encounter a problem and decide that the LCE equation will be your in the problem. **However, once that choice is made, you must use it consistently throughout the problem!** Thus, for a given mass in an LCE problem, choose the most conven-ient selection for 'ground-level'—and stick with that choice for the duration of yo-ur problem-solving.

2. *You can use different ground-levels for different masses.*

Correct—for each mass you see in a problem, you have the freedom to choose a 'ground-level' that is convenient for you. The ground-level of each mass may be separate as long as you consistently use each ground-level throughout the prob-lem. The illustration below should help: Consider ropes. If someone pulls on M_3 with a force of 100 Newtons at 30°above the horiz— the ontal, all of the blocks will be set into motion. Eventually, M_1 and M_2 will fall down the nearby ramps. If we were asked to solve for the velocity of the three blocks after they have traveled 2 m, the LCE equation would be the best choice of a problem-solving strategy. When using the LCE equation, you could set the ground-level for each of the three blocks at different locations

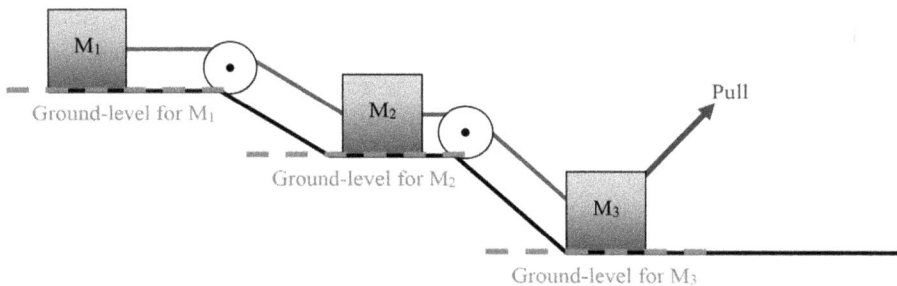

2.4.4.2 Work done by 'Other' forces
3. *When W_{other} is involved in a problem, there is no easy way to calculate its value—you have to perform a vector 'Dot'-product. Also, we often have to use Newton's laws of motion just to determine how to write the forces that are contributing to W_{other}.*

Yes—remember that W_{other} comes from the work done by forces other than gravity or a spring.. Thus W_{other} is the work done by a push or pull. tension. , or friction:

$$\underbrace{\underbrace{\left(\vec{P}\bullet\vec{d}\right)}_{\text{term 1}}+\underbrace{\left(\vec{T}\bullet\vec{d}\right)}_{\text{term 2}}+\underbrace{\left(\vec{N}\bullet\vec{d}\right)}_{\text{term 4}}+\underbrace{\left(\vec{f}\bullet\vec{d}\right)}_{\text{term 5}}}_{W_{other}}+\underbrace{\left(\vec{F}_{sp}\bullet\vec{d}\right)}_{\text{term 3}}+\underbrace{\left(\vec{W}\bullet\vec{d}\right)}_{\text{term 6}} = \underbrace{m\cdot\vec{a}\bullet\vec{d}}_{\text{term 7}}$$

Because the work done by each of these forces can **ONLY** be calculated using the vector 'Dot'-product, we have to write each force as a vector. **We must resort to any problem—solving strategy that will allow us to write these forces as vectors Indeed,** we often have to perform a separate Newton's laws calculation, just to be able to write the forces as vectors—before we even begin thinking about computing a vector 'Dot'-product.

IOP Publishing

Simplified Classical Mechanics, Volume 2 (Second Edition)

Gravity and the conservation laws

Gregory A DiLisi

Chapter 3

Rotational motion

This chapter introduces the concept of rotational motion. Instead of limiting motion to a straight line, objects under analysis can now rotate, spin, and tumble. The motion of spinning objects can get fairly sophisticated because both linear and rotational motion are often coupled. The quantities of center-of-mass, moment of inertia, and torque are defined and discussed. Emphasis is placed on the vector 'Cross-Product,' a method of multiplying the perpendicular components of two vectors. Linear parameters, such as displacement, velocity, and acceleration, are transformed into analogous rotational parameters. This transformation, from linear to rotational parameters, allows for the development of four problem-solving techniques to determine the rotational motion of an object about some axis of rotation. Problem-solving strategies are modeled throughout the chapter.

After you understand about the sun and the stars and the rotation of the earth, you may still miss the radiance of the sunset.

—Alfred North Whitehead

Joy is the mainspring in the whole
Of endless Nature's calm rotation.
Joy moves the dazzling wheels that roll
In the great Time-piece of Creation.

—Friedrich Schiller

I shall now recall to mind that the motion of the heavenly bodies is circular, since the motion appropriate to a sphere is rotation in a circle.

—Nicolaus Copernicus

doi:10.1088/978-0-7503-6402-7ch3

The lazy manage to keep up with the earth's rotation just as well as the industrious.

—Mason Cooley

In the rotation of crops there was a recognized season for wild oats; but they were not to be sown more than once.

—Edith Wharton

Faces come and faces go in circular rotation. But something yearns within to grow beyond infatuation.

—Don McLean

Squaring numbers is a symmetrical process that I like very much. And when I divide one number by another, I see a spiral rotating downwards in larger and larger loops that seem to warp and curve. The shapes coalesce into the right number.

—Daniel Tammet

As long as the world is turning and spinning, we're gonna be dizzy and we're gonna make mistakes.

—Mel Brooks

In order to share one's true brilliance one initially has to risk looking like a fool: genius is like a wheel that spins so fast, it at first glance appears to be sitting still.

—Criss Jami

In a constantly revolving circle every point is simultaneously a point of departure and a point of return. If we interrupt the rotation, not every point of departure is a point of return.

—Karl Marx

As long as the world keeps spinning, I'll keep riffing.

—Baron Vaughn

It's not magic! It's physics. The speed of the turn is what keeps you upright. It's like a spinning top.

—Deborah Bull

A circle is the reflection of eternity. It has no beginning and it has no end - and if you put several circles over each other, then you get a spiral.

—Maynard James Keenan

I grew up very strongly with this sense of time being circular: that it constantly returned upon itself.

—Richard Flanagan

We live on a spinning planet in a world of spin.

—Christopher Buckley

The same thing is to be understood of all bodies, revolved in any orbits. They all endeavour to recede from the centres of their orbits, and were it not for the opposition of a contrary force which restrains them to and detains them in their orbits, which I therefore call Centripetal, would fly off in right lines with a uniform motion.

—Issac Newton

Nature is ever at work building and pulling down, creating and destroying, keeping everything whirling and flowing, allowing no rest but in rhythmical motion, chasing everything in endless song out of one beautiful form into another.

—John Muir

The moon gravitates towards the earth and by the force of gravity is continually drawn off from a rectilinear motion and retained in its orbit.

—Issac Newton

Your dreams come crushing down when you tow the wrong path by looking at what others are doing. The Milky Way Galaxy would have been crushed down by now if each planet had left its own orbit to revolve elsewhere!

—Israelmore Ayivor

3.1 Motivation

In volume 1, chapter 1, we developed some mathematical tools. The focus of that chapter was to develop our conventions for scalar notation and vector notation.

In volume 1, chapter 2, we defined four kinematic quantities (i.e., speed, velocity, acceleration, and jerk) that will be used throughout this series to describe how objects move.

In volume 1, chapter 3, we remained focused on describing **how** objects move and developed the UAM technique for solving problems. We saw that this technique involved three vector equations and is most applicable to 'ideal situations,' short-term accelerations, and free-falling objects (i.e., projectile motion). This was our first technique allowing for numerical analysis of motion.

In volume 1, chapters 4 and 5, we tackled the question of **why** objects move. Thanks to Isaac Newton and his three laws of motion, we saw that objects move because a net external force exists on them. This was our second technique allowing for numerical analysis of motion.

In volume 1, chapter 6, we adapted Newton's laws of motion to the special situation of 'Uniform Circular Motion' or 'UCM' In a sense, UCM is a sub-strategy of Newton's laws of motion. The focus of volume 1, chapters 4 and 5 was the general motion of objects when they are subjected to various common forces while volume 1, chapter 6 was a specific case of objects moving in a circle (or partial circle).

In volume 2, chapter 1, we digressed a little bit from our problem-solving strategies to examine more closely the origin of the gravitational field force that was introduced earlier in the series. We found that the formula for 'weight' that we had been using in the past was really a simplification of a much more elaborate law; namely, Newton's 'Universal Law of Gravitation' (abbreviated as the 'ULG'). The ULG determines the gravitational interaction of any two bodies—whether they are on the surface of the Earth or not.

In volume 2, chapter 2, we developed an entirely new approach to problem-solving. We defined two new physical quantities called 'work' (or 'energy') and 'linear momentum' and saw that these quantities were conserved; that is to say, they remained fixed over time. In certain situations, we were able to use principles of conservation to analyze a new set of everyday motions. For example, situations involving only kinetic, gravitational potential, and elastic potential energies could be solved using the Law of Conservation of Energy. On the other hand, collisions and explosions were particularly suited for analysis using the Law of Conservation of Momentum.

In volume 2, chapter 3, we will essentially *double* the number of techniques we have in our arsenal of problem-solving recipes. We will take the same approach that has paid us dividends in the past—we will define some new physical quantities and explore how these quantities allow us to tackle different types of motion. In the process of defining these new physical quantities, we also need to develop one last new mathematical operation—the vector 'Cross'-product method for multiplying two vectors.

3.2 Getting ready

3.2.1 Anticipatory set

Today's class is held at the local ice rink. Your favorite physics professor enters the rink and skates to center-ice where she performs an acrobatic triple-axel. Upon landing, she immediately goes into a high speed, one-toe spin. When she starts to rotate, you notice that she brings her arms over her head. Without pushing on the ice, without anything to grab, she magically increases her rate of rotation—spinning faster and faster at a dizzying rate.

After class, you see a clumsy cat named 'Tom' sitting on a high tree branch. Suddenly, he darts after a clever mouse named 'Jerry.' Unfortunately, our poor cat stumbles over some branches and falls to the ground. With almost certain disaster waiting for him when he hits the ground, our cat is able to right himself and land softly on his feet. Unable to push on the air, without anything to grab, he magically begins to rotate until he lands upright. As the saying goes, the cat has once again 'landed on all fours.'

Without grabbing onto anything, your favorite physics professor somehow spins faster and faster.

Ice Skater Image Credit: Victoria VIAR PRO/Shutterstock.com.

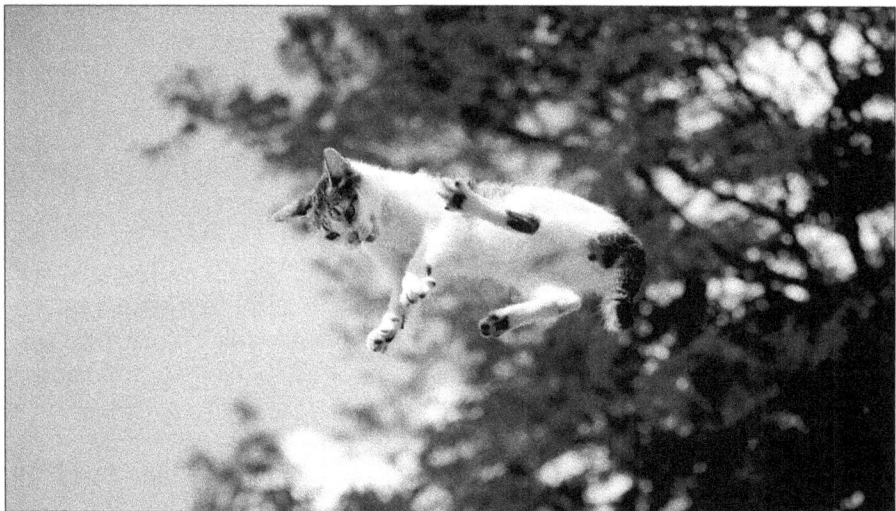

Without grabbing onto anything, "Tom" is able to right himself in mid-air and land on his feet.

Cat Image Credit: yordanka caridad almaguer/Shutterstock.com.

The previous two examples involve motion we have not previously considered in this series on *Classical Mechanics*. In the first example, the professor is not moving linearly. Essentially, she remains in one place (center-ice), yet we certainly would not describe her as 'motionless.' None of our previous problem-solving strategies can describe her dizzying spinning motion. Likewise in our second example, the motion of the cat seems to defy the problem-solving strategies we have so far developed. Overall, the cat does move linearly (i.e., from the tree branch to the ground), yet we certainly would not describe his motion as following a simple free-falling trajectory. In addition to free-falling, the cat is performing some very complex rotational motion. Something about the cat's spinning motion confounds the problem. *How do we describe rotational motion about a fixed point? How do we describe rotational motion about a point that is moving linearly? How is the professor able to spin faster and faster, even though she is unable to grab something and exert a force to change her motion? How is the cat able to rotate his body and land on his feet, even though he is unable to grab something and exert a force to change his motion?*

3.2.2 Objective

By the end of this chapter, you will be able to:
- Define the 'The Center-of-Mass' of an object.
- Define the 'Moment of Inertia' of an object about an axis of rotation.
- Multiply two vectors using the vector 'Cross'-product.
- Interpret the physical meaning of the vector 'Cross'-product.
- Define the 'Torque' exerted by a force on an object.
- Define the angular (or rotational) parameters of a point (P.) on an object: 'Angular Displacement' $(\vec{\theta})$, 'Angular Velocity' $(\vec{\omega})$, and 'Angular Acceleration' $(\vec{\alpha})$.
- Transform the *linear parameters* (i.e., linear displacement $[\vec{r}$ or $\vec{x}]$, linear velocity $[\vec{v}]$, and linear acceleration $[\vec{a}]$) of a point (P.) on an object to the *rotational parameters* (i.e., angular displacement $[\vec{\theta}]$, angular velocity $[\vec{\omega}]$, and angular acceleration $[\vec{\alpha}]$) of that same point on the object.
- Develop four problem-solving techniques to determine the rotational motion of an object about some axis of rotation.

3.2.3 Purpose

This information is needed:
- Because calculating the 'Center-of-Mass' of an object enables us to determine the point about which an object spins or rotates—it is the point at which all of an object's mass acts as if it were located.
- Because the 'Moment of Inertia' measures the ease or difficulty associated with spinning an object about an axis of rotation.
- Because calculating the 'Torque' exerted by a force on an object measures the extent to which that force can cause the object to rotate.

- Because the vector 'Cross'-product will give us additional practice in using our vector notation (which will help us gain speed and accuracy in manipulating vectors); but more importantly, the vector 'Cross'-product will now enable us to calculate how much of one vector is *perpendicular* to another vector (i.e., what fraction of one vector is perpendicular to the direction of another vector).

- Because once we transform the linear parameters of a point (P.) on an object to the rotational parameters of that same point on the object, the number of problem-solving techniques at our disposal will double. In the previous chapters in this series on *Classical Mechanics*, we were limited to analyzing the linear motion of an object located at a single point. Now we will be able to analyze more sophisticated motion; namely, the linear motion of an object located at a single point coupled with the rotational motion of that object about that point. Thus, in addition to objects moving along a straight line, we will be able to analyze objects that spin as they move.

3.3 Giving information

3.3.1 Instructional input

3.3.1.1 The center-of-mass

Let's dive right in and define a new physical quantity. The purpose of this new physical quantity will become apparent after we compute it in the context of a few practice problems.

Consider an irregularly-shaped object of mass, M_{total}. This object can be a child's toy, a car, a person walking down the street, or a simple blob of clay. Let's start with a simple blob of clay. Place a coordinate system near the object—you can place the coordinate system anywhere you like, but once you choose your coordinate system, it has to remain fixed because all of your upcoming computations will be made relative to your chosen coordinate system:

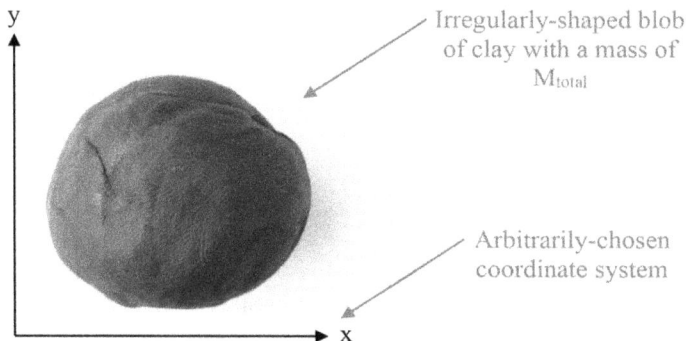

Irregularly-shaped blob of clay with a mass of M_{total}

Arbitrarily-chosen coordinate system

Blue Plasticine Image Credit: sergua/Shutterstock.com.

We are now going to compute the 'Center-of-Mass' of the blob (abbreviated from this point forward at the 'COM'). To calculate the COM of any object, a precise procedure must be followed. First, imagine you have a pair of tweezers. Next, use the tweezers to remove the smallest-sized chunk of mass you can imagine from the object. Now weigh that chunk (designated as m_1) and multiply its mass by its location (designated as $\vec{r_1} = (x_1, y_1)$).

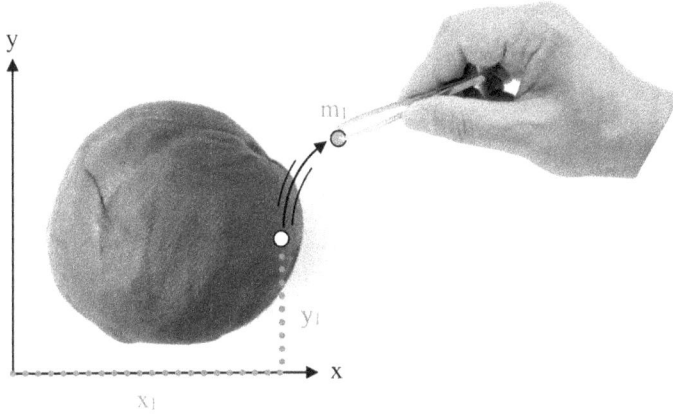

Blue Plasticine Image Credit: sergua/Shutterstock.com. Hand Image Credit: Ju Jae-young/Shutterstock.com.

Repeat the process with another small chunk of mass. Namely, use the tweezers to remove another small-sized chunk of mass from the object. Next, weigh that chunk (designated as m_2). Finally, multiply the mass of the chunk by its location (designated as $\vec{r_2} = (x_2, y_2)$).

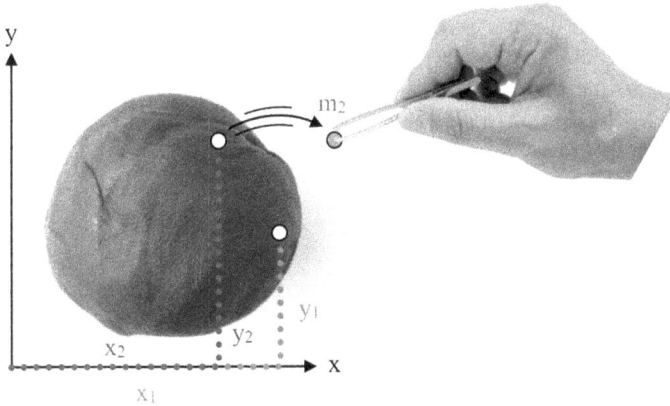

Blue Plasticine Image Credit: sergua/Shutterstock.com. Hand Image Credit: Ju Jae-young/Shutterstock.com.

Repeat this process over and over until you have removed all of the object's mass. This process may need to be repeated 50 times, 278 times, 1038 times, etc, depending on the number of small chunks you need to remove.

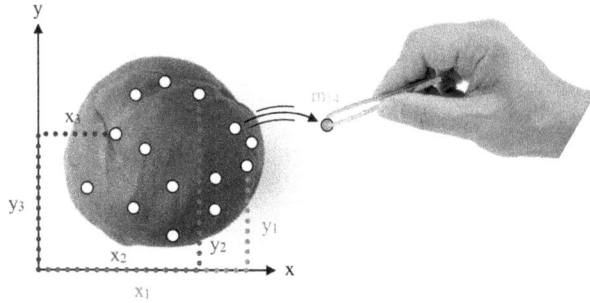

Blue Plasticine Image Credit: sergua/Shutterstock.com. Hand Image Credit: Ju Jae-young/Shutterstock.com.

Once you complete this tedious, but straightforward process, we define the COM as:

▶ **Definition of the Center-of-Mass of an object:**

$$\vec{r}_{COM} = \frac{(m_1 \times \vec{r_1}) + (m_2 \times \vec{r_2}) + (m_3 \times \vec{r_3}) + \cdots}{M_{total}} \text{ or}$$

$$\vec{r}_{COM} = \frac{m_1 \times \binom{x_1}{y_1} + m_2 \times \binom{x_2}{y_2} + m_3 \times \binom{x_3}{y_3} + \cdots}{M_{total}}$$

For convenience, we abbreviate our definition as follows:

$$\vec{r}_{COM} = \frac{\sum\limits_{i=1}^{n} m_i * \vec{r_i}}{M_{total}},$$ where n is the number of pieces into which the object has been cut.

This simple definition incorporates a number of features that are worth itemizing and clarifying. As usual, we'll check these features when we actually start calculating the COM in example problems:

1. The COM is a point in space ... it is a location ... it is a coordinate. When all is said and done, your answer for the COM should be a coordinate: $\vec{r}_{COM} = (x_{COM}, y_{COM})$.
2. An object only has one COM.
3. The COM of an object may or may not be on the object itself. For example, the COM of a typical boomerang is not actually on the boomerang:

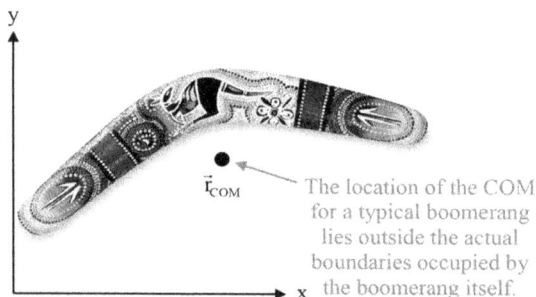

$\vec{r}_{COM}$ — The location of the COM for a typical boomerang lies outside the actual boundaries occupied by the boomerang itself.

Boomerang Image Credit: Sanit Fuangnakhon/Shutterstock.com.

4. The COM is a vector! As mentioned above, we are actually calculating a point in space, relative to our arbitrarily-chosen coordinate system. Therefore, we need to keep *x*- and *y*-components separate.

5. The units of the COM are 'meters.' To calculate the COM of an object, we first take each small chunk of mass and multiply it by its location—thus, the numerator in our expression for $\vec{r}_{COM}$ has units of (kg · m). These products are then added together and divided by the total mass, M_{total}. Therefore, the denominator in our expression for $\vec{r}_{COM}$ has units of kg. Dividing the numerator by the denominator leads to units of (kg · m) kg^{-1}. Thus, the remaining units are simply 'meters.'

6. One of the first questions you might be asking is: *'Into how many pieces do I cut the object?'* In other words, *'How small are the pieces that I cut with the tweezers?'* Your determination of the COM improves as you cut smaller and smaller pieces out of the object. You have to imagine using the tweezers to pull the smallest size chunks of mass out of the object as possible. The smaller the chunks of mass that you pull out of the object, the more accurate your determination of the COM will be. In short, use your tweezers to remove the *smallest-sized* chunks of mass that you possibly can from the object! Removing many, many, small masses will result in a better determination of the COM than removing only a few, large masses.

7. Next, you might be wondering what the COM of an object represents. In other words, *'What is the **physical interpretation** of the COM of an object?'* Notice that during our procedure for determining the COM of the blob of clay, we sliced tiny pieces of clay off of the blob until there was no clay left to slice. We then mathematically weighted the location of every small chunk of clay by the mass at that location. That is to say, each point in the blob of clay was multiplied by the amount of mass at its location. Thus, the COM of an object is physically interpreted at the 'mass—weighted location of an object.' Think of it this way—heavier (i.e., more massive) parts of an object are more important than less massive parts and should therefore be given more emphasis. Therefore, these heavier points are given more emphasis by multiplying their locations by the masses at their locations. To illustrate the physical interpretation of the COM of an object, let's consider an object that looks relatively uniform on the outside, but actually contains many different masses on the inside. For example, a human being looks relatively smooth and uniform on the outside, but is actually made up of different organs, of different masses, on the inside. A suitcase aboard a passenger airliner looks relatively smooth and uniform on the outside but may actually have heavy (i.e., books, alarm clocks, and shoes) and/or light (i.e., clothes, socks, and sweatshirts) objects scattered throughout its interior. My favorite example is a candy bar. From the outside, the candy bar looks smooth and uniform … like a simple bar of chocolate. However, on the inside, the candy bar may actually contain peanuts, peanut butter, caramel, chewy nougat, or maybe even wafers. Consider the complicated, but delicious-looking candy bar shown below:

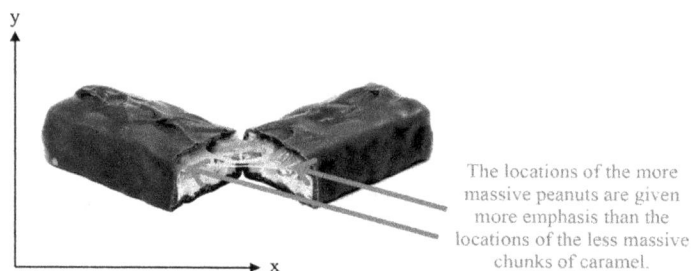

The locations of the more massive peanuts are given more emphasis than the locations of the less massive chunks of caramel.

This Snickers-broken image has been obtained by the author from the Wikimedia website, where it is stated to have been released into the public domain. It is included within this article on that basis.

The procedure for calculating the COM of the candy bar ensures that more emphasis is given to each location of a peanut than to each location of a piece of caramel. Since each peanut weighs more than each chunk of caramel, the location of each peanut is mathematically weighted by the amount of mass at that location.

As a final example of the physical interpretation of the COM of an object, consider the three unevenly weighted pencils shown below. Because the location of the green weight shifts from pencil to pencil, the COM shifts to a corresponding location where more of the weight is distributed. Remember, the COM of an object is the 'mass-weighted location of an object'—meaning, the location of the COM is where most of the object's mass is located.

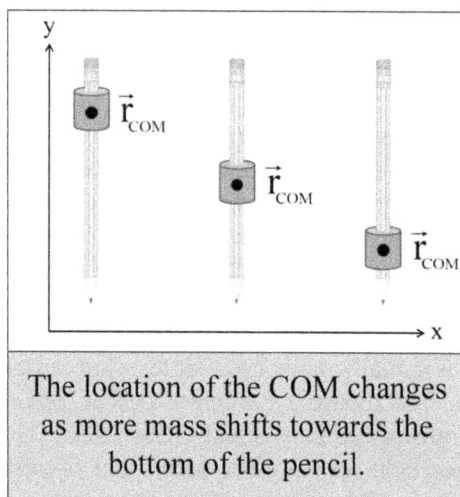

The location of the COM changes as more mass shifts towards the bottom of the pencil.

Pencil Image Credit: Pixabay by Clker-Free-Vector-Images.

8. Last, you most certainly must be wondering, 'WHY did we define the COM of an object?' How does knowing the location of an object's COM help us solve

problems? To answer these questions, let's calculate the velocity and acceleration of the COM of a body. Let's start with the velocity. To calculate the velocity of the COM, just calculate the change in location of the COM over time:

$$\vec{v}_{COM} = \frac{\Delta(\vec{r}_{total})}{\Delta t}$$

The COM is just the mass-weighted location of the object. Use the short-hand notation for the definition of the COM and note that the total mass of the object is constant.

$$\vec{v}_{COM} = \frac{\Delta\left(\dfrac{\sum\limits_{i=1}^{n} m_i * \vec{r}_i}{M_{total}}\right)}{\Delta t} = \frac{\Delta\left(\sum\limits_{i=1}^{n} m_i * \vec{r}_i\right)}{M_{total} \cdot \Delta t}$$

Each individual mass is also constant so only the <u>location</u> of each individual mass changes.

$$\vec{v}_{COM} = \frac{\left(\sum\limits_{i=1}^{n} m_i * \dfrac{\Delta\vec{r}_i}{\Delta t}\right)}{M_{total}} = \frac{\left(\sum\limits_{i=1}^{n} m_i * \vec{v}_i\right)}{M_{total}}$$

$$\vec{v}_{COM} = \frac{\vec{p}_{total}}{M_{total}}$$

The numerator is just the total linear momentum of all of the masses. If you consider the object to be a "system" of many individual masses, then the numerator represents the total linear momentum of that system.

Now we can compute the acceleration of the COM by calculating the change in its velocity over time:

$$\vec{a}_{COM} = \frac{\Delta(\vec{v}_{total})}{\Delta t}$$

Plug in our result for the velocity of the COM from above.

$$\vec{a}_{COM} = \frac{\Delta\left(\dfrac{\vec{p}_{total}}{M_{total}}\right)}{\Delta t} = \frac{\Delta\vec{p}_{total}}{M_{total} \cdot \Delta t}$$

However, we know from the original formulation of Newton's second law of motion, that the change in total momentum of a system (with respect to time) is just the total force on the system.

$$\vec{a}_{COM} = \frac{\vec{F}_{total}}{M_{total}} \quad \text{or} \quad \vec{F}_{total} = M_{total} * \vec{a}_{COM}$$

This is a surprising, but fantastic and super-helpful result! All of the forces acting on an object act as if all of its mass were concentrated at the COM. The shape of the object is unimportant. The size of the object is unimportant. The object acts as if all of its mass were at a single, special point … the Center−of−Mass. For example, consider the flight of a boomerang. Suppose we place a green point of the boomerang's tip and a black dot on its COM. We now track the motion of the two dots as the boomerang flies through the air. We see that the green dot follows a complicated, erratic path through the air as the boomerang spins in flight. The path is so complicated that we don't even bother trying to describe nor analyze it. However, the path of the COM is easy to describe and analyze—the path is that of projectile motion near the surface of the earth. The entire boomerang acts as if all of its mass were located at the COM. The boomerang can rotate, spin, and tumble all it wants, but the object ultimately behaves as if all the mass were at its COM. This surprising result is going to be very helpful for problem-solving as it will soon allow us to double the number of techniques in our problem-solving arsenal.

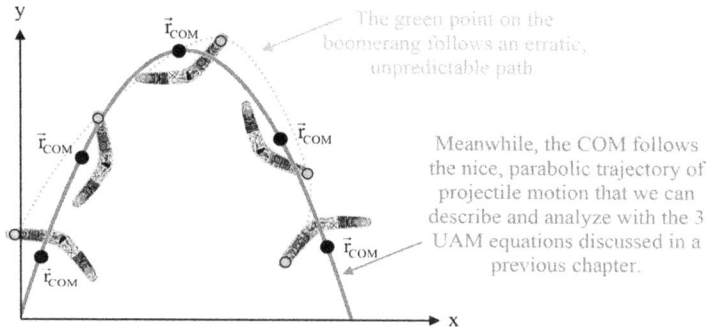

Boomerang Image Credit: Sanit Fuangnakhon/Shutterstock.com.

3.3.2 Modeling

❶ **EX 1:** Four point-masses (masses that are so small, they essentially occupy only a point in space ... like marbles or ball-bearings) are connected by massless sticks (sticks that are so lightweight, they essentially have no mass ... like toothpicks or plastic straws) into the shape shown below:

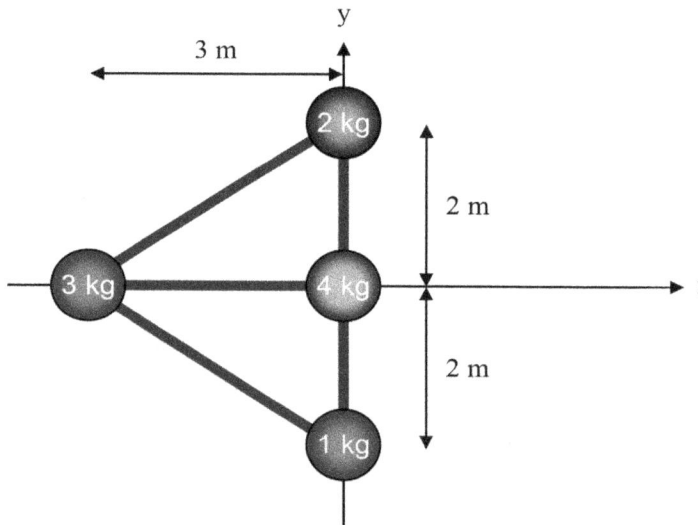

Calculate the location of the shape's COM.

☑ **ANSWER:** Imagine using our tweezers to remove the four point-masses one at a time. After removing each point-mass, we multiply its position by the mass of that we have just removed:

$$\vec{r}_{COM} = \frac{1 \times \begin{pmatrix} 0 \\ -2 \end{pmatrix} + 2 \times \begin{pmatrix} 0 \\ 2 \end{pmatrix} + 3 \times \begin{pmatrix} -3 \\ 0 \end{pmatrix} + 4 \times \begin{pmatrix} 0 \\ 0 \end{pmatrix}}{10} = \begin{pmatrix} -0.9 \\ 0.2 \end{pmatrix} \text{ mor}(-0.9, 0.2) \text{ m.}$$

☺ **COMMENTS:** Let's check some of the features we expect to find when calculating a COM:

1. ✓ The result is a location or point in space—yes, our answer is a coordinate.
2. ✓ The result may or may not be on the object itself—yes, the COM is actually not located on the object itself. The COM is just below and to the left of the origin.
3. ✓ The COM is a vector—yes, we were careful to keep x- and y-components separate in our calculations.
4. ✓ The result has units of meters—yes.
5. ✓ Last, let's remind ourselves of the physical interpretation of the COM—the COM is the 'mass-weighted location of the object.' The COM is the point at which the object acts as if all of its mass were located. If we were to put a small black dot on the object's COM and throw the object across the room, we would see the object spin, tumble, flip, and flop ... but its COM would follow the parabolic path of a simple point-like mass of 10 kg undergoing projectile motion through the air.

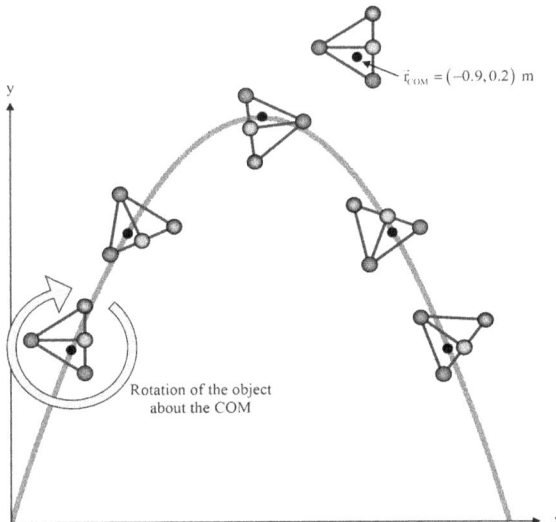

$\vec{r}_{COM} = (-0.9, 0.2)$ m

Rotation of the object about the COM

❷ **EX 2:** Four point-masses are connected by massless sticks into the shape shown below:

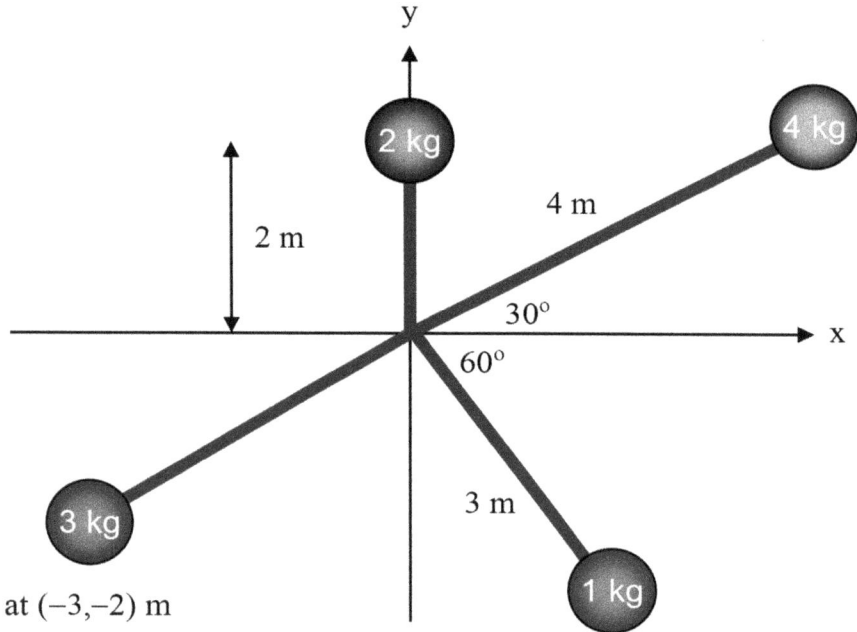

at $(-3, -2)$ m

Calculate the location of the shape's COM.

☑ **ANSWER:** Imagine using our tweezers to remove the four point-masses one at a time. After removing each point-mass, we multiply its position by the mass of that we have just removed:

$$\vec{r}_{COM} = \frac{1 \times \begin{pmatrix} 3 \cdot \cos(60^o) \\ -3 \cdot \sin(60^o) \end{pmatrix} + 2 \times \begin{pmatrix} 0 \\ 2 \end{pmatrix} + 3 \times \begin{pmatrix} -3 \\ -2 \end{pmatrix} + 4 \times \begin{pmatrix} 4 \cdot \cos(30^o) \\ 4 \cdot \sin(30^o) \end{pmatrix}}{10}$$

$$= \begin{pmatrix} 0.635 \\ 0.34 \end{pmatrix} \text{mor}(0.635, 0.34) \text{ m.}$$

☺ **COMMENTS:** Let's check some of the features we expect to find when calculating a COM:
1. ✓ The result is a location or point in space—yes, our answer is a coordinate.
2. ✓ The result may or may not be on the object itself—yes, the COM is actually not located on the object itself.
3. ✓ The COM is a vector—yes, we were careful to keep x- and y-components separate in our calculations.
4. ✓ The result has units of meters—yes.

3.3.3 Checking for understanding

3.3.3.1 Where is the human body's center-of-mass?

Before wrapping up our discussion of the 'Center-of-Mass,' let's calculate the COM for a typical human being (both males and females) and perform some amusing demonstrations that highlight how useful the COM can be for analyzing motion. The goal of these demonstrations is to highlight our discovery, stated a few previous pages back, that 'an object acts as if all of its mass were located at the COM.'

To calculate the COM for a typical human being, we have to first consider the human body to be a system of discrete masses (i.e., organs, limbs, bones, etc) located at various positions throughout the body. First, imagine we have x-ray eyes and can see the various discrete masses that make up a human body:

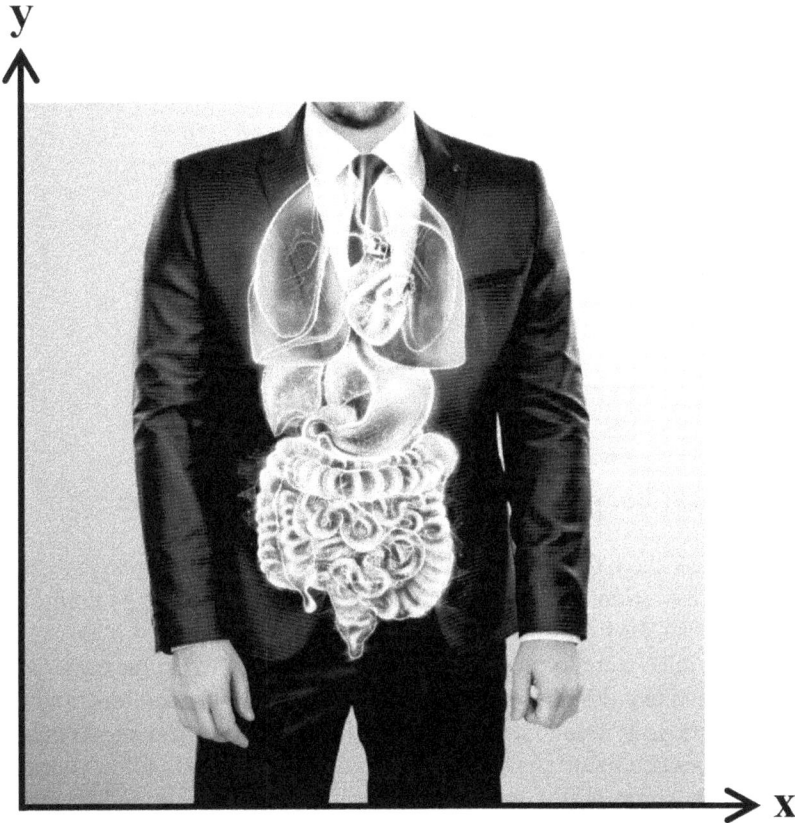

X-ray Image Credit: Marko Aliaksandr/Shutterstock.com.

Next, we would perform the following calculation using all the parts of the human body:

$$\vec{r}_{\substack{\text{COM} \\ \text{Human} \\ \text{Body}}} = \frac{m_{\substack{\text{Right} \\ \text{Kidney}}} \times \begin{pmatrix} x_{\substack{\text{Right} \\ \text{Kidney}}} \\ y_{\substack{\text{Right} \\ \text{Kidney}}} \end{pmatrix} + m_{\substack{\text{Left} \\ \text{Kidney}}} \times \begin{pmatrix} x_{\substack{\text{Left} \\ \text{Kidney}}} \\ y_{\substack{\text{Left} \\ \text{Kidney}}} \end{pmatrix} + m_{\substack{\text{Right} \\ \text{Lung}}} \times \begin{pmatrix} x_{\substack{\text{Right} \\ \text{Lung}}} \\ y_{\substack{\text{Right} \\ \text{Lung}}} \end{pmatrix} + m_{\substack{\text{Left} \\ \text{Lung}}} \times \begin{pmatrix} x_{\substack{\text{Left} \\ \text{Lung}}} \\ y_{\substack{\text{Left} \\ \text{Lung}}} \end{pmatrix} + \cdots}{M_{\text{total}}}$$

Remember, our determination of the COM improves as we cut the object into smaller and smaller pieces; so even the right kidney can be cut into smaller pieces … even the left kidney can be cut into smaller pieces … etc. Finally, after cutting the typical human body into small, discrete masses, the previous calculation puts the COM along the long-axis of the body, somewhere near the pelvis.

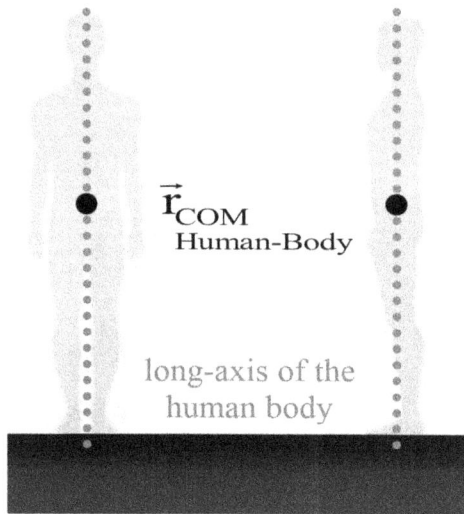

Human Body Image Credit: Net Vector/Shutterstock.com.

Knowing where the COM is located for a typical human body, you can perform an interesting demonstration to show dramatically that an object acts as if all of its mass were located at its COM. In other words, you can demonstrate that a person acts as if all of his or her mass were located at the COM. This demonstration works identically for both males and females (unlike some of the other upcoming demonstrations). Stand in a doorway with your nose touching the door jamb and your feet straddling it. Try to stand on your toes (see figure (A) below). Normally, when you stand on your toes, your body instinctively compensates for the imbalance in forces by leaning forward to move its COM over your toes. If you are standing with your nose against a door jamb, you cannot lean forward to place your COM over your toes. The result is that you can no longer stand on your toes. Repeat the demonstration, but this time, hold a heavy weight or some heavy textbooks in each hand with your arms outstretched forward on opposite sides of the door jamb (see figure (B) below). The added masses shift your COM over your toes and you can therefore stand on your toes.

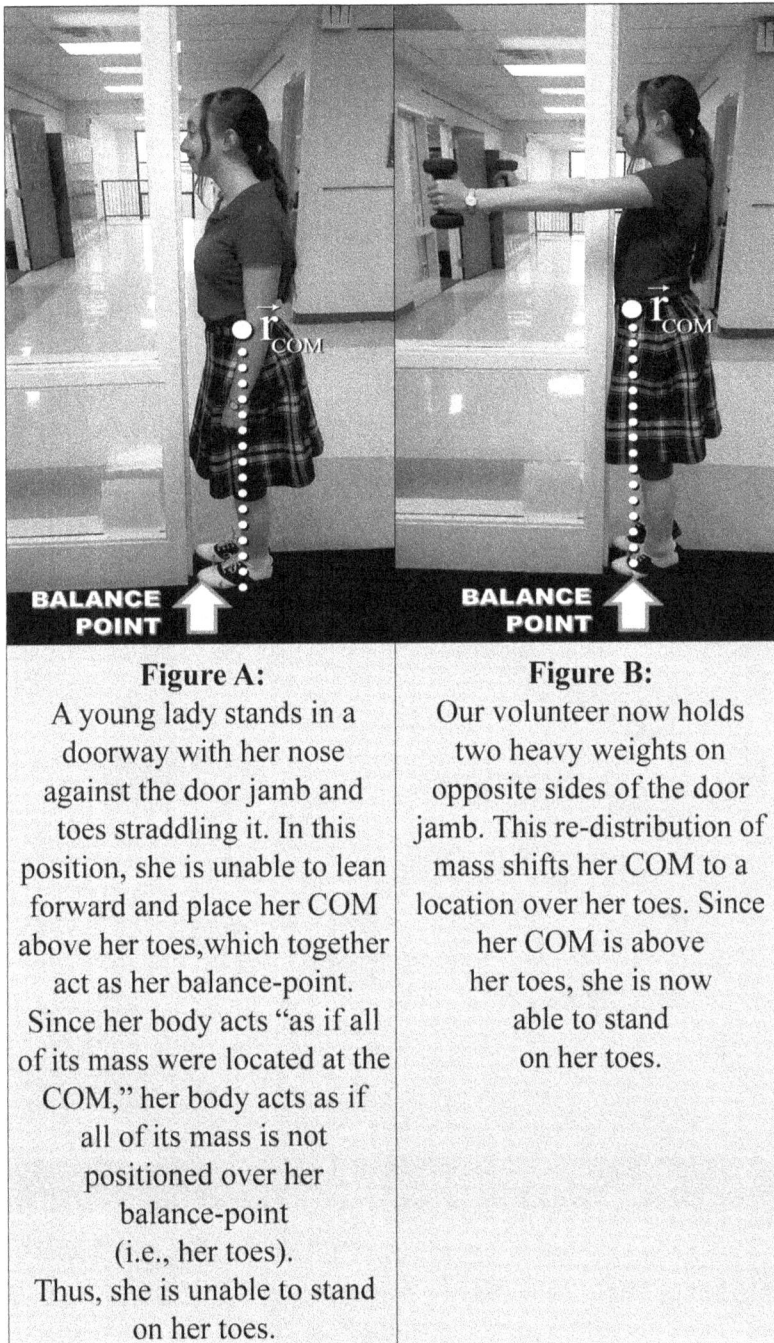

Figure A:
A young lady stands in a doorway with her nose against the door jamb and toes straddling it. In this position, she is unable to lean forward and place her COM above her toes, which together act as her balance-point. Since her body acts "as if all of its mass were located at the COM," her body acts as if all of its mass is not positioned over her balance-point (i.e., her toes). Thus, she is unable to stand on her toes.

Figure B:
Our volunteer now holds two heavy weights on opposite sides of the door jamb. This re-distribution of mass shifts her COM to a location over her toes. Since her COM is above her toes, she is now able to stand on her toes.

Image Credit: Author.

This effect also explains why a pregnant woman tends to lean backwards when walking. In the latter stages of pregnancy, a woman's COM shifts forward and slightly upward, requiring her to lean backwards on her heels to maintain equilibrium. This strains the back and other muscles unaccustomed to this posture.

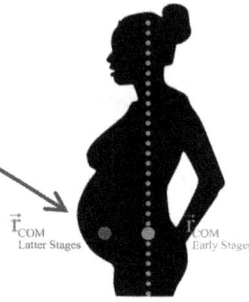

In the early stages of pregnancy, the COM is located in its usual position along the long-axis of the body. However, in the latter stages of pregnancy (as more mass accumulates near a woman's belly) the COM shifts forward. To maintain balance, a woman must lean backwards to keep her COM over her heels.

$\vec{r}_{COM}$ Latter Stages $\vec{r}_{COM}$ Early Stages

Pregnant Woman Image Credit: Pixabay by waldryano.

Some additional interesting demonstrations are possible when you compare the location of the COM of a typical human male versus the location of the COM of a typical human female. The typical male is essentially 'apple-shaped' with dense muscle in the chest and shoulders. Meanwhile, the typical female is essentially 'pear-shaped' with more mass distributed in the pelvis. The difference in the distribution of mass between males and females puts the location of their respective COMs in slightly different locations. Since more mass is distributed in the chest of typical males, his COM is still along the long-axis of the body but is slightly higher along the axis than the COM of a female.

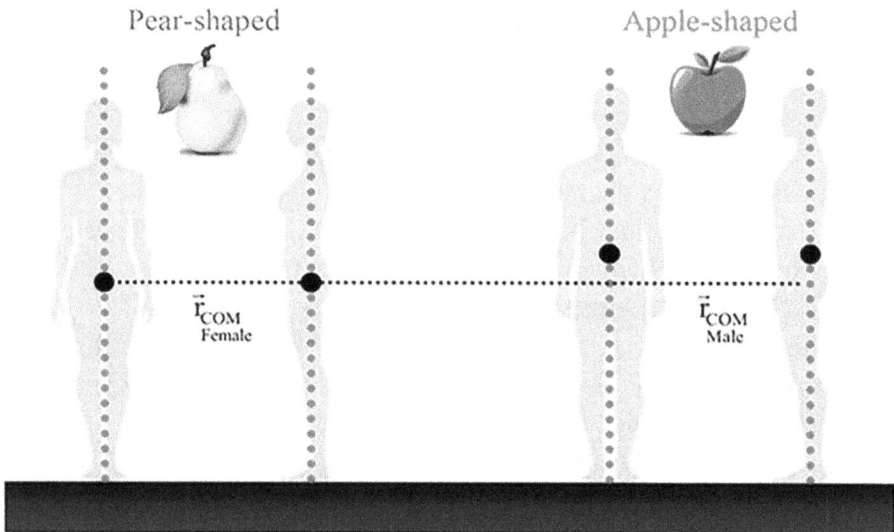

Pear-shaped   Apple-shaped

$\vec{r}_{COM}$ Female $\vec{r}_{COM}$ Male

The COM of a female is generally lower than the COM of a male.

Apple Image Credit: Pixabay by DesignerRiya. Human Body Image Credit: Net Vector/Shutterstock.com. Pear Image Credit: Pixabay by OpenClipart-Vectors.

To see the difference in the location of the COM between males and females, try the following two demonstrations. First, measure the length of your foot from heel to toe. Stand on the floor with the tip of your toes at a distance of twice this length from a wall. For example, if your foot is 10 inches in length, face a wall with the tips of your toes exactly 20 inches from the wall. Next, bend at the waist and place your head against the wall. Hold 3–4 heavy textbooks against your chest and try to stand upright by bending at the waist (see figure (A) below). Do not use your hands to stabilize yourself!

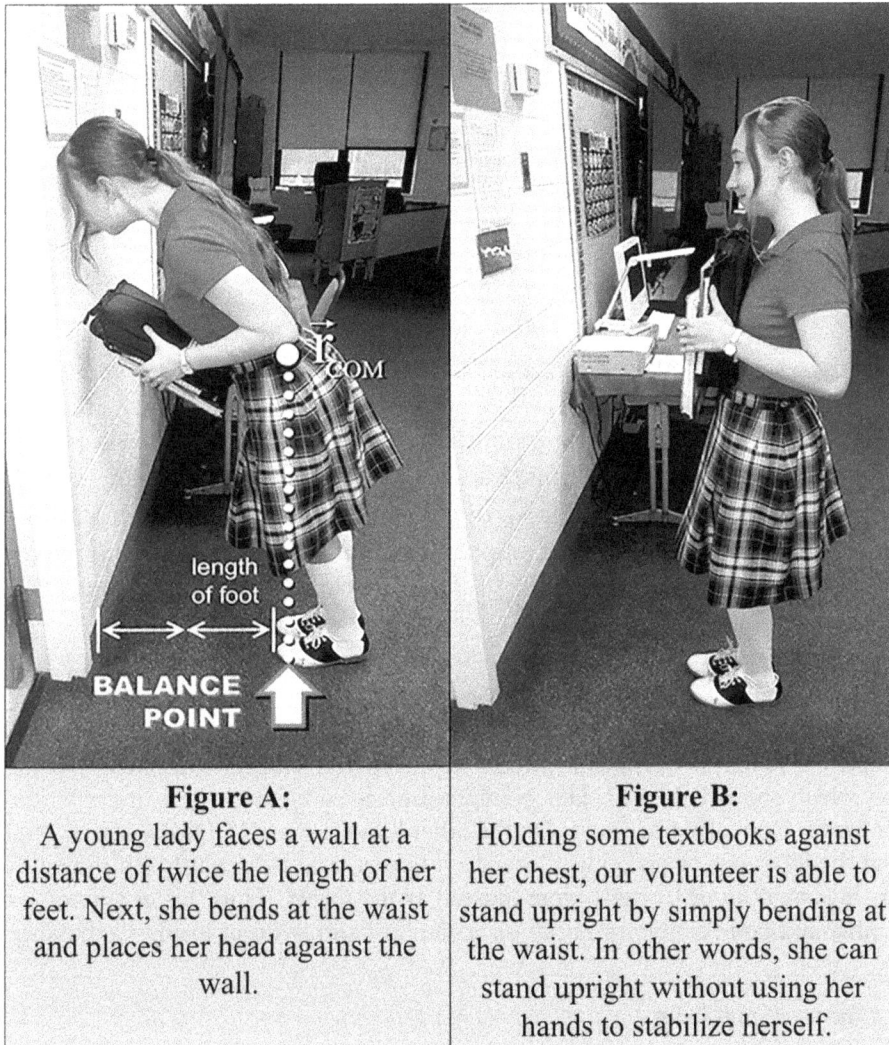

Figure A:	Figure B:
A young lady faces a wall at a distance of twice the length of her feet. Next, she bends at the waist and places her head against the wall.	Holding some textbooks against her chest, our volunteer is able to stand upright by simply bending at the waist. In other words, she can stand upright without using her hands to stabilize herself.

Image Credit: Author.

A typical female is able to stand upright while a typical male is unable to do so (see figure (B) above). Since a female's COM is in her pelvis (i.e., lower along the long-axis of the body)—and since we know that an object acts as if all of its mass were located at the COM—her COM remains over her feet. Even while holding several large textbooks against her chest, a female's COM remains near her pelvis. Her configuration remains stable and she can stand upright. On the other hand, a male's higher COM is shifted even higher into his torso by holding the textbooks against his chest. When leaning against the wall, his COM is actually in front of his feet. His configuration is unstable and he cannot stand upright.

Next, kneel on the ground and place your elbows against your knees. Stretch your fingers and place a small object (a stick of lip balm, glue stick, bottle of correction fluid, or butane lighter work perfectly) at the location of the tips of your outstretched fingers (see figure (A) below). Clasp your hands behind your back and try to knock over the object with your nose without losing your balance (see figure (B) below).

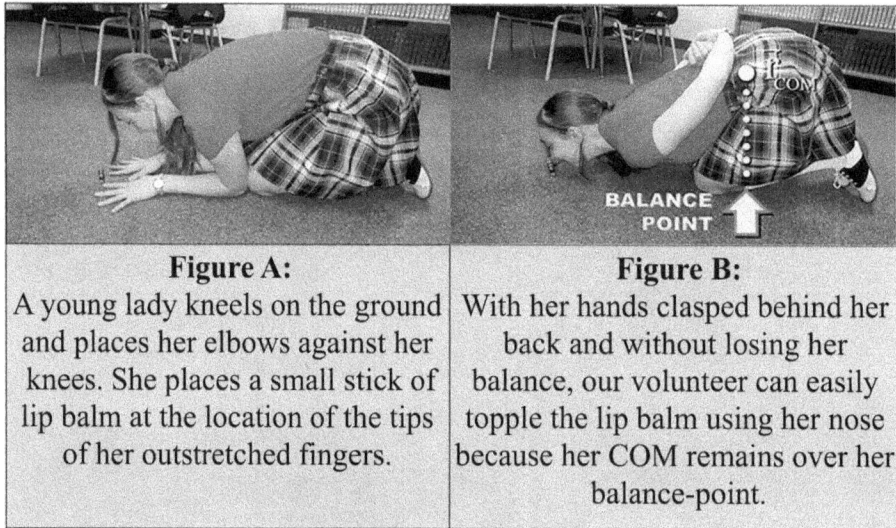

Figure A:	**Figure B:**
A young lady kneels on the ground and places her elbows against her knees. She places a small stick of lip balm at the location of the tips of her outstretched fingers.	With her hands clasped behind her back and without losing her balance, our volunteer can easily topple the lip balm using her nose because her COM remains over her balance-point.

Image Credit: Author.

Again, a female's COM is located near her pelvis and remains over her knees, upon which she is balanced. Her configuration is stable and she can easily knock over the lip balm using her nose. On the other hand, a male's COM is slightly higher along the long-axis of his body, closer to his chest, and is forward of his knees. His body acts as if all of its mass were located at the COM. Thus, his configuration is unstable and he is unable to knock over the lip balm without losing his balance.

3.3.4 Instructional input

3.3.4.1 The moment of inertia about an axis of rotation
Now that we have defined the 'Center-of-Mass' of an object and practiced calculating it, let's move forward and define another new physical quantity. Like the COM, the utility of this new physical quantity will become apparent after we compute it in the context of a few practice problems. Eventually, these new physical quantities will assemble into a new set of problem-solving strategies that help us determine how objects rotate and spin.

Consider our irregularly-shaped object. Like before, this object can be a child's toy, a car, a person walking down the street, or a simple blob of clay. Let's continue to use our favorite blob of clay. Stick an imaginary thumbtack (or nail ... or pin ... or stake) somewhere in (or near) the object—this thumbtack will designate the 'axis of rotation.' The 'axis of rotation' is the *imaginary axis, or line,* about which we will spin the object. You can place the thumbtack anywhere you like, but once you choose your axis of rotation, it has to remain fixed because all of your upcoming computations will be made relative to your chosen axis. Place your coordinate system so that the axis of rotation passes through the origin of the coordinate system:

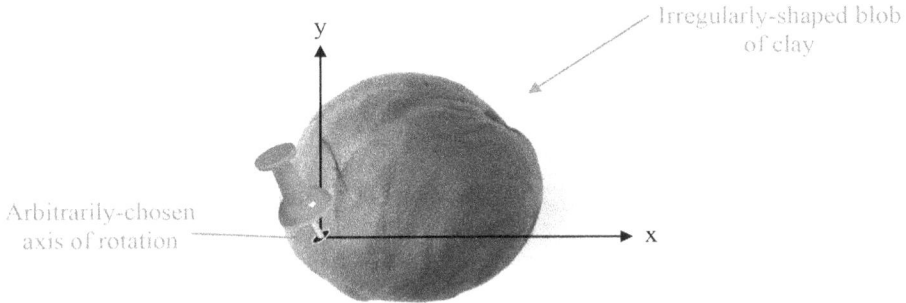

Blue Plasticine Image Credit: sergua/Shutterstock.com. Thumbtack Image Credit: Pixabay by OpenClipart-Vectors.

We are now going to compute the 'Moment of Inertia' of the blob relative to the axis of rotation (abbreviated from this point forward at the 'MOI'). Similar to the procedure for determining an object's COM, to calculate the MOI of any object, a precise procedure must be followed. First, imagine you have the same trusty pair of tweezers you used when determining an object's COM. Use the tweezers to remove the smallest-sized chunk of mass you can imagine from the object. Next, weigh that chunk (designated as m_1) and multiply its mass by the square of the *perpendicular distance* from the chunk to the axis of rotation (designated as $r_{\perp 1}$).

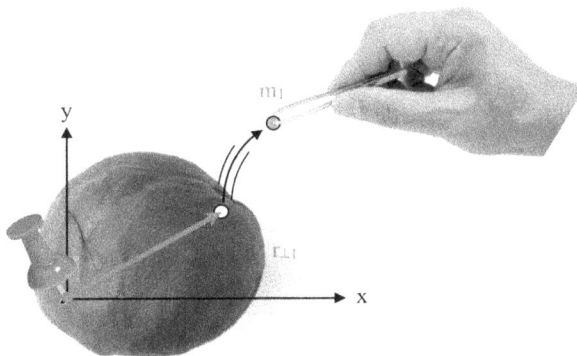

Blue Plasticine Image Credit: sergua/Shutterstock.com. Hand Image Credit: Ju Jae-young/Shutterstock.com. Thumbtack Image Credit: Pixabay by OpenClipart-Vectors.

Repeat the process with another small chunk of mass. Namely, use the tweezers to remove another small-sized chunk of mass from the object. Next, weigh that chunk (designated as m_2) and multiply its mass by the square of the *perpendicular distance* from the chunk to the axis of rotation (designated as $r_{\perp 2}$).

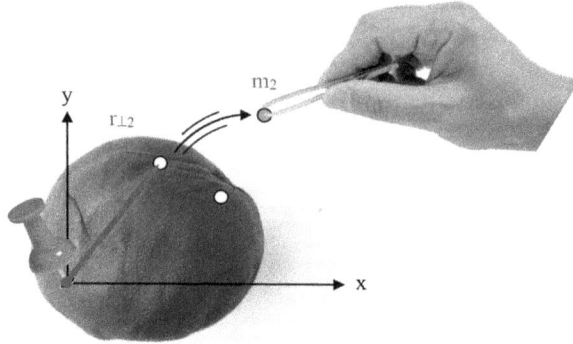

Blue Plasticine Image Credit: sergua/Shutterstock.com. Hand Image Credit: Ju Jae-young/Shutterstock.com. Thumbtack Image Credit: Pixabay by OpenClipart-Vectors.

Repeat this process over and over until you have removed all of the object's mass. This process may need to be repeated 50 times, 278 times, 1038 times, etc, depending on the number of small chunks you need to remove. The process is similar, *but not identical*, to the procedure used to determine an object's COM. Once you complete this tedious, but straightforward process, we define the MOI as:

▶ **Definition of the Moment of Inertia of an object about an axis of rotation:**

$$I = (m_1 \times r_{\perp 1}^2) + (m_2 \times r_{\perp 2}^2) + (m_3 \times r_{\perp 3}^2) + \cdots$$

For convenience, we abbreviate our definition as follows:

$I = \sum_{i=1}^{n} m_i * r_{\perp i}^2$, where n is the number of pieces into which the object has been cut.

As usual, this simple definition incorporates a number of features that are worth itemizing and clarifying. We'll check these features when we actually start calculating the MOI in example problems:

1. A single object has an *infinite number* of Moments of Inertia because we calculate the MOI relative to an axis of rotation. An object's MOI depends on where you put the thumbtack! If you move the thumbtack, you change the object's MOI. Since the thumbtack can be placed at an infinite number of locations (in the object, near the object, on Mars, on Pluto … anywhere you can imagine), an object has an infinite number of Moments of Inertia. In summary, you can only calculate the MOI relative to an axis of rotation. You must place the thumbtack before any calculations can proceed.

2. The axis of rotation need not pass through the object itself. In other words, the thumbtack need not touch the object. For example, we can calculate the MOI of a tetherball as it spins around the top of a pole, even though our thumbtack is not actually passing through the tetherball. Since the tetherball

is spinning around the pole (with the help of the rope), it still must have a MOI relative to the pole.

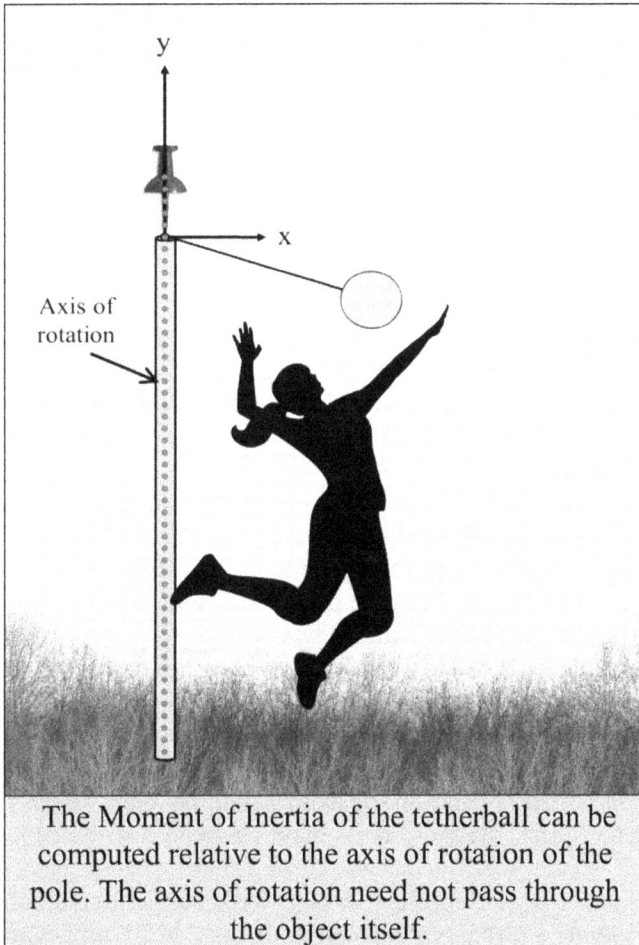

The Moment of Inertia of the tetherball can be computed relative to the axis of rotation of the pole. The axis of rotation need not pass through the object itself.

Background Image Credit: Pixabay by Ridderhof. Tetherball Image Credit: Enola99d/ Shutterstock.com. Thumbtack Image Credit: Pixabay by TheDigitalArtist. Volleyball Player Image Credit: Pixabay by Mohamed_hassan.

3. The MOI is a scalar! Therefore, we need not keep x- and y-components separate.
4. To calculate an object's COM, you must compute a ratio, using a denominator of M_{total}. However, to calculate an object's MOI relative to an axis of rotation, you do not compute a ratio. In other words, when calculating a MOI, do not divide your summation of $\sum_{i=1}^{n} m_i \times r_{\perp i}^2$ by M_{total}.

Making this unnecessary and incorrect calculation is a very common mistake that students typically make. Unfortunately, since COMs and MOIs are often computed in the same problem, students often confuse the formulas. When calculating the MOI, you do not need to divide it by M_{total}!

5. The units of the MOI are 'kg · m^2' To calculate the MOI of an object, we first take each small chunk of mass and multiply it by the square of its perpendicular distance to the axis of rotation—thus, the units are simply 'kg' times 'meters^2.'

6. In the same way you improve your determination of an object's COM by cutting smaller and smaller masses out of the object, so too will you improve your determination of the MOI by cutting smaller and smaller masses out of the object. You have to imagine using the tweezers to pull the smallest-sized chunks of mass out of the object as possible. The smaller the chunks of mass that you pull out of the object, the more accurate your determination of the MOI will be. In short, use your tweezers to remove the *smallest-sized* chunks of mass that you possibly can from the object! Removing many, many, small masses will result in a better determination of the MOI than removing only a few, large masses.

7. Next, you might be wondering what the MOI of an object represents. In other words, *'What is the **physical interpretation** of the MOI of an object?'* Notice that during our procedure for determining the MOI of the blob of clay, we sliced tiny pieces of clay from the blob until there was no clay left to slice. We then multiplied the mass of these chunks by the square of the perpendicular distance to the axis of rotation. Thus, we are building a physical parameter that increases as mass is moved farther and farther from the axis of rotation. In short, the MOI of an object measures the object's distribution of mass relative to the axis of rotation—the larger the MOI of an object, the more of its mass has been distributed *far* from the axis of rotation. Conversely, the smaller the MOI of an object, the more of its mass has been distributed *near* the axis of rotation. Thus, the MOI measures the difficulty associated with rotating an object about an axis of rotation. A large MOI implies that an object is *difficult* to rotate about a given axis of rotation. A small MOI implies an object is *easy* to rotate about a given axis of rotation.

8. Next, we should say something about the phrase, 'the *perpendicular distance* to the axis of rotation.' Another common mistake that students make when computing the MOI of an object is in determining the values of $r_{\perp 1}, r_{\perp 2}, r_{\perp 3}$, etc. Students often think that these distances must lie along the object itself … or maybe they must lie along the mechanism that connects an object to its axis of rotation. Not true! Instead, when computing $r_{\perp 1}, r_{\perp 2}, r_{\perp 3}$, etc, simply use the path that forms a 90° angle with respect to the axis of rotation. This 90° path is considered the 'perpendicular distance' and is labeled the *'moment arm'* (The symbol '$\perp$' is used as the symbol for 'perpendicular' while '$\parallel$' is used as the symbol for 'parallel'). Consider the situation of a tetherball rotating around a pole. Clearly, the axis of rotation is the pole—after all, the pole is the axis around which the ball is rotating. When calculating the MOI

for the tetherball, you need *not* find the distance from the tetherball to the pole along the rope. Instead, you calculate the MOI by using the perpendicular distance (i.e., the *'moment arm'*) between the tetherball and the pole. Notice that r makes a 90° angle with respect to the pole.

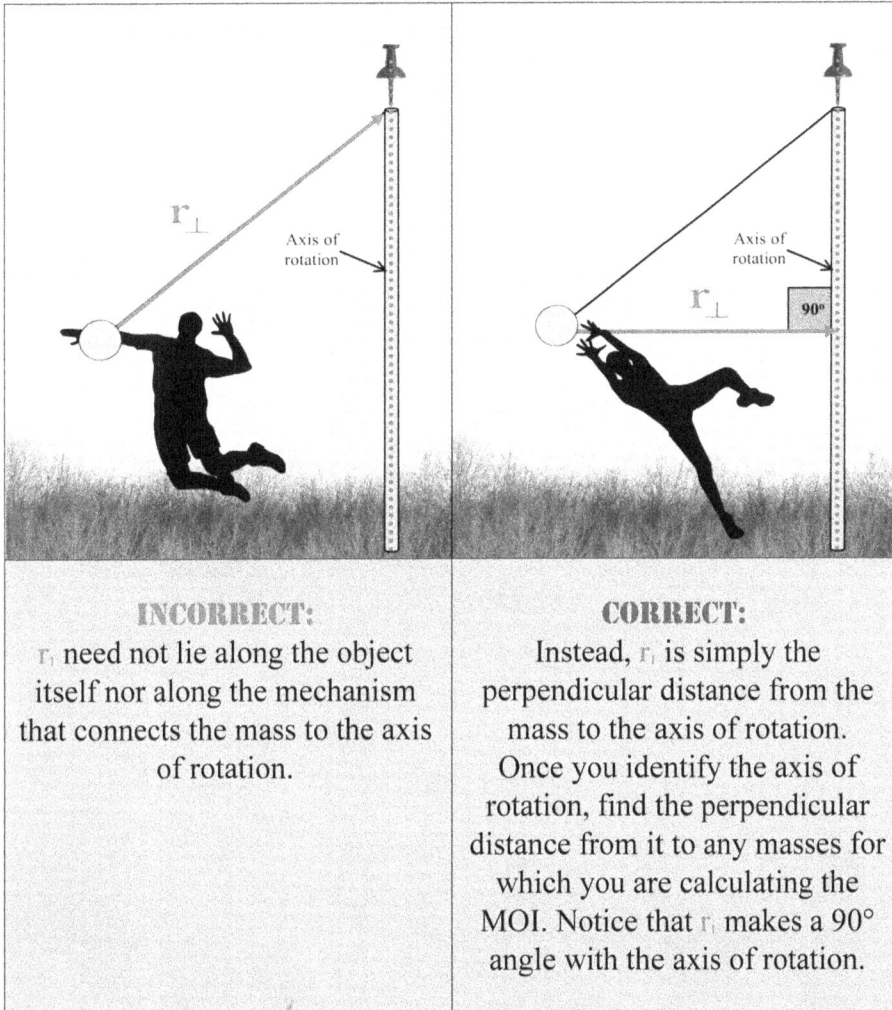

INCORRECT:	CORRECT:
$r_{\perp}$ need not lie along the object itself nor along the mechanism that connects the mass to the axis of rotation.	Instead, $r_{\perp}$ is simply the perpendicular distance from the mass to the axis of rotation. Once you identify the axis of rotation, find the perpendicular distance from it to any masses for which you are calculating the MOI. Notice that $r_{\perp}$ makes a 90° angle with the axis of rotation.

Background Image Credit: Pixabay by Ridderhof. Tetherball Image Credit: Enola99d/ Shutterstock.com. Thumbtack Image Credit: Pixabay by TheDigitalArtist. Volleyball Player Image Credit: Pixabay by Clker-Free-Vector-Images.

9. Last, you might be wondering about the MOI of some common, everyday objects. *'Do we really have to calculate the MOI for objects we encounter over and over again?'* For example, when calculating the motion of a can of soup rolling across your kitchen countertop, has some great physicist already

calculated the MOI for a solid drum rotating about its central axis? When analyzing the motion of a solid marble rolling across the floor, has some world-famous engineer already calculated the MOI of a solid sphere rotating about its diameter? Thankfully, the answer to both of these questions is 'Yes.' The MOI of some common, everyday objects (with commonly-encountered axes of rotation) have already been calculated and compiled for your use. For example, if you tell me the radius and mass of a solid drum (i.e., a can of soup), I can quickly tell you its MOI about a central axis using the following table (The answer is: $I_{\text{Solid Drum}} = 1/2\ M \times R^2$). If you tell me the radius and mass of a solid ball (i.e., a marble), I can quickly tell you its MOI about any diameter using the following table (The answer is: $I_{\text{Solid Sphere}} = 2/5\ M \times R^2$).

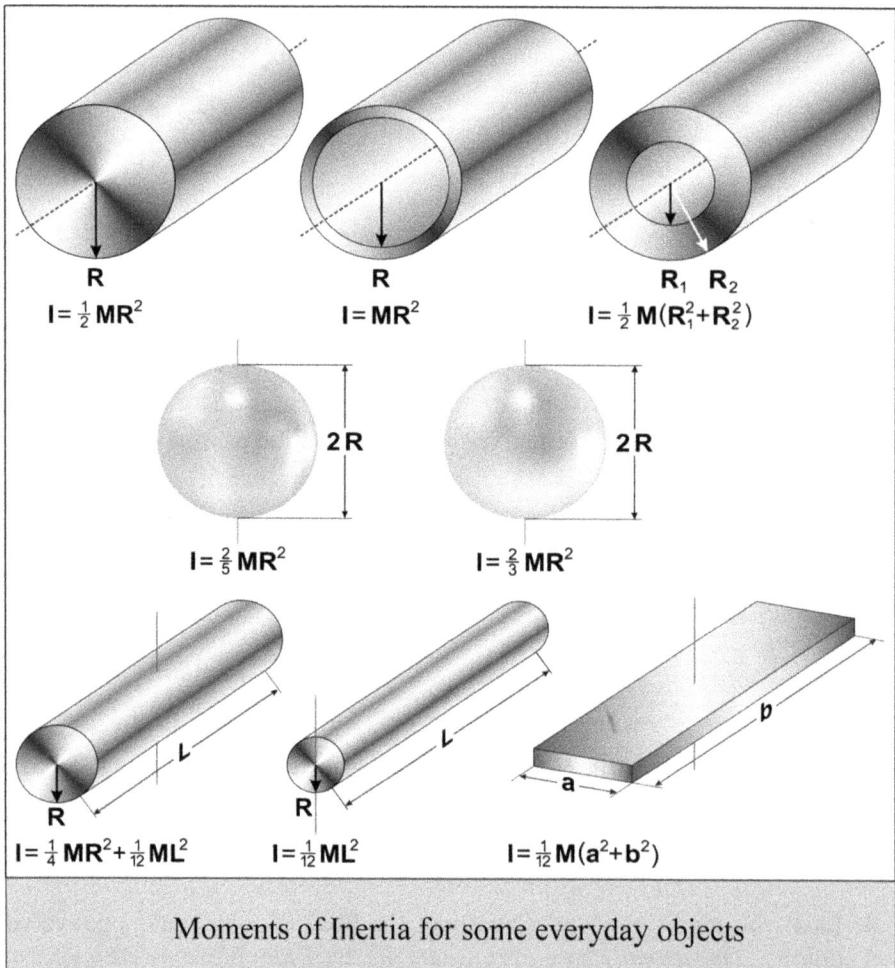

$I = \frac{1}{2}MR^2$

$I = MR^2$

$I = \frac{1}{2}M(R_1^2 + R_2^2)$

$2R$ $I = \frac{2}{5}MR^2$

$2R$ $I = \frac{2}{3}MR^2$

$I = \frac{1}{4}MR^2 + \frac{1}{12}ML^2$

$I = \frac{1}{12}ML^2$

$I = \frac{1}{12}M(a^2 + b^2)$

Moments of Inertia for some everyday objects

Moments of Inertia Image Credit: Fouad A. Saad/Shutterstock.com.

3.3.5 Modeling

❶ **EX 1:** Let's re-examine the strange object from a previous example but instead of calculating its COM, let's calculate its MOI about various axes of rotations. Consider the object made of four point-masses (masses that are so small, they essentially occupy only a point in space ... like marbles or ball-bearings) that are connected by massless sticks (sticks that are so lightweight, they essentially have no mass ... like toothpicks or plastic straws) into the shape shown below. Determine the object's MOI relative to an axis of rotation that is perpendicular to the xy-plane and passes through:

a) The 1 kg mass located at $(0,-2)$ m.
b) The 2 kg mass located at $(0,+2)$ m.
c) The 3 kg mass located at $(-3,0)$ m.
d) The 4 kg mass located at $(0,0)$ m.

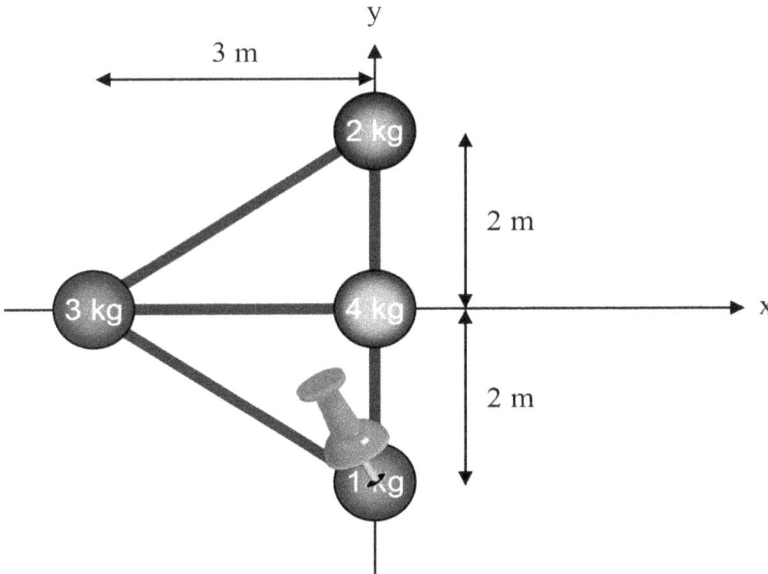

Thumbtack Image Credit: Pixabay by OpenClipart-Vectors.

☑ **ANSWER:** To start, imagine that we stick a thumbtack into the page, directly through the 1-kg point-mass. Note that this thumbtack is directed into the page, so the axis of rotation is directed perpendicular to the plane of the xy-axes, not in the plane of the xy-axes. This designates our axis of rotation. Next, imagine using our tweezers to remove the four point-masses one at a time. After removing each point-mass, we multiply its mass by the square of the perpendicular distance to the axis of rotation. Notice that we have to use the Pythagorean theorem to determine that the 3-kg point-mass is $\sqrt{13}$ m (or 3.6 m) from the axis of rotation.

a) $I_{\text{axis } a} = 1*(0^2) + 2*(4^2) + 3*(3.6^2) + 4*(2^2) = 87 \text{ kg} \cdot \text{m}^2$

We then move the thumbtack to the other three masses and repeat the process:

b) $I_{\text{axis } b} = 1*(4^2) + 2*(0^2) + 3*(3.6^2) + 4*(2^2) = 71 \text{ kg} \cdot \text{m}^2$

c) $I_{\text{axis } c} = 1*(3.6^2) + 2*(3.6^2) + 3*(0^2) + 4*(3^2) = 75 \text{ kg} \cdot \text{m}^2$

d) $I_{\text{axis } d} = 1*(2^2) + 2*(2^2) + 3*(3^2) + 4*(0^2) = 39 \text{ kg} \cdot \text{m}^2$

☺ **COMMENTS:** Let's check some of the features we expect to find when calculating a MOI:

1. ✓ The MOIs are scalars—yes.
2. ✓ The results have units of kg · m²—yes.
3. ✓ Last, let's remind ourselves of the physical interpretation of the MOI—the MOI measures how easily the object can be rotated. Comparing the MOI values we just calculated, we see that the object is hardest to rotate if the thumbtack is placed into the 1-kg point-mass (since $I_{\text{axis } a}$ is the largest value). Conversely, the object is easiest to rotate if the thumbtack is placed into the 4-kg point-mass (since $I_{\text{axis } d}$ is the smallest value). Finally, the object rotates with roughly the same ease/difficulty whether the thumbtack is placed in the 2- or 3-kg point-mass (since $I_{\text{axis } b} \approx I_{\text{axis } c}$).

❷ **EX 2:** Four point-masses are connected by massless sticks into the shape shown below:

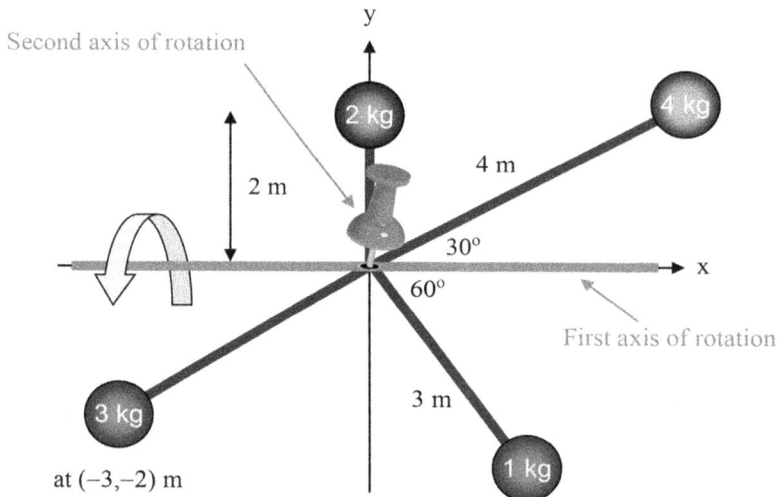

Thumbtack Image Credit: Pixabay by OpenClipart-Vectors.

Would someone have more difficulty rotating this object: (a) about an axis of rotation lying along the x-axis, or (b) about an axis of rotation that is perpendicular to the xy-plane and passes through the origin?

☑ **ANSWER:** The method by which we answer this question is to calculate the MOI for both axes of rotation. The larger MOI will tell us the axis of rotation about which the object is harder to rotate.

a) Note that the first choice of the axis of rotation implies that the object can rotate in and out of the xy-plane (as indicated by the circulating arrow). To determine the MOI, we first need to use some trigonometric functions to determine the *perpendicular distance* between each point-mass and the axis of rotation; however, once we determine these distances, calculating the MOI should be straightforward:

$$I_{\text{axis a}} = 1*[(3* \sin(60°))^2] + 2*(2^2) + 3*(2^2) + 4*[(4* \sin(30°))^2] = 42.75 \text{ kg} \cdot \text{m}^2$$

b) Next, we drive our trusty thumbtack to the origin and calculate the MOI as follows:

$$I_{\text{axis } b} = 1 \times (3^2) + 2 \times (2^2) + 3 \times (3.6^2) + 4 \times (4^2) = 120 \text{ kg} \cdot \text{m}^2$$

☺ **COMMENTS:** Let's check some of the features we expect to find when calculating a MOI:

1. ✓ The MOI is a scalar—yes.
2. ✓ The result has units of kg · m²—yes.
3. ✓ Last, let's remind ourselves of the physical interpretation of the MOI—the MOI measures how easy (or difficult) the object can be rotated. Clearly, rotating the object about the second axis of rotation (with $I_{\text{axis } b} = 120 \text{ kg} \cdot \text{m}^2$) is more difficult than rotating it about the first axis of rotation (with $I_{\text{axis } a} = 42.75 \text{ kg} \cdot \text{m}^2$).

3.3.6 Instructional input

3.3.6.1 Multiplying vectors by vectors

3.3.6.1.1 The vector 'cross'-product
In the first chapter of this series, we spent a great deal of time discussing vector notation and the rules for vector addition and subtraction. Later, when we introduced the conservation laws, we spent a great deal of time discussing the rules for vector multiplication. We said there were four instances in which vectors could be multiplied:

▶ **CASE 1—Multiplying a vector by a scalar:** The simplest case of vector multiplication involves multiplying a vector by a scalar. If $\overrightarrow{A} = (A_x, A_y)$ and k is a scalar, then $k \cdot \overrightarrow{A} = (k \cdot A_x, k \cdot A_y)$. This is nothing new—we simply multiply each of the components of $\overrightarrow{A}$ by the scalar k, making sure we keep x- and y-components separate. We have been doing so throughout all the chapters in this series:

❶ **EX 1:** If acceleration is given by the vector: $\overrightarrow{a} = (1.5, 4.4)$m s^{-2}, what is $(20 \text{ kg}) \cdot \overrightarrow{a}$?

☑ **ANSWER:** Simply multiply each component of the acceleration by the scalar 20 kg: $(20 \text{ kg}) \cdot \overrightarrow{a} = (20 \text{ kg}) \cdot (1.5, 4.4)$m s$^{-2} = (30, 88)$ Newtons.

▶ **CASE 2—Dividing a vector by a scalar:** Another simple case of vector multiplication involves dividing a vector by a scalar. If $\overrightarrow{A} = (A_x, A_y)$ and k is a scalar, then $\overrightarrow{A} \div k = (A_x/k, A_y/k)$. Again, the methodology is nothing new. We simply divide each of the components of $\overrightarrow{A}$ by the scalar k, making sure we keep x- and y-components separate.

❷ **EX 2:** If force is given by the vector: $\overrightarrow{F} = (50.8, 44.6)$ Newtons, what is $\overrightarrow{a} = \frac{\overrightarrow{F}}{(2 \text{ kg})}$?

☑ **ANSWER:** Simply divide each component of the force by the scalar 2 kg:

$$\overrightarrow{a} = \frac{(50.8, 44.6) \text{ Newtons}}{(2 \text{ kg})} = \left(\frac{50.8}{2}, \frac{44.6}{2}\right) \text{ m s}^{-2} = (25.4, 22.3) \text{ m s}^{-2}.$$

▶ **CASE 3—The 'Dot'-product—Multiplying a vector by a vector:** When calculating 'work' or 'energy' in the context of conservation laws, we developed a new mathematical procedure for multiplying a vector by another vector. This procedure was called the vector 'Dot'-product. If $\overrightarrow{A} = (A_x, A_y)$ and $\overrightarrow{B} = (B_x, B_y)$, then:

$$\overrightarrow{A} \bullet \overrightarrow{B} = (A_x, A_y) \bullet (B_x, B_y) = A_x \cdot B_x + A_y \cdot B_y.$$

The methodology begins to get tricky here. For starters, when you 'dot' two vectors, the result is a scalar. Next, our procedure for calculating a 'Dot'-product involves multiplying only parallel components. In other words, the product of the x-component of $\overrightarrow{A}$ and the x-component of $\overrightarrow{B}$ is added to the product of the y-component of $\overrightarrow{A}$ and the y-component of $\overrightarrow{B}$. Finally, we interpret the vector 'Dot'-product as telling us the degree to which two vectors are *parallel*. For instance, if two vectors are completely perpendicular, their 'Dot'-product is zero. The more aligned two vectors are, the larger their 'Dot'-product will be.

❸ **EX 3:** If force is given by the vector: $\overrightarrow{F} = (50.8, 44.6)$ Newtons and displacement is given by vector $\overrightarrow{d} = (10, 3)$m, what is $\overrightarrow{F} \bullet \overrightarrow{d}$?

☑ **ANSWER:** Simply follow the procedure for 'dotting' two vectors:

$$\overrightarrow{F} \cdot \overrightarrow{d} = \begin{pmatrix} 50.8 \\ 44.6 \end{pmatrix} \cdot \begin{pmatrix} 10 \\ 3 \end{pmatrix} = (50.8 \times 10) + (44.6 \times 3) = 508 + 133.8 = 641.8 \text{ Joules}$$

► **CASE 4—The 'Cross'-product—Multiplying a vector by a vector:** Recall from our discussion of conservation laws that when multiplying two vectors, the symbol between the two vectors is extremely important. The 'DOT' (i.e., symbolized by • or •) and the 'CROSS' (i.e., symbolized by × or ×) are not interchangeable symbols for multiplying two vectors. In other words, $\overrightarrow{A} \cdot \overrightarrow{B} \neq \overrightarrow{A} \times \overrightarrow{B}$. If you see a 'DOT' between two vectors, follow the rules for 'dotting' two vectors that were outlined above; however, if you see a 'CROSS' between two vectors, follow the rules for 'crossing' two vectors as outlined below:

If two vectors are written in vector notation, $\overrightarrow{A} = (A_x, A_y)$ and $\overrightarrow{B} = (B_x, B_y)$:

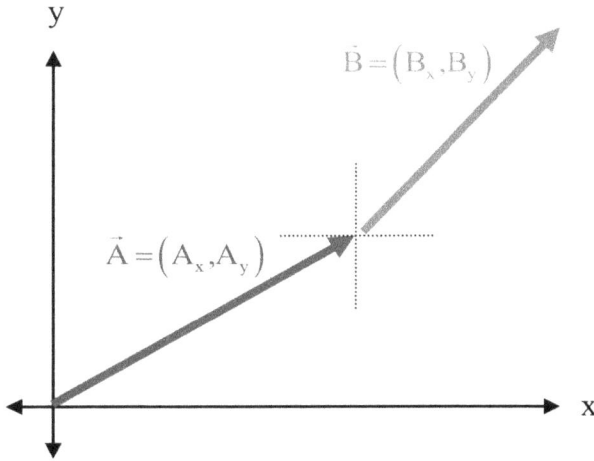

then the vector 'Cross'-product is defined as:

$$\vec{A} \times \vec{B} = \begin{pmatrix} A_x \\ A_y \end{pmatrix} \times \begin{pmatrix} B_x \\ B_y \end{pmatrix} = \left(A_x \cdot B_y \right) - \left(A_y \cdot B_x \right) \text{ with } \begin{cases} \odot \text{ or CCW or } \circlearrowleft, \text{ if positive} \\ \otimes \text{ or CW or } \circlearrowright, \text{ if negative} \end{cases}$$

Notice the answer is a vector—but the resulting direction is given by new symbols. These symbols represent the directions of "out" or "in."

Computing the vector 'Cross'-product is *much more complicated* than computing the relatively straightforward vector 'Dot'-product. I've been teaching physics for decades now and can tell you that some of the most commonly-occurring errors made by students in a *Classical Mechanics* course involve calculating the vector 'Cross'-product. The reason the vector 'Cross'-product is so difficult to calculate is because several small steps need to be done correctly. Additionally, the vector 'Cross'-product results in a third vector that points out of the plane of the two vectors that were crossed. In other words, the 'Cross'-product of vectors is a three-dimensional procedure. You have to envision vectors and arrows pointing in and out of the book you are reading! Hopefully, these aspects of the vector 'Cross'-product will become clear as we work our way through several examples. Let's examine some of the features of the 'Cross'-product more closely:

1. The units of the 'Cross'-product are the units of vector $\vec{A}$, times the units of vector $\vec{B}$. For instance, if $\vec{A}$ has units of 'Cats' and $\vec{B}$ has units of 'Dogs,' then $\vec{A} \times \vec{B}$ has units of 'Cats · Dogs.'

2. The 'Cross'-product is <u>not</u> 'commutative.' In other words, the order of vectors is important: $\vec{A} \times \vec{B}$ is <u>not</u> identical to $\vec{B} \times \vec{A}$. In fact, the 'Cross'-product is 'anti-commutative,' meaning that if you reverse the order of the two vectors, you will end up with the negative, or opposite, result. Thus, $(\vec{A} \times \vec{B}) = -(\vec{B} \times \vec{A})$. Mixing up the order of vectors is a very common error when computing vector 'Cross'-products.

3. The vector 'Dot'-product tells us the degree to which two vectors are parallel while the vector 'Cross'-product tells us the degree to which two vectors are perpendicular. Physically, the 'Cross'−product multiplies perpendicular components of two vectors. In other words, the x-component of $\vec{A}$ multiplies the y-component of $\vec{B}$, while the y-component of $\vec{A}$ multiplies the x-component of $\vec{B}$. We will need to explore this in more detail in the upcoming examples, but we will now interpret the 'Cross'-product as a method for multiplying the perpendicular components of two vectors. If two vectors are parallel, their vector 'Cross'-product will be zero. If two vectors are perpendicular, namely at 90° to one another, their vector 'Cross'-product will be maximized.

4. Notice that the definition of the vector 'Cross'-product involves a minus-sign. The product of $A_y \cdot B_x$ is subtracted from the product of $A_x \cdot B_y$. Here lies another common source of error when computing the vector 'Cross'-product.

5. The vector 'Cross'-product takes two vectors, multiplies them, and **results in a vector!** However, the resulting vector no longer lies in the plane of vectors $\vec{A}$ and $\vec{B}$. The resulting 'Cross'-product is actually 'in' or 'out' of the plane of vectors $\vec{A}$ and $\vec{B}$. This result is very difficult for students to grasp and is clearly the most obvious source of

confusion when trying to compute the vector 'Cross'-product. Here is the problem—we multiply two two-dimensional vectors and end up with a vector that points in the third dimension. So, if we draw two vectors on our sheet of paper, the vector 'Cross'-product actually points 'in' or 'out' of the sheet of paper. *How do we draw such an arrow in three-dimensions? How can we visualize what this vector 'Cross'-product looks like if it is actually pointing in or out of the book you are reading?* To address these concerns, physicists and mathematicians have adopted a set of symbols to help us visualize vectors in three-dimensions. More importantly, these symbols will help us draw three-dimensional vectors on a two-dimensional sheet of paper; namely, the book you are currently reading.

3.3.6.2 Representing vectors 'out of' and 'into' this book

Imagine a simple arrow. The tip has a nice, sharp, shiny bullet-shaped point. The back-end has four feathers functioning as its fletching to stabilize the arrow in flight.

Etymology:
The fin-shaped aerodynamic device on the end of an arrow is called the 'fletching,' from the French word 'flèche,' meaning 'arrow.'

The tip of the arrow is a bullet-shaped point.	The back end of the arrow has four feathers, or fins, directed out of it

Left image: Bullet Image Credit: Pixabay by BRAIN_PAIN. Right image: This DFRArrow (1) image has been obtained by the author from the Wikimedia website, where it is stated to have been released into the public domain. It is included within this article on that basis.

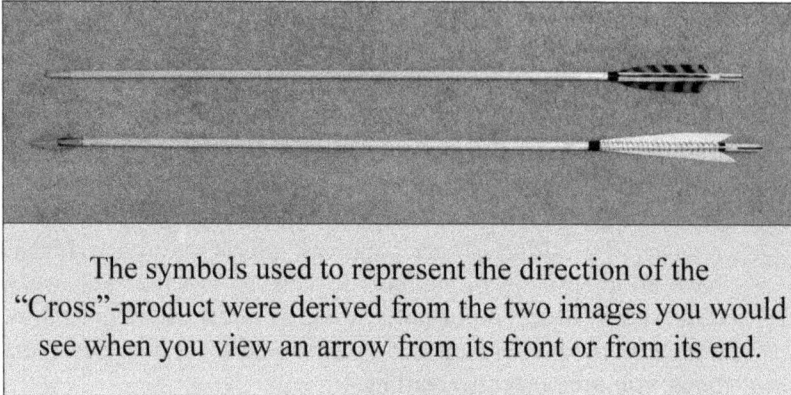

The symbols used to represent the direction of the "Cross"-product were derived from the two images you would see when you view an arrow from its front or from its end.

This Two arrows image has been obtained by the author from the Wikimedia website, where it is stated to have been released into the public domain. It is included within this article on that basis.

► **To represent a vector pointing OUT OF this book:**

Three symbols are used to represent a vector pointing out of this book. The three symbols look nothing alike but they are three completely equivalent representations to a physicist or mathematician.

Image a simple arrow pointing out of this book, directly at your face. *What would you see?* You would see only the tip of the arrow—essentially, a small circle outlining the outer-edge of the bullet-shaped tip, with a small dot in the center indicating where the actual point of the arrow is.

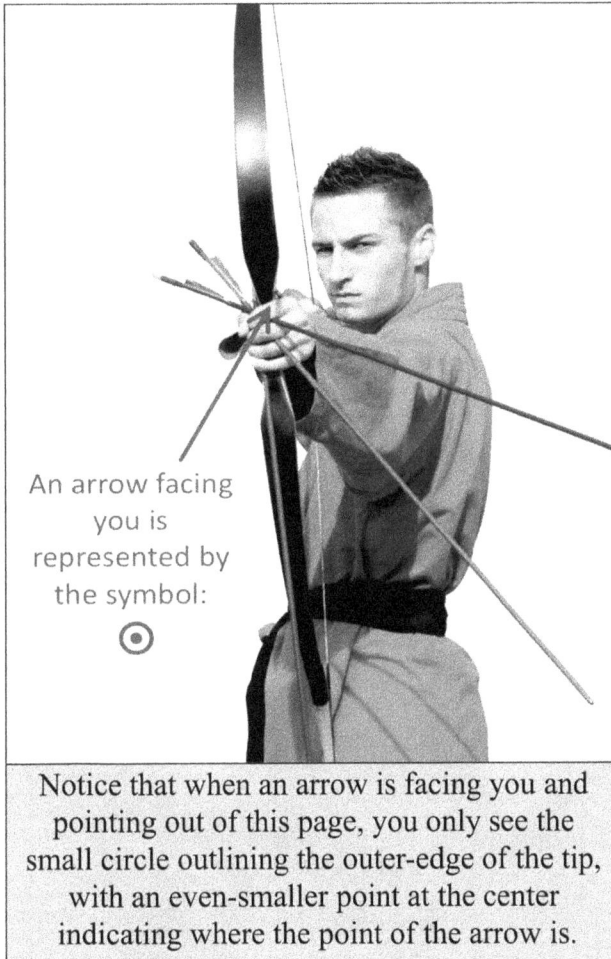

An arrow facing you is represented by the symbol:

⊙

Notice that when an arrow is facing you and pointing out of this page, you only see the small circle outlining the outer-edge of the tip, with an even-smaller point at the center indicating where the point of the arrow is.

Archer Credit Image: Gergo Orban/Shutterstock.com.

Thus, one symbol for an arrow pointing at you is '⊙'.

Likewise, imagine that, *using your right hand*, you point your thumb out of this book, directly at your face. *What would you see?* You would see your fingers 'curling' or 'twirling' in a Counter-Clockwise direction.

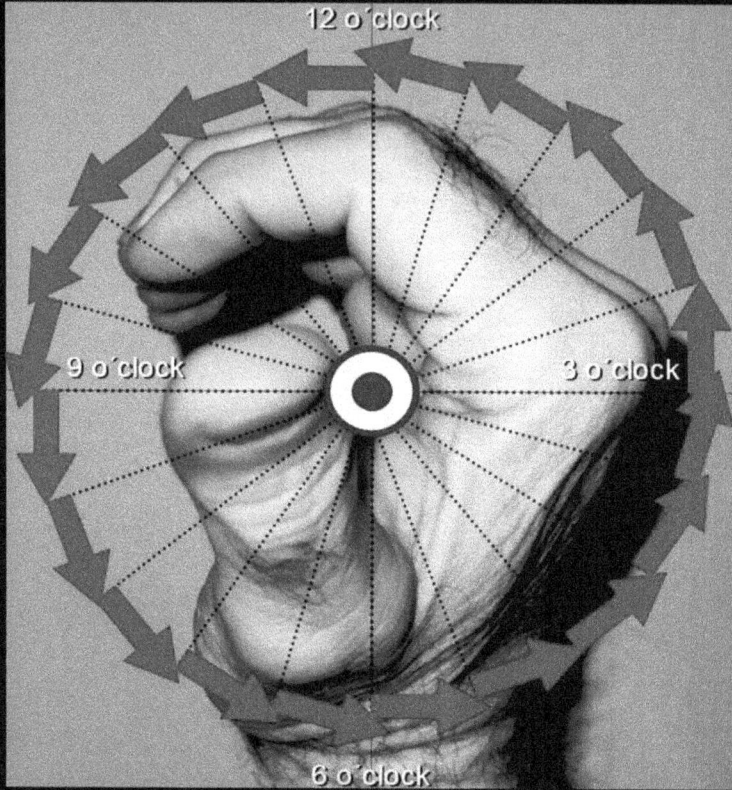

YOUR RIGHT THUMB FACING TOWARD YOU

12 o'clock

9 o'clock

3 o'clock

6 o'clock

If you direct the thumb on your right hand out of the book, your fingers naturally "curl" in a Counter-Clockwise direction.

Image Credit: Author.

Thus, if you imagine that the thumb on your right hand is an arrow point out of the plane of this book, two other symbols for an arrow pointing at you are '↺' or 'CCW' (as an abbreviation for 'Counter-Clockwise').

▶ **To represent a vector pointing INTO this book:**

Similarly, three symbols are used to represent a vector pointing into this book. The three symbols look nothing alike but they are three completely equivalent representations to a physicist or mathematician.

Now imagine the simple arrow pointing into this book, directly away from your face. *What would you see?* You would see only the fletching (i.e., the feathers or fins) of the arrow—essentially, a cluster of four feathers, in the shape of an '×'.

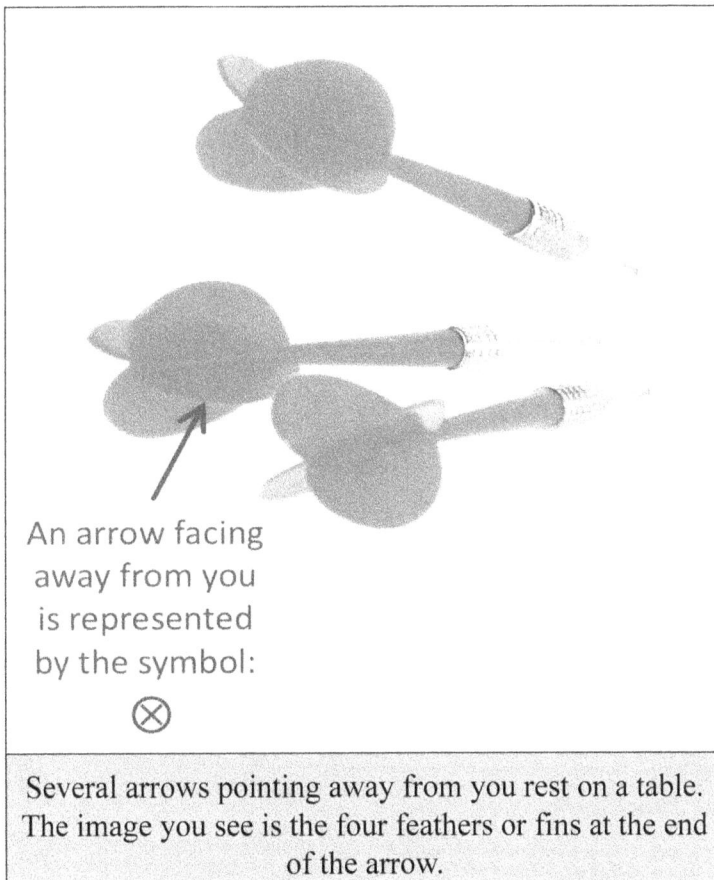

An arrow facing away from you is represented by the symbol:

$\otimes$

Several arrows pointing away from you rest on a table. The image you see is the four feathers or fins at the end of the arrow.

Darts Image Credit: Africa Studio/Shutterstock.com.

Thus, one symbol for an arrow point at you is '$\otimes$'.

Likewise, imagine that, *using your right hand*, you point your thumb into this book, directly away from your face. *What would you see?* You would see your fingers 'curling' or 'twirling' in a Clockwise direction.

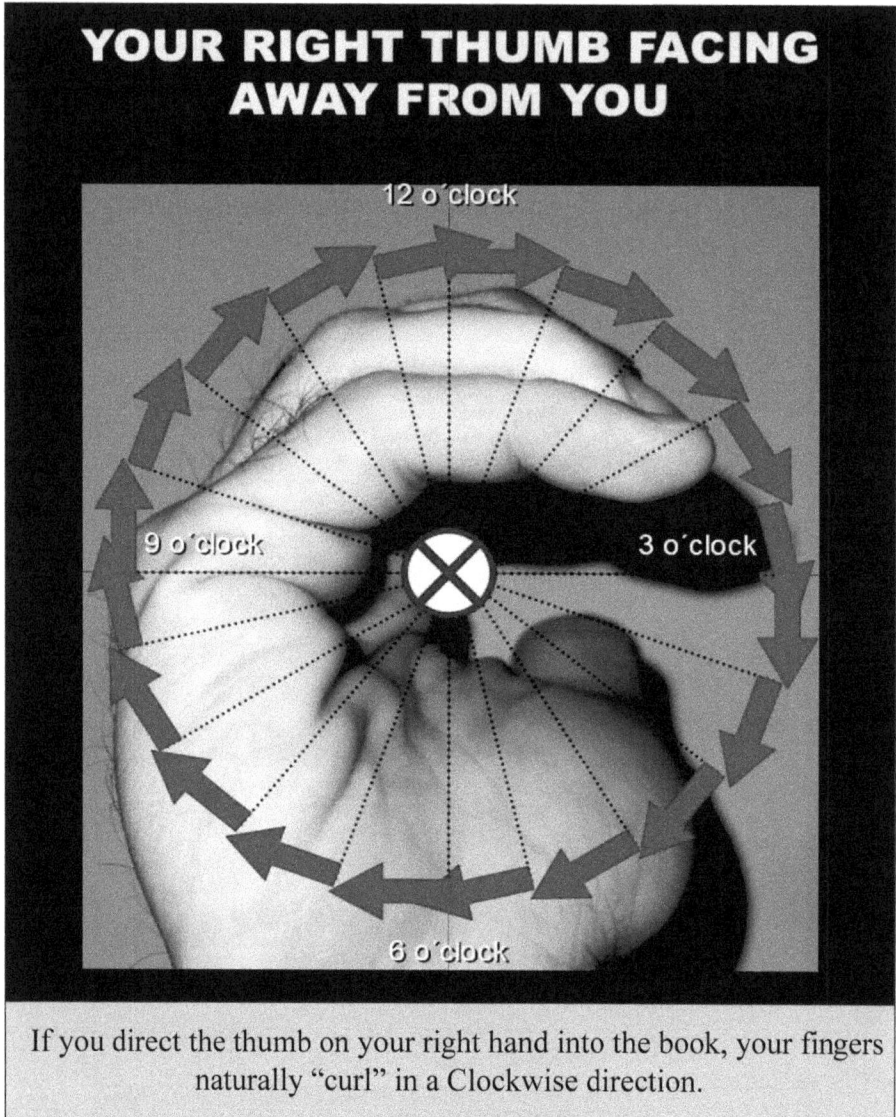

YOUR RIGHT THUMB FACING AWAY FROM YOU

12 o'clock

9 o'clock

3 o'clock

6 o'clock

If you direct the thumb on your right hand into the book, your fingers naturally "curl" in a Clockwise direction.

Image Credit: Author.

Thus, if you imagine that the thumb on your right hand is an arrow point into the plane of this book, two other symbols for an arrow point away from you are '↻' or 'CW' (as an abbreviation for 'Clockwise').

Let's try some examples and check that we have mastered the art of calculating a vector 'Cross'-product and representing vectors that point in and out of this book.

3.3.7 Modeling

❶ **EX 1:** Two vectors are given: $\overrightarrow{A} = (3, 4)$ Newtons and $\overrightarrow{B} = (5, 6)$ m. What are the vector 'Cross'-products $\overrightarrow{A} \times \overrightarrow{B}$ and $\overrightarrow{B} \times \overrightarrow{A}$?

☑ **ANSWER:** Let's follow the multiplication rules stated above. Also, for reasons that will become clear in a moment, let's write our vectors as column-vectors rather than row-vectors:

$$\overrightarrow{A} \times \overrightarrow{B} = \begin{pmatrix} 3 \\ 4 \end{pmatrix} \times \begin{pmatrix} 5 \\ 6 \end{pmatrix} = (3 \cdot 6) - (4 \cdot 5) = 18 - 20 = -2 \otimes \text{Newton} \cdot \text{m}.$$

You would state your answer as '2 Newton · m into the page.'

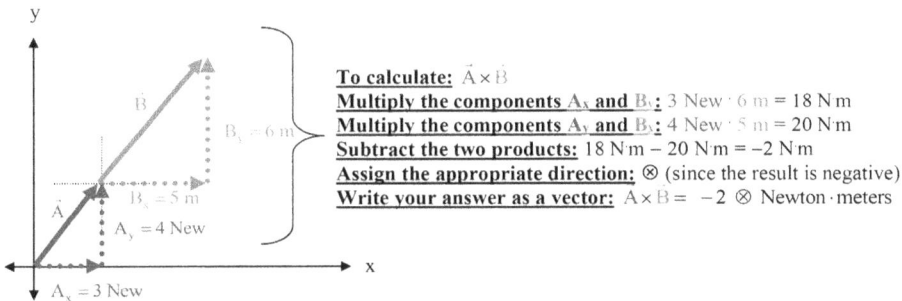

To calculate: $\vec{A} \times \vec{B}$
Multiply the components A_x and B_y: 3 New · 6 m = 18 N·m
Multiply the components A_y and B_x: 4 New · 5 m = 20 N·m
Subtract the two products: 18 N·m − 20 N·m = −2 N·m
Assign the appropriate direction: ⊗ (since the result is negative)
Write your answer as a vector: $A \times B = -2 \otimes$ Newton·meters

To calculate the second 'Cross'-product, just reverse the order of the operation of subtraction:

$$\overrightarrow{B} \times \overrightarrow{A} = \begin{pmatrix} 5 \\ 6 \end{pmatrix} \times \begin{pmatrix} 3 \\ 4 \end{pmatrix} = (5 \cdot 4) - (6 \cdot 3) = 20 - 18 = +2 \odot \text{Newton} \cdot \text{m}.$$

You would state your answer as '2 Newton · m out of the page.'

1. ☺ **COMMENTS:** Let's check each of the five features of vector 'Cross'-products:

   1. ✓ The result has units of vector $\overrightarrow{A}$ (which are 'Newtons'), times the units of vector $\overrightarrow{B}$ (which are 'meters').
   2. ✓ The procedure is anti-commutative. Our two answers demonstrate that $(\overrightarrow{A} \times \overrightarrow{B}) = -(\overrightarrow{B} \times \overrightarrow{A})$.
   3. ✓ The result multiplies perpendicular components of the two vectors. Notice that vector notation keeps x- and y-components separate. Then performing the 'Cross'-product, we multiply x-components with y-components.
   4. ✓ The vector 'Cross'-product involves a subtraction. When calculating $\overrightarrow{A} \times \overrightarrow{B}$, the product of $A_y \cdot B_x$ is subtracted from the product of $A_x \cdot B_y$. When calculating $\overrightarrow{B} \times \overrightarrow{A}$, the product of $B_y \cdot A_x$ is subtracted from the product of $B_x \cdot A_y$.

5. ✓ The result is a vector—yes, our answer was not written with the usual vector notation involving a set of parentheses; however, we agreed that the symbols ⦿ and ⊗ could be used to represent vectors that point 'out' or 'in' from the plane of this book. Remember, vectors are quantities that carry two, linked pieces of information. Each of these answers contains two linked pieces of information—a magnitude (of 2 Newton · m) and a direction (⊗ or ⦿). $\overrightarrow{A} \times \overrightarrow{B}$ points *into* the plane of this book while $\overrightarrow{B} \times \overrightarrow{A}$ points *out of* the plane of this book.

If you look over this first example of a vector 'Cross'-product, you'll notice that we preferred to write the vectors as column-vectors rather than row-vectors (although writing them as row-vectors is perfectly fine). The reason for this preference is because a neat and useful mnemonic device results. When performing the 'Cross'-product and using col-umn-vectors, you'll notice that the x- and y-components of the two vectors are aligned side-by-side. When multiplying the side-by-side components, you will naturally form the 'crisscross' symbol of an ×:

$$\vec{A} \times \vec{B} = \begin{pmatrix} A_x \\ A_y \end{pmatrix} \begin{pmatrix} B_x \\ B_y \end{pmatrix} = A_x \cdot B_y - A_y \cdot B_x$$

or

$$\vec{A} \times \vec{B} = \times$$

This is a very useful mnemonic device because it constantly reminds us that the 'Cross'-product physically multiplies the perpendicular components of two vectors. This mnemonic reminds us that if we ever encounter a physical situation in which we must ensure that two vectors are perpendicular or if we need to check whether or not two vectors are perpendicular, then the vector 'Cross'-product is the way to go!

❷ **EX 2:** Four vectors are given: $\overrightarrow{A} = (3, 4)$ Newtons, $\overrightarrow{B} = (5, 6)$m, $\overrightarrow{C} = (-7, 8)$Cats, and $\overrightarrow{D} = (-4, 3)$Dogs. Calculate the following vector 'Cross'-products:

a) $\overrightarrow{A} \times \overrightarrow{B}$
b) $\overrightarrow{A} \times \overrightarrow{C}$
c) $\overrightarrow{A} \times \overrightarrow{D}$
d) $\overrightarrow{C} \times \overrightarrow{D}$
e) $\overrightarrow{C} \times \overrightarrow{A}$

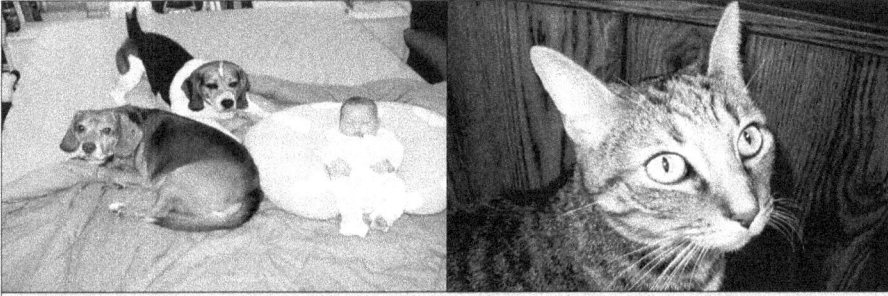

Dogs and a Cat–and a Baby–Again! This was the same set of vectors we tackled in volume 2, chapter 2 when we were learning to perform the vector "Dot"-product. Now, we are using the same four vectors to practice performing the vector "Cross"-product.

Image Credit: Author.

☑ **ANSWER:** Let's follow the multiplication rules stated above and we'll write our vectors as column-vectors rather than row-vectors:

a) $\vec{A} \times \vec{B} = \begin{pmatrix} 3 \\ 4 \end{pmatrix} \times \begin{pmatrix} 5 \\ 6 \end{pmatrix} = (3 \cdot 6) - (4 \cdot 5) = 18 - 20 = -2 \otimes$ Newton · m

b) $\vec{A} \times \vec{C} = \begin{pmatrix} 3 \\ 4 \end{pmatrix} \times \begin{pmatrix} -7 \\ 8 \end{pmatrix} = (3 \cdot 8) - (4 \cdot (-7)) = 24 + 28 = 52 \odot$ Newton · Cats

c) $\vec{A} \times \vec{D} = \begin{pmatrix} 3 \\ 4 \end{pmatrix} \times \begin{pmatrix} -4 \\ 3 \end{pmatrix} = (3 \cdot 3) - (4 \cdot (-4)) = 9 + 16 = 25 \odot$ Newton · Dogs

d) $\vec{C} \times \vec{D} = \begin{pmatrix} -7 \\ 8 \end{pmatrix} \times \begin{pmatrix} -4 \\ 3 \end{pmatrix} = ((-7) \cdot 3) - (8 \cdot (-4)) = -28 + 32 = 11 \odot$ Cats · Dogs

e) $\vec{C} \times \vec{A} = \begin{pmatrix} -7 \\ 8 \end{pmatrix} \times \begin{pmatrix} 3 \\ 4 \end{pmatrix} = ((-7) \cdot 4) - (8 \cdot 3) = -28 - 24 = -52 \otimes$ Newton · Cats

☺ **COMMENTS:** Let's check each of the five features:

1. ✓ The results have units of vector $\vec{A}$ times the units of vector $\vec{B}$.
2. ✓ The procedure is anti-commutative—notice that answers (b) and (e) have the exact same numeric values, but opposite signs. We could have skipped a calculation in part (e) because we know immediately that it must be the negative as the answer in part (a). As we know, order is very important when 'Crossing' two vectors.
3. ✓ The results multiply perpendicular components of two vectors. Notice that part (c) demonstrates the maximum possible value of multiplying these two vectors. This should not surprise us because

vectors $\overrightarrow{A}$ and $\overrightarrow{D}$ **are perpendicular**—and the 'Cross'-product only multiplies perpendicular components. Even if we didn't realize that $\overrightarrow{A}$ and $\overrightarrow{D}$ *are perpendicular*, the 'Cross'-product would tell us, 'Vector $\overrightarrow{A}$ is perpendicular (or has perpendicular components) to vector $\overrightarrow{D}$.'

PART (C) ILLUSTRATES HOW THE 'CROSS'–PRODUCT SHOULD BE   PHYSICALLY INTERPRETED—IT TELLS HOW MUCH OF ONE VECTOR IS PERPENDICULAR TO ANOTHER VECTOR!

4. ✓ The operation of subtraction was used in each calculation.
5. ✓ The results are all vectors, with the directions symbolized by ◉ or ⊗!

❸ **EX 3:** Two vectors are given: $\overrightarrow{A} = (3, 4)$ Newtons and $\overrightarrow{B} = (5, 6)$ meters. What is the vector 'Cross'-product $-5\overrightarrow{A} \times 2\overrightarrow{B}$?

☑ **ANSWER:** Let's just follow the conventions we've used in the previous two examples. Also, an alternative way to do this problem is to multiply the two vectors by the appropriate scalars (i.e., $-5\overrightarrow{A}$ and $2\overrightarrow{B}$), before using them in a 'Cross'-product.

$$-5\overrightarrow{A} \times 2\overrightarrow{B} = -5\begin{pmatrix}3\\4\end{pmatrix} \times 2\begin{pmatrix}5\\6\end{pmatrix} = \begin{pmatrix}-15\\-20\end{pmatrix} \times \begin{pmatrix}10\\12\end{pmatrix} = (-15 \times 12) - (-20 \times 10)$$
$$= -180 + 200 = 20 \odot \text{ Newton} \cdot \text{m}$$

☺ **COMMENTS:** Let's check each of the five features:

1. ✓ The result has units of vector $\overrightarrow{A}$ (which are 'Newtons'), times the units of vector $\overrightarrow{B}$ (which are 'meters').
2. ✓ The procedure is anti-commutative—we could have calculated $(2\overrightarrow{B}) \times (-5\overrightarrow{A})$ and ended up with the exact opposite answer.
3. ✓ The result multiplies perpendicular components of the two vectors.
4. ✓ The operation of subtraction was used in each calculation.
5. ✓ The results are all vectors, with the directions symbolized by ◉ or ⊗!

3.3.8 Checking for understanding

3.3.8.1 All vector operations

Before moving to our next topic, review the vector operations below and perform the additions, subtractions, and multiplications on the following page.

SUMMARY OF VECTOR OPERATIONS

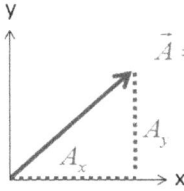

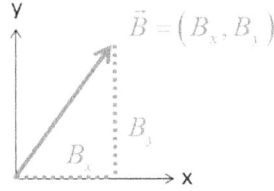

$$\vec{A} = \left(A_x, A_y \right)$$

$$\vec{B} = \left(B_x, B_y \right)$$

1. **VECTOR ADDITION:**

$$\vec{A} + \vec{B} = \left(A_x + B_x, A_y + B_y \right) = \begin{pmatrix} A_x + B_x \\ A_y + B_y \end{pmatrix}$$

2. **VECTOR SUBTRACTION:**

$$\vec{A} - \vec{B} = \left(A_x - B_x, A_y - B_y \right) = \begin{pmatrix} A_x - B_x \\ A_y - B_y \end{pmatrix}$$

3. **MULTIPLICATION BY A SCALAR:**

$$k\vec{A} = \left(k * A_x, k * A_y \right) = \begin{pmatrix} k * A_x \\ k * A_y \end{pmatrix}$$

4. **MULTIPLICATION – THE VECTOR *"DOT"*-PRODUCT:**

$$\vec{A} \bullet \vec{B} = \left(A_x * B_x \right) + \left(A_y * B_y \right)$$

5. **MULTIPLICATION – THE VECTOR *"CROSS"*-PRODUCT:**

$$\vec{A} \times \vec{B} = \left(A_x * B_y - A_y * B_x \right) \text{ with } \left\{ \begin{matrix} \bigcirc \text{ or } CCW \text{ or } \circlearrowleft, \text{ if positive} \\ \otimes \text{ or } CW \text{ or } \circlearrowright, \text{ if negative} \end{matrix} \right\}$$

$$= \begin{pmatrix} 0 \\ 0 \\ A_x * B_y - A_y * B_x \end{pmatrix}$$

EXAMPLES AND APPLICATIONS

$\vec{A} = (3,0)\ m$

$\vec{B} = (3,4)\ m$

$\vec{C} = (0,4)\ m$

$\vec{D} = (-2,0)\ m$

1. **VECTOR ADDITION:**
   $\vec{A} + \vec{B} = (3+3, 0+4) = (6,4)\ m$

2. **VECTOR SUBTRACTION:**
   $\vec{A} - \vec{B} = (3-3, 0-4) = (0,-4)\ m$
   Why? Adds or subtracts to get a resultant
   Applications: Conservation of Momentum: $\vec{p}_{total} = \vec{p}_1 + \vec{p}_2 + \vec{p}_3 + \ldots$

3. **MULTIPLICATION BY A SCALAR:**
   $10\vec{A} = (10*3, 10*0) = (30,0)\ m$
   Why? *"Scales"* or stretches a vector
   Applications: Newton's 2nd Law: $\sum \vec{F}_{ext} = m\vec{a}_{net}$

4. **MULTIPLICATION – THE VECTOR *"DOT"*-PRODUCT:**
   $\vec{A} \bullet \vec{B} = (3*3) + (0*4) = 9\ m^2$
   $\vec{A} \bullet \vec{C} = (3*0) + (0*4) = 0$
   $\vec{A} \bullet \vec{D} = (3*-2) + (0*0) = -6\ m^2$
   Why? Checks if two vectors are PARALLEL
   Applications: Energy: $U = \vec{F} \bullet \vec{d}$

5. **MULTIPLICATION – THE VECTOR *"CROSS"*-PRODUCT:**
   $\vec{A} \times \vec{B} = (3*4 - 0*3) = (0,0,12)\ m^2$ or $12\ \odot\ m^2$
   $\vec{A} \times \vec{C} = (3*4 - 0*0) = (0,0,12)\ m^2$ or $12\ \odot\ m^2$
   $\vec{A} \times \vec{D} = (3*0 - 0*-2) = (0,0,0)\ m^2$ or 0
   Why? Checks if two vectors are PERPENDICULAR
   Applications: Rotational Dynamics

3.3.9 Instructional input

3.3.9.1 Torque

In this chapter, we have defined and practiced calculating two new physical quantities; namely an object's 'Center-of-Mass' and its 'Moment of Inertia' about an axis of rotation. In addition, we have defined and practiced computing a new vector operation; namely, the vector 'Cross'-product. By defining one last physical quantity, we will have the necessary tools to generate a new set of problem-solving strategies that help us determine how objects rotate and spin.

Imagine we are building a new house and need to tighten a bolt. We pull downward on our trusty crescent wrench and hope the bolt turns. As physicists, we would like to find a way to measure the extent to which our downward applied force causes the bolt to tighten. *How likely is our pull, exerted at the end of a crescent wrench, going to result in tightening the bolt?* In a general sense, we would like to define a physical quantity that measures if (or by how much) a force can cause an object to rotate about an axis.

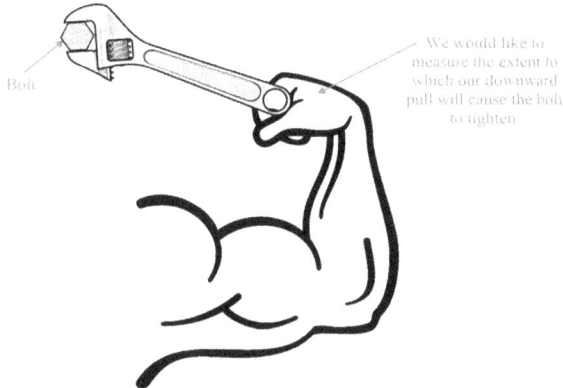

Bolt Image Credit: Pixabay by Clker-Free-Vector-Images. Strong Arm Image Credit: Net Vector/Shutterstock.com.

Like before, we begin our analysis by sticking an imaginary thumbtack (or nail … or pin … or stake) somewhere in (or near) the object—this thumbtack will designate the 'axis of rotation.' The 'axis of rotation' is the *imaginary line* about which we will spin the object. You can place the thumbtack anywhere you like, but once you choose your axis of rotation, it has to remain fixed because all of your upcoming computations will be made relative to your chosen axis. Similar to our procedure for calculating an object's COM or its MOI about an axis of rotation, place your coordinate system so that the axis of rotation passes through the origin of the coordinate system.

Next, draw and label the vector $\vec{F}$, that represents the force that is causing the object to rotate. In our example, the force is the downward applied force exerted by our muscles at the end of the crescent wrench. Finally, draw and label the vector $\vec{r}$, that represents the 'lever arm.' The 'lever arm' is the vector **from** the axis of rotation **to** the point at which the force is attached. In our example, the 'lever arm' is the crescent wrench since it represents the vector from the bolt to the point at which our hand grabs the crescent wrench to apply our downward pull.

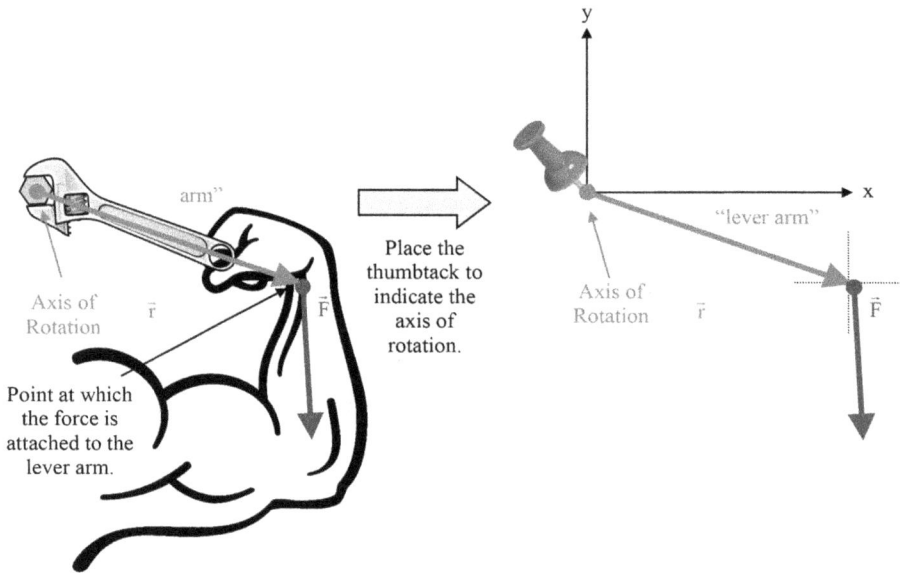

Bolt Image Credit: Pixabay by Clker-Free-Vector-Images. Strong Arm Image Credit: Net Vector/Shutterstock.com. Thumbtack Image Credit: Pixabay by OpenClipart-Vectors.

We define the 'Torque' that the force $\overrightarrow{F}$ exerts on an object about an axis of rotation as follows:

▶ **Definition of the Torque that a force exerts on an object about an axis of rotation:**

$\overrightarrow{\tau} = \overrightarrow{r} \times \overrightarrow{F}$, where $\overrightarrow{r}$, is the 'lever arm' from the axis of rotation to the point at which the force is acting on the object.

Etymology:
> The word 'torque' comes from the Latin word 'torquere' that translates into 'to twist.'

As usual, this simple definition incorporates a number of features that are worth itemizing and clarifying. We'll check these features when we actually start calculating the torque in example problems:

1. Torque is a vector calculated using the vector 'Cross'-product! However, our answer will not be written using the standard vector notation involving a set of parentheses. Instead, since torque is computed using the vector operation of 'Cross'-product multiplication, it will be written using the symbols ⊙ and ⊗ to indicate direction. More importantly, since '↺(CCW)' or '↻(CW)' were also acceptable symbols for the

direction of a 'Cross'-product, torque will determine if a force causes an object to rotate in a counter-clockwise or clockwise direction.

2. The units of torque are 'Newton · m.'

3. Order is important! Be sure $\overrightarrow{r}$ is the first vector in your 'Cross'-product. Since the vector 'Cross'-product is anti-commutative, you have to be extremely careful in ordering the vectors $\overrightarrow{r}$ and $\overrightarrow{F}$ before 'crossing' them; otherwise, you will compute an answer with the opposite sign. In other words: $\overrightarrow{\tau} = \overrightarrow{r} \times \overrightarrow{F}$, not $\overrightarrow{F} \times \overrightarrow{r}$.

4. Last, let's discuss the physical interpretation of torque. We started this section by trying to define a physical quantity that measures if (or by how much) a force can cause an object to rotate about an axis. Let's see if our definition of torque accomplishes that goal. Consider a hinged door as seen from above. You want to open the door, as easily as possible, by simply applying a push somewhere to it. Let's consider three questions to help us determine how to formulate our definition of torque:

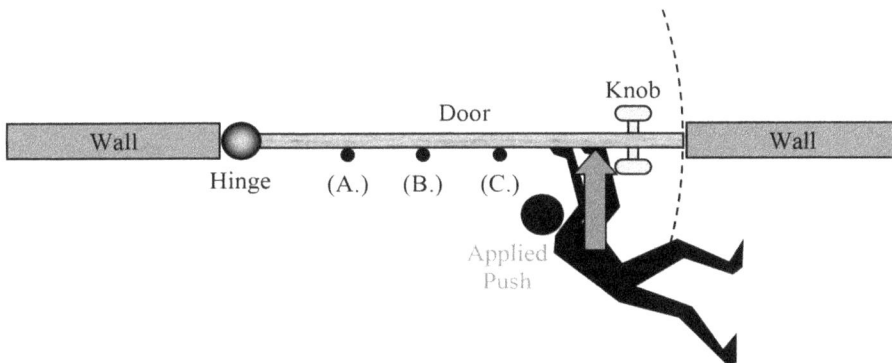

Silhouette of Man Image Credit: Pixabay by OpenClipart-Vectors.

- Question 1: Should torque be proportional to the applied push? Do you think the door will open more easily if you apply 100 lbs of force to the door versus only 2 lbs of force? I think we can all agree that the door opens in proportion to the force applied to it. In other words, the door more easily swings if a large force is applied to it. Thus, our definition for torque should be proportional to force: $\tau \propto \overrightarrow{F}$ ✓

- Question 2: Should torque be proportional to the length of the lever arm? Look at the drawing above. We are looking at a hinged door from the vantage point of the ceiling—the hinge is on the left while the doorknob is on the right. The door is free to swing along the path indicated by the dashed line. Do you think the door will open more easily if the green person applies 100 lbs. of force to point (A), (B), or (C)? Clearly, the door opens more easily the farther from the hinge the force is applied. I would certainly try to open a door by pushing on point (C) rather than point (A)—that's why the doorknob is located as far away from the hinge as possible! In fact, if you try to open a door by applying a force exactly at the hinge (i.e., the axis of rotation), the door will not swing at all. Imagine trying to open the door by pushing directly on the hinge—the door would be impossible to move! In other words, the door more easily swings if a force is applied to it far from the hinge. Thus, our definition for torque should be proportional to length of the lever arm: $\tau \propto \vec{r}$ ✓

- Question 3: Last, should torque be proportional to the direction at which the force is applied? Look at the drawing below. Again, we are looking at a hinged door from the vantage point of the ceiling—the hinge is on the left while the doorknob is on the right. The door is free to swing along the path indicated by the dashed line. Do you think the door will open more easily if our green friend applies 100 lbs. of force in a direction *perpendicular* to the length of the door ($\vec{P_A}$) or in a direction somewhat *parallel* to the length of the door ($\vec{P_B}$)?

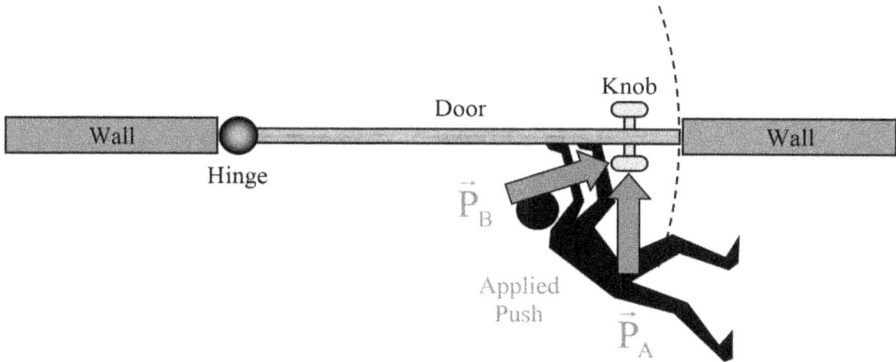

Silhouette of Man Image Credit: Pixabay by OpenClipart-Vectors.

Clearly, the door opens more easily as the more perpendicular force is applied. In fact, if you try to open a door by applying a force exactly parallel to the length of the door, the door will not swing at all. Imagine trying to open the door by pushing completely along the length of the door—the door would be impossible to move! In other words, the door more easily swings if a force is applied perpendicularly to the lever arm. Thus, our definition for torque should somehow ensure that the force and lever arm are at 90° with respect to one another: $\tau \propto \vec{r} \perp \vec{F}$ ✓

Combining these three arguments, we are left with our definition of torque: $\vec{\tau} = \vec{r} \times \vec{F}$. We have built a physical quantity that:

- increases as force increases;
- increases as the length of the lever arm increases; and
- increases as the force and lever arm become more perpendicularly-aligned (ensured by the procedures involved in the vector 'Cross'-product).

Thus, torque measures if (or by how much) a force can cause an object to rotate about an axis. A large torque implies that a force will likely cause an object to rotate about a given axis of rotation. A small torque implies that a force will less likely cause an object to rotate about a given axis of rotation.

3.3.10 Modeling

❶ **EX 1:** A rope pulls on a wheel as shown below:

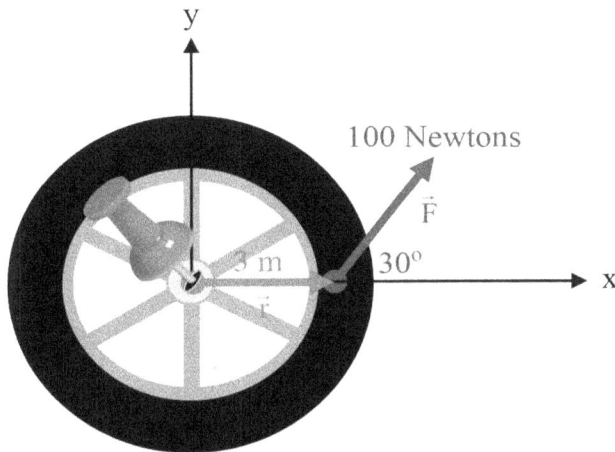

Thumbtack Image Credit: Pixabay by OpenClipart-Vectors.

Calculate the torque exerted by the rope on the wheel.

☑ **ANSWER:** Torque is just the 'Cross'-product of the lever arm and applied force.

$$\vec{\tau} = \vec{r} \times \vec{F} = \begin{pmatrix} 3 \\ 0 \end{pmatrix} \times \begin{pmatrix} 100 \cdot \cos(30^\circ) \\ 100 \cdot \sin(30^\circ) \end{pmatrix} = \begin{pmatrix} 3 \\ 0 \end{pmatrix} \times \begin{pmatrix} 86.6 \\ 50 \end{pmatrix}$$

$$= (3 \cdot 50) - (0 \cdot 86.6) = 150 - 0 = 150 \circlearrowleft \text{Newton} \cdot \text{meters}$$

☺ **COMMENTS:** Let's check each of the four features of torque:
    1. ✓ The result is a vector, with the direction symbolized by ↺. This tells us that the rope causes the wheel to spin in a Counter-Clockwise direction. The conclusion that the rope causes the wheel to spin in a Counter-Clockwise direction seems obvious from the drawing, but

we can sleep well knowing that our mathematical procedure agrees with our intuition. In upcoming problems, the direction of resulting motion will not always be so intuitively obvious. We can trust in our procedures to give us the correct results.

2. ✓ The result has units of vector $\vec{r}$ (which are 'meters'), times the units of vector $\vec{F}$ (which are 'Newtons').

3. ✓ Order is important—we ordered $\vec{r}$ first, followed by $\vec{F}$, before 'crossing' them.

4. ✓ The computation of torque multiplies perpendicular components of the two vectors. We interpret our answer for torque as a measure of the extent to which the rope will cause the wheel to spin in the Counter-Clockwise direction.

❷ **EX 2:** Calculate the torques exerted by the four forces shown below:

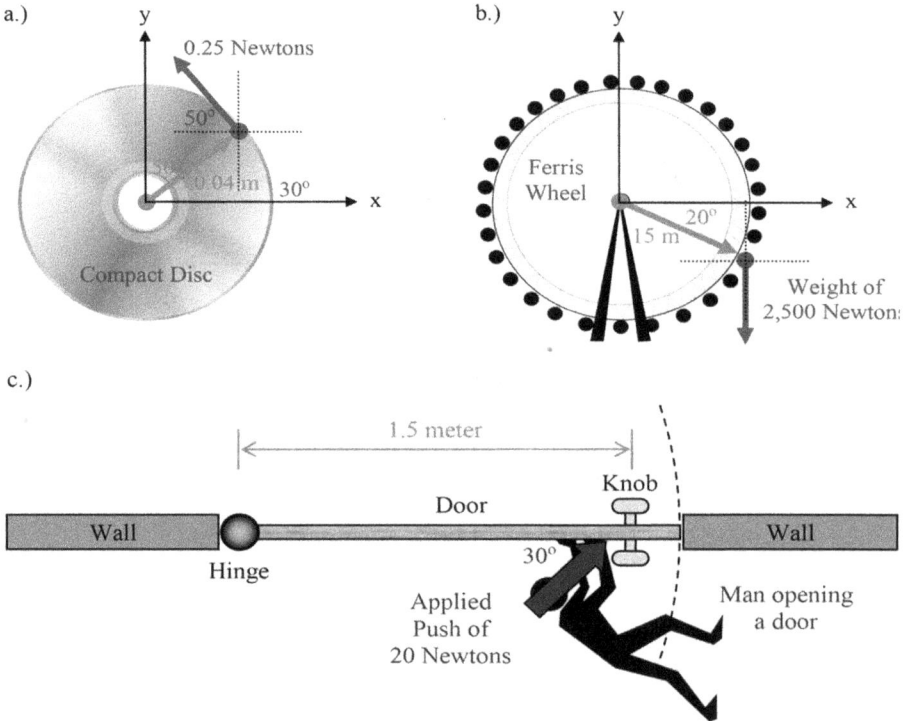

a.)

b.)

c.)

d.)

Compact Disc Image Credit: Pixabay by Clker-Free-Vector-Images.
Ferris Wheel Image Credit: Pixabay by Clker-Free-Vector-Images. Plane
Image Credit: Pixabay by nurbs999. Silhouette of Man Image Credit:
Pixabay by OpenClipart-Vectors.

☑ **ANSWER:** The key to calculating torque is to carefully write each force and lever arm as a vector, then carefully compute the 'Cross'-products:

a)

$$\vec{\tau} = \vec{r} \times \vec{F} = \begin{pmatrix} 0.04 \cdot \cos(40°) \\ 0.04 \cdot \sin(40°) \end{pmatrix} \times \begin{pmatrix} -0.25 \cdot \cos(50°) \\ 0.25 \cdot \sin(50°) \end{pmatrix} = \begin{pmatrix} 0.031 \\ 0.026 \end{pmatrix} \times \begin{pmatrix} -0.161 \\ 0.192 \end{pmatrix}$$

$$= (0.031 \cdot 0.192) - (0.026 \cdot (-0.161))$$

$$= 0.0060 + 0.0042 = 0.01 \circlearrowleft \text{Newton} \cdot \text{meters}.$$

Interpret this result as telling you that the applied force will cause the Compact Disk to rotate Counter-Clockwise.

b)

$$\vec{\tau} = \vec{r} \times \vec{F} = \begin{pmatrix} 15 \cdot \cos(20°) \\ -15 \cdot \sin(20°) \end{pmatrix} \times \begin{pmatrix} 0 \\ -2,500 \end{pmatrix} = \begin{pmatrix} 14.1 \\ -5.13 \end{pmatrix} \times \begin{pmatrix} 0 \\ -2,500 \end{pmatrix}$$

$$= (14.1 \cdot (-2,500)) - (-5.13 \cdot 0)$$

$$= -35,250 + 0 = -35,250 \text{ Newton} \cdot \text{meters}.$$

Interpret this result as telling you that the weight of the bucket will cause the Ferris Wheel to rotate Clockwise.

c)

$$\vec{\tau} = \vec{r} \times \vec{F} = \begin{pmatrix} 1.5 \\ 0 \end{pmatrix} \times \begin{pmatrix} 20 \cdot \cos(30°) \\ 20 \cdot \sin(30°) \end{pmatrix} = \begin{pmatrix} 1.5 \\ 0 \end{pmatrix} \times \begin{pmatrix} 17.3 \\ 10 \end{pmatrix} = (1.5 \cdot 10) - (0 \cdot 17.3)$$

$$= 15 - 0 = 15 \circlearrowleft \text{Newton} \cdot \text{meters}.$$

Interpret this result as telling you that the applied force of the man's hands will cause the door to rotate Counter-Clockwise.

d)

$$\vec{\tau} = \vec{r} \times \vec{F} = \begin{pmatrix} -1.5 \cdot \cos(30°) \\ 1.5 \cdot \sin(30°) \end{pmatrix} \times \begin{pmatrix} -100 \cdot \cos(75°) \\ -100 \cdot \sin(75°) \end{pmatrix} = \begin{pmatrix} -1.3 \\ 0.75 \end{pmatrix} \times \begin{pmatrix} -25.9 \\ -96.6 \end{pmatrix}$$
$$= (-1.3 \cdot (-96.6)) - (0.75 \cdot (-25.9))$$
$$= 125.58 + 19.42 = 145 \circlearrowleft \text{Newton} \cdot \text{meters}.$$

Interpret this result as telling you that the applied force of the woman's hands will cause the propeller to rotate Counter-Clockwise.

☺ **COMMENTS:** Let's check each answer for the four features of torque:
1. ✓ The results are vectors, with the direction symbolized by ↻ or ↺.
2. ✓ The results have units of 'Newton · m'.
3. ✓ We ordered $\vec{r}$ first, followed by $\vec{F}$, before 'crossing' all vectors.
4. ✓ We interpret our answers for torque as measures of the extent to which each force will cause the objects to spin.

❸ **EX 3:** Three forces are applied at the center of the three petals of a *Fidget Spinner* as shown below. The distance from the center of the fidget (i.e., the axis around which it spins) to the center of its three petals is 2 cm, or 0.02 m. In what direction does the fidget spin?

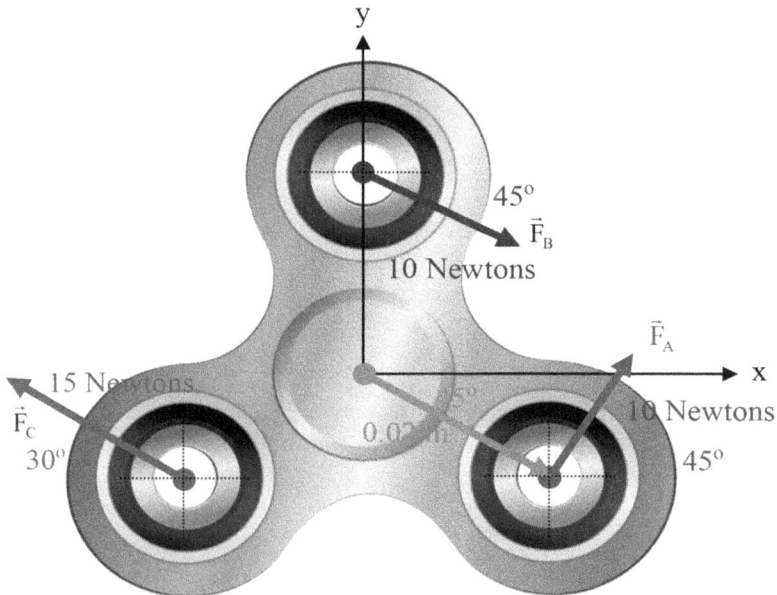

Fidget Spinner Image Credit: Pixabay by TonesB.

☑ **ANSWER:** In this example, you will begin to see the advantages of our formal definition of torque. Looking at the three forces, the direction of the net motion is not obvious. Just looking at the picture, we cannot easily tell if the spinner will rotate Clockwise or Counter-Clockwise. To answer the question, simply calculate the torques from each forces and add them as vectors to get a resultant torque.

a)

$$\vec{\tau_A} = \vec{r} \times \vec{F_A} = \begin{pmatrix} 0.02 \cdot \cos(45^\circ) \\ -0.02 \cdot \sin(45^\circ) \end{pmatrix} \times \begin{pmatrix} 10 \cdot \cos(45^\circ) \\ 10 \cdot \sin(45^\circ) \end{pmatrix} = \begin{pmatrix} 0.014 \\ -0.014 \end{pmatrix} \times \begin{pmatrix} 7.07 \\ 7.07 \end{pmatrix}$$

$$= (0.014 \cdot 7.07) - ((-0.014) \cdot 7.07) = 0.099 + 0.099 = 0.198 \circlearrowleft \text{Newton} \cdot \text{meters}.$$

b)

$$\vec{\tau_B} = \vec{r} \times \vec{F_B} = \begin{pmatrix} 0 \\ 0.02 \end{pmatrix} \times \begin{pmatrix} 10 \cdot \cos(30^\circ) \\ -10 \cdot \sin(30^\circ) \end{pmatrix} = \begin{pmatrix} 0 \\ 0.02 \end{pmatrix} \times \begin{pmatrix} 8.66 \\ -0.5 \end{pmatrix}$$

$$= (0 \cdot (-0.5)) - ((0.02) \cdot 8.66) = 0 - 0.173 = -0.173 \circlearrowright \text{Newton} \cdot \text{meters}.$$

c)

$$\vec{\tau_C} = \vec{r} \times \vec{F_C} = \begin{pmatrix} -0.02 \cdot \cos(45^\circ) \\ -0.02 \cdot \sin(45^\circ) \end{pmatrix} \times \begin{pmatrix} -15 \cdot \cos(30^\circ) \\ 15 \cdot \sin(30^\circ) \end{pmatrix} = \begin{pmatrix} -0.014 \\ -0.014 \end{pmatrix} \times \begin{pmatrix} -13 \\ 7.5 \end{pmatrix}$$

$$= (-0.014 \cdot 7.5) - ((-0.014) \cdot (-13)) = -0.105 - 0.182 = -0.287 \circlearrowright \text{Newton} \cdot \text{meters}.$$

Now calculate the net torque:

$$\vec{\tau}_{Net} = \vec{\tau_A} + \vec{\tau_B} + \vec{\tau_C} = (+0.198 \circlearrowleft) + (-0.173 \circlearrowright) + (-0.287 \circlearrowright) = -0.262 \circlearrowright \text{Newton} \cdot \text{meters}.$$

☺ **COMMENTS:** Torque is a vector and can be added just like any other vector, but instead of keeping x- and y-components separate, keep Clockwise- and Counter-Clockwise-components separate. Since our result is negative, the spinner will rotate Clockwise!

3.3.11 Instructional input

3.3.11.1 Angular displacement
At this point, we have defined three new physical quantities, one of which involves a new mathematical procedure to multiply two vectors (i.e., the vector 'Cross'-product).

Physical quantity	Definition	Interpretation
Center-of-Mass	$\vec{r}_{COM} = \dfrac{\sum\limits_{i=1}^{n} m_i \times \vec{r_i}}{M_{total}}$, where n is the number of pieces into which an object has been cut.	The COM is the point at which an object acts as if all of its mass were located.
Moment of Inertia (about an axis of rotation)	$I = \sum\limits_{i=1}^{n} m_i \times r_i^2$, where n is the number of pieces into which an object has been cut.	The MOI measures the difficulty associated with rotating an object about an axis of rotation.
Torque	$\vec{\tau} = \vec{r} \times \vec{F}$, where $\vec{r}$, is the 'lever arm' from the axis of rotation to the point at which the force is acting on an object.	Torque measures the ease with which a force can cause an object to rotate about an axis.

Armed with these three quantities, we can now tackle rotational motion. In a few short pages, we will double the number of techniques in our arsenal of problem-solving strategies.

To start, consider a rigid body (a purple blob of clay) that rotates about a fixed axis. In its 'initial' location, we place a point (P.) on the blob at a distance $\vec{r}$ from the axis of rotation. Next, we rotate the blob to a new, 'final' location and measure the distance 'x' that the point (P.) traveled. This distance x is an arc length:

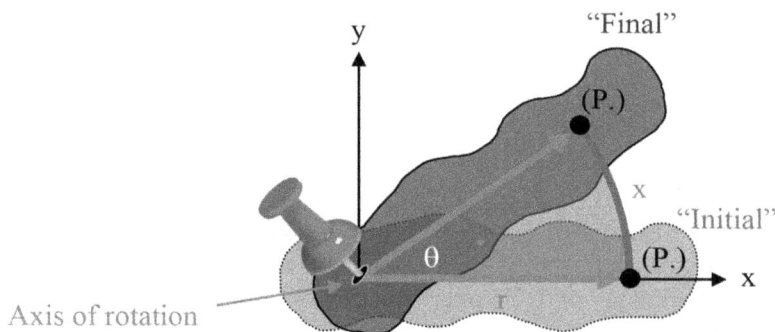

Thumbtack Image Credit: Pixabay by OpenClipart-Vectors.

Until now, we could describe the motion of point (P.) in terms of 'x,' the linear displacement (or arc length) that the point moved as a result of rotating the rigid body. However, we now seek to define the location of point (P.), not in terms of its linear displacement, but in terms of its angular displacement. Knowing the angle through which the body rotated may prove to be just as informative as knowing the linear distance that it moved. We simply need to relate the arc length that the point (P.) moved to the angle through which the body rotated. The parameter that connects these two quantities is r, the distance from (P.) to the axis of rotation. Thus, we define the 'Angular Displacement' of the point (P.) to be the ratio of the distance traveled by (P.) to the distance from point (P.) to the axis of rotation:

▶ **Definition of the Angular Displacement:**

$$\vec{\theta} = \frac{x}{r} \text{ with } \left\{ \begin{array}{l} \odot \text{ or } CCW \text{ or }, \text{ if positive} \\ \otimes \text{ or } CW \text{ or }, \text{ if negative} \end{array} \right\}, \text{ where } x \text{ is the linear distance}$$

traveled by a point (P.) and r is the distance from (P.) to the axis of rotation

---

Etymology:
The word 'angle' comes from the Latin word 'angulus' that translates into 'corner.'

As usual, this simple definition incorporates a number of features (some very confusing) that are worth itemizing and clarifying. We'll check these features when we actually start calculating angular displacements in example problems:

1. Even though angular displacement is the ratio of two lengths and is therefore a *dimensionless* quantity, we measure it in units called '**radians**,' abbreviated as '**rad**.' One rad is equal to the angular displacement that subtends an arc length equal to the radius of a given circle.

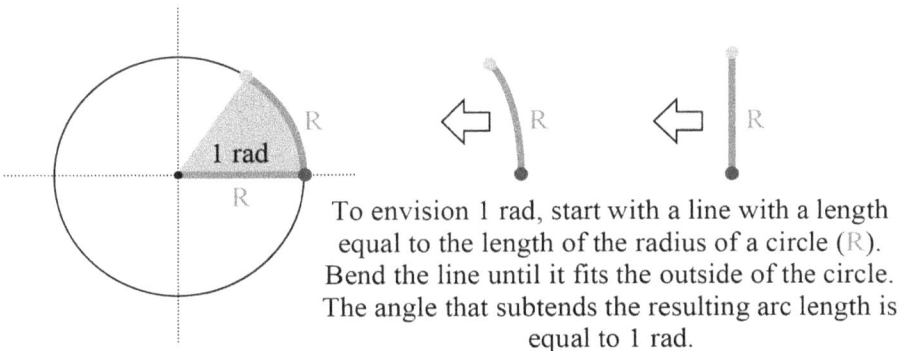

To envision 1 rad, start with a line with a length equal to the length of the radius of a circle (R). Bend the line until it fits the outside of the circle. The angle that subtends the resulting arc length is equal to 1 rad.

Think of the 'rad' as a *reminder* that angular displacement is dimensionless. Even though a rad is dimensionless, it is considered the SI (*Système International*) unit of angular displacement. I agree that this definition can be very confusing—a unit that is dimensionless? However, most physics textbooks use rads to measure angles.

2. Another way to envision the size of a rad is to imagine a circle of radius R. Note that the circle's circumference is $2\pi \cdot R$. Therefore, a point traveling one complete revolution of the circle will cover a linear distance equal to the circumference of the circle, $x = 2\pi \cdot R$. Therefore, 1 revolution is 2π rad, or ≈ 6.283 rad.

3. One final way to envision the size of a rad is to note that one revolution of a circle is indeed equal to 2π rads, but is also equal to $360°$. Therefore, $1\text{rad} = 360°/2\pi = 57.296°$. However, 'radians' are not to be confused with 'degrees.' Again, in most physics applications and textbooks, angles are measured in 'radians,' never 'degrees.'

4. Last, angular displacement is a vector. If positive, the vector is symbolized by '⊙,' '↺,' or 'CCW' and indicates a Counter-Clockwise displacement. If negative, the vector is symbolized by '⊗,' '↻,' or 'CW' and indicates a Clockwise displacement.

With our simple definition of angular displacement ($\vec{\theta} = x/r$ with a symbol of ⊙ or ⊗ to indicate direction), we have accomplished an amazing feat! We can now transform between linear and angular parameters. In other words, we can take all of our problem−solving techniques that employ linear parameters and map them into comparable problem−solving techniques that employ angular parameters. Let's briefly re-visit ALL of our problem-solving techniques covered by previous chapters in the series on *Classical Mechanics* and transform their governing equations from linear parameters to angular parameters. Once we have the new sets of equations, we can practice solving various problems.

3.3.11.2 Linear-to-angular transformations

To keep the notation clean and simple, we will temporarily suspend vector notation with each of our transformations. Just realize that all of the linear parameters are vectors to be written using sets of parentheses (x-component, y-component), while all of the angular parameters are vectors to be written with the appropriate symbol ⊙ or ⊗ to represent Clockwise or Counter-Clockwise motion.

▶ **Kinematic Definitions:**

I. Displacement: Our first transformation comes directly from our definition of angular displacement.

> **Transformation:**
> $$x = r \cdot \theta$$

II. Velocity: From our kinematic definition of linear velocity, we can derive an expression for angular velocity:
$$v = \frac{\Delta x}{\Delta t} = \frac{\Delta(r \cdot \theta)}{\Delta t} = r \cdot \frac{\Delta(\theta)}{\Delta t} = r \cdot \omega.$$

> **Transformation:**
> $$v = r \cdot \omega$$

where ω is called the 'Angular Velocity' and is measured in rad s^{-1}.

III. Acceleration: From our kinematic definition of linear acceleration, we can derive an expression for angular acceleration: $a = \frac{\Delta v}{\Delta t} = \frac{\Delta (r \cdot \omega)}{\Delta t} = r \cdot \frac{\Delta(\omega)}{\Delta t} = r \cdot \alpha$.

Transformation:
$$a = r \cdot \alpha$$

---

where α is called the 'Angular Acceleration' and is measured in rad s^{-2}.

▶ **Uniformly Accelerated Rotational Motion (α is fixed and abbreviated as _U.α._ _M._):**

I. _UαM #1_: This transformation comes directly from our definition of angular acceleration. If we assume that the angular acceleration is constant (i.e., α is fixed) and that the initial time is labeled as $t_{\text{initial}} = 0$, then: $\alpha = \frac{\Delta \omega}{\Delta t} = \frac{\omega_{\text{final}} - \omega_{\text{initial}}}{t_{\text{final}} - t_{\text{initial}}} = \frac{\omega - \omega_o}{t - 0} = \frac{\omega - \omega_o}{t}$, or $\omega = \omega_o + \alpha \cdot t$.

Transformation:
$$v = v_o + a \cdot t$$
‖tranforms
$$\omega = \omega_o + \alpha \cdot t$$

II. _UαM #2_: We derive the second _UαM_ equation using the same approach we employed to derive the second UAM equation. First, we calculate the average angular velocity using our kinematic definition: $\overline{\omega} = \frac{\Delta \theta}{\Delta t} = \frac{\theta_{\text{final}} - \theta_{\text{initial}}}{t_{\text{final}} - t_{\text{final}}} = \frac{\theta - \theta_o}{t}$, or $\overline{\omega} = \frac{\theta - \theta_o}{t}$. Second, we note that since the angular acceleration is constant, the average angular velocity is just the simple average of the initial and final angular velocities: $\overline{\omega} = (\omega - \omega_o)/2$. Equating the two expressions for $\overline{\omega}$ gives us:

Transformation:
$$r = r_o + (v_o \cdot t) + \tfrac{1}{2} \cdot a \cdot t^2$$
‖tranforms
$$\theta = \theta_o + (\omega_o \cdot t) + \tfrac{1}{2} \cdot \alpha \cdot t^2$$

III. _UαM #3_: Simply combine _UαM #1_ and _UαM #2_ to derive _UαM #3_:

Transformation:

$$v^2 = v_o^2 + 2 \cdot a \cdot (r - r_o)$$

$$\Downarrow \text{tranforms}$$

$$\omega^2 = \omega_o^2 + 2 \cdot \alpha \cdot (\theta - \theta_o)$$

▶ **Newton's second law of motion and Radial Acceleration:**

    I. <u>Newton's second law of motion</u>: To write Newton's second law of motion for purely rotational motion requires a slightly more involved derivation. Consider a rigid body rotating with angular acceleration α and angular velocity ω. The net force and acceleration on the point (P.) are broken into components that are parallel (also called 'radial') and perpendicular (also called 'tangential') to $\vec{r}$.

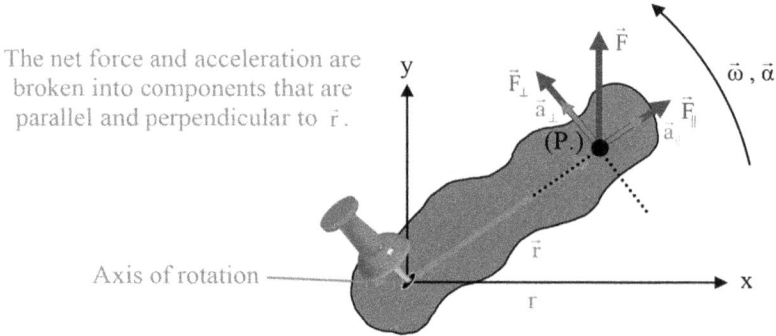

The net force and acceleration are broken into components that are parallel and perpendicular to $\vec{r}$.

Axis of rotation

Thumbtack Image Credit: Pixabay by OpenClipart-Vectors.

Start with Newton's second law of motion:

$$\sum \vec{F}_{\text{external}} = m \cdot \vec{a}$$

Break the force and acceleration into components that are parallel and perpendicular to $\vec{r}$.

$$\sum \left(\vec{F}_{\parallel} + \vec{F}_{\perp} \right)_{\text{external}} = m \cdot \left(\vec{a}_{\parallel} + \vec{a}_{\perp} \right)$$

Take the vector "Cross"-product of both sides with respect to $\vec{r}$.

$$\sum \vec{r} \times \left(\vec{F}_{\parallel} + \vec{F}_{\perp} \right)_{\text{external}} = m \cdot \vec{r} \times \left(\vec{a}_{\parallel} + \vec{a}_{\perp} \right)$$

$$\sum \left(\left(\vec{r} \times \vec{F}_{\parallel} \right) + \left(\vec{r} \times \vec{F}_{\perp} \right) \right)_{\text{external}} = m \cdot \left(\left(\vec{r} \times \vec{a}_{\parallel} \right) + \left(\vec{r} \times \vec{a}_{\perp} \right) \right)$$

Now recall that the vector 'Cross'-product physically multiplies perpendicular components of two vectors. Thus, the cross products of $\vec{r} \times \overrightarrow{F_{\parallel}}$ and $\vec{r} \times \overrightarrow{a_{\parallel}}$ are zero.

$$\sum\left(\vec{r} \times \vec{F}_{\perp}\right)_{\text{external}} = m \cdot \vec{r} \times \vec{a}_{\perp}$$

Recall from our kinematic transformations that the linear acceleration can be written as an angular acceleration $a = r \cdot \alpha$.

$$\sum\left(\vec{r} \times \vec{F}_{\perp}\right)_{\text{external}} = m \cdot \vec{r} \times (r \cdot \vec{\alpha})$$

Although this last equation looks complicated, it greatly simplifies when you realize that $(\vec{r} \times \overrightarrow{F_{\perp}})_{\text{external}} = \tau_{\text{external}}$ (the definition of torque) and $m \cdot r \times r = I$ (the definition of the moment of inertia). Thus, Newton's second law of motion for purely rotational motion simplifies to:

> **Transformation:**
> $$\sum \overrightarrow{F}_{\text{external}} = m \cdot \overrightarrow{a}$$
> $\parallel$ tranforms
> $$\sum \overrightarrow{\tau}_{\text{external}} = I \cdot \overrightarrow{\alpha}$$

II. Radial acceleration: Our expression for radial acceleration was derived in the context of Uniform Circular Motion (that we abbreviated as UCM). Our expression, derived in volume 1, chapter 6, was: $a_{\text{radial}} = v^2/R$, where v was the constant speed of the moving object and R was the radius of the circle upon which the object moved. To derive a comparable angular expression, simply use our transformation from linear velocity to angular velocity: $v = R \cdot \omega$

> **Transformation:**
> $$a_{\text{radial}} = \frac{v^2}{R}$$
> $\parallel$ tranforms
> $$a_{\text{radial}} = R \cdot \omega^2$$

We should pause here and discuss the difference between the two formulas we have just derived for rotational accelerations because at first, a contradiction seems to exist. A few pages ago, when we transformed the formula for linear acceleration into angular acceleration, we arrived at the formula: $a = r \cdot \alpha$. Now, when we transform the equation for radial acceleration into angular acceleration, we arrive at: $a_{\text{radial}} = R \cdot \omega^2$. So ... which formula is correct? As you probably have guessed, both formulas are correct because they describe different situations.

Situation 1—Uniform Circular Motion: Recall that for Uniform Circular Motion, an object was assumed to be moving along a circular path, **but at constant speed**. In such a situation, an object's instantaneous velocity is indeed changing because as it moves around the circle, only its *direction* changes. As a result, its velocity necessarily has to change even though the magnitude of the velocity is fixed. In such a situation, the object does not speed up as it moves along the circle, so it has a fixed angular velocity (i.e., $\Delta\omega/\Delta t = 0$) and therefore has no angular acceleration (i.e., $\alpha = 0$). However, it still has 'centripetal'—or 'radial'—acceleration (given by $a_{\text{radial}} = v^2/R = R \cdot \omega^2$) because of the simple fact that it is moving on a circular path. In summary, an object moving on a circular path is, by definition, accelerating toward the center of the circle even though it is not changing its speed along its path of motion.

Situation 2—An object accelerates along a circular path: In addition to simple uniform circular motion, more sophisticated motion is possible along a circular path. Namely, an object can speed up as it travels along its circular path. In such a situation, it still experiences 'centripetal'—or 'radial'—acceleration (given by $a_{\text{radial}} = v^2/R = R \cdot \omega^2$) because of the simple fact that it is moving on a circular path. However, it also experiences 'tangential' acceleration perpendicular to the radius of the circle and given by the formula: $a_{\text{tangential}} = r \cdot \alpha$. This is where our expression for linear acceleration comes into play. The tangential acceleration arises from the changing value of angular velocity with respect to time (i.e., $\alpha = \Delta\omega/\Delta t \neq 0$). Of course, the total acceleration experienced by the object is the vector sum of the tangential and radial accelerations:

$$a_{\text{total}} = \sqrt{(a_{\text{radial}})^2 + (a_{\text{tangential}})^2}\,.$$

Situations 1 and 2 are nicely summarized in the graphic below:

SITUATION 1: Uniform Circular Motion	SITUATION 2: Accelerated
THE OBJECT IS NOT SPEEDING UP, NOR SLOWING DOWN	THE OBJECT IS SPEEDING UP OR SLOWING DOWN

- Velocity involves direction. Therefore, since the object's direction constantly changes as it moves around the circle (in both situations), instantaneous velocity is not constant in either situation.
- Radial Component of Acceleration: In situation 1, a_{rad} remains constant at v^2/R or $\omega^2 * R$, but the direction changes.
- In situation 1, this remains constant since v is fixed for Uniform Circular Motion.
- Tangential Component of Acceleration: This is zero in situation 1. However, in situation 2, the acceleration of the object along the circular path means that a tangential component of acceleration exists.
- This changes in both situations because the object's direction is always changing. In situation 1, only the direction changes but the magnitude stays constant at $\omega * R$. In situation 2, both the magnitude and direction change.

The difference between "radial" and "tangential" accelerations.

▶ **Conservation Laws:**

    I. Law of Conservation of Energy (LCE): The transformations for LCE are very simple:

- Gravitational Potential Energy ($\Delta U_g = -m \cdot g \cdot \Delta y$) depends only on the location of an object's Center-of-Mass, so is not impacted by rotational motion.
- Elastic Potential Energy ($\Delta U_e = -\frac{K}{2} \times \Delta x$) only depends on the deformation of springs so it is not impacted by rotational motion.
- Kinetic Energy ($\Delta KE = \frac{1}{2} \times m \times v^2$) depends on the velocity of a moving object so certainly it will be impacted by the possible extra energy needed to rotate an object as it moves. To transform our expression for Kinetic Energy, simply insert the kinematic transformation for linear-to-angular velocity ($v = r \cdot \omega$) into our expression for KE and note that mr^2 is nothing more than our definition of the Moment of Inertia of an object about an axis of rotation.

Transformation:

$$\Delta KE = \Delta(\tfrac{1}{2} \cdot m \cdot v^2)$$

⇃tranforms

$$\Delta KE = \Delta(\tfrac{1}{2} \cdot I \cdot \omega^2)$$

- **Law of Conservation of Momentum (LCL):** like LCE, the transformation for LCP is very straightforward. If you will recall from a previous chapter in this series, the Law of Conservation of Linear Momentum can be stated as follows:

$$\text{If } \sum \overrightarrow{F}_{\text{external}} = 0 \text{ then } \sum \overrightarrow{p}_{\text{initial}} = \sum \overrightarrow{p}_{\text{final}},$$

where $\overrightarrow{p}$ is 'Linear Momentum,' define as $\overrightarrow{p} = m \cdot \overrightarrow{v}$.

We have already seen in all of the previous transformations that when transforming from linear-to-angular motion, torque replaces force, moment of inertia replaces mass, and angular velocity replaces linear velocity. Thus, the Law of Conservation of Angular Momentum can be stated as follows:

Transformation:

$$\text{If } \sum \overrightarrow{\tau}_{\text{external}} = 0,$$
$$\text{then}$$
$$\sum \overrightarrow{L}_{\text{initial}} = \sum \overrightarrow{L}_{\text{final}},$$

where $\overrightarrow{L}$ is 'Angular Momentum,' define as $\overrightarrow{L} = I \cdot \overrightarrow{\omega}$.

3.3.12 Checking for understanding

Let's compile a summary of all of our problem-solving techniques for linear motion and see how they transform into equivalent techniques for rotational motion. This table will be very helpful for the upcoming problems we will tackle.

Name	Governing equations	Applications
LINEAR MOTION: Motion along a line		
Kinematics	**Speed:** $v = \frac{\Delta r}{\Delta t} = \frac{r_{final} - r_{initial}}{t_{final} - t_{initial}}$ **Velocity:** $\vec{v} = \frac{\Delta \vec{r}}{\Delta t} = \frac{\vec{r}_{final} - \vec{r}_{initial}}{t_{final} - t_{initial}}$ **Acceleration:** $\vec{a} = \frac{\Delta \vec{v}}{\Delta t} = \frac{\vec{v}_{final} - \vec{v}_{initial}}{t_{final} - t_{initial}}$	These kinematic definitions state the physical quantities which we are interested in computing in order to describe linear motion.
Uniformly Accelerated Motion	**UAM #1:** $\vec{v} = \vec{v}_o + \vec{a} \cdot t$ **UAM #2:** $\vec{r} = \vec{r}_o + (\vec{v}_o \cdot t) + \frac{1}{2} \cdot \vec{a} \cdot t^2$ **UAM #3:** $\vec{v}^2 = \vec{v}_o^2 + 2 \cdot \vec{a} \cdot (\vec{r} - \vec{r}_o)$	The UAM equations describe how objects move when subjected to a constant linear acceleration. These equations are most often used to describe free-falling objects near the surface of the Earth but can also be used to describe 'idealized situations' or short-term accelerations.
Newton's laws of motion	**First Law:** An 'inertial reference frame' is any frame in which a body with zero net force acting upon it does not accelerate. **Second Law:** $\sum \vec{F}_{external} = m \cdot \vec{a}$ **Third Law:** Forces are pairwise interactions between two bodies.	Newton's three laws of motion are the hallmark of any course on *Classical Mechanics*. The first law defines an 'inertial reference frame.' The second law solves problems numerically with the help of Free Body Diagrams. The third law defines a force.
Uniform Circular Motion	$a_{radial} = v^2/R$, directed towards the center of the circle.	In the case of uniform circular motion, the net centripetal acceleration has a specific form.
Law of Conservation of Energy (LCE):	$W_{other} + KE_{initial} + U_{g_{initial}} + U_{e_{initial}}$ $= KE_{final} + U_{g_{final}} + U_{e_{final}}$ with $KE = \frac{1}{2} \cdot m \cdot v^2$, $U_g = m \cdot g \cdot y$, and $U_e = \frac{1}{2} \cdot K \cdot (\Delta x)^2$	The total energy of a system is 'conserved'—that is to say, fixed over time. This principle is always valid but is most often used to describe situations in which an object is moving, acted upon by gravity, acted upon by springs, or acted upon by 'conservative forces.'
Law of Conservation of Linear Momentum (LCP):	If $\sum \vec{F}_{external} = 0$, then $\sum \vec{p}_{initial} = \sum \vec{p}_{final}$, where $\vec{p}$ is 'Linear Momentum,' define as $\vec{p} = m \cdot \vec{v}$.	In the absence of external forces, the total linear momentum of a system is 'conserved'—that is to say, fixed over time. This principle is most often used to analyze 'collisions' and 'explosions.'

TRANSFORMATIONS: Convert from linear motion of a point (P.) on a rotating object to the angular motion of point (P.)

Center-of-Mass	$$\vec{r}_{COM} = \frac{\sum_{i=1}^{n} m_i \times \vec{r_i}}{M_{total}}, \text{where } n \text{ is the}$$ number of pieces into which an object has been cut.	The COM is the point at which an object acts as if all of its mass were located.
Moment of Inertia about an axis of rotation	$I = \sum_{i=1}^{n} m_i * r_{\perp i}^2$, where n is the number of pieces into which an object has been cut.	The MOI measures the difficulty associated with rotating an object about an axis of rotation.
Torque	$\vec{\tau} = \vec{r} \times \vec{F}$, where $\vec{r}$, is the 'lever arm' from the axis of rotation to the point at which the force is acting on an object.	Torque measures the extent to which a force can cause an object to rotate about an axis.
Angular Displacement	**Angular Displacement:** $x = r \cdot \theta$ **Angular Velocity:** $v = r \cdot \omega$ **Angular Acceleration (Radial):** $a_{radial} = \omega^2 \cdot r$ **Angular Acceleration (Tangential):** $a_{tangential} = r \cdot \alpha$	As a rigid body rotates, the linear displacement (x) of a point ($P.$) on the body is related to the angle through which the body rotated (θ).

Transformations	**Quantity:**	**Linear:**	**Angular:**	Any of the equations governing linear motion can be written as angular analog equations by simply making the substitutions listed to the left.
	Displacement:	$\vec{r}$ or $\vec{x}$	$\vec{\theta}$	
	Velocity:	$\vec{v}$	$\vec{\omega}$	
	Acceleration:	$\vec{a}$	$\vec{\alpha}$	
	Inertia:	m	I	
	Force:	$\vec{F}$	$\vec{\tau}$	
	Momentum:	$\vec{p}$	$\vec{L}$	

ANGULAR OR ROTATIONAL MOTION: Motion about a fixed axis

Kinematics	**Velocity:** $\vec{\omega} = \frac{\Delta \vec{\theta}}{\Delta t} = \frac{\vec{\theta}_{final} - \vec{\theta}_{initial}}{t_{final} - t_{initial}}$ **Acceleration:** $\vec{\alpha} = \frac{\Delta \vec{\omega}}{\Delta t} = \frac{\vec{\omega}_{final} - \vec{\omega}_{initial}}{t_{final} - t_{initial}}$	These kinematic definitions state the physical quantities which we are interested in computing in order to describe rotational motion.
Uniformly Accelerated Motion	**UαM #1:** $\vec{\omega} = \vec{\omega}_o + \vec{\alpha} \cdot t$ **UαM #2:** $\vec{\theta} = \vec{\theta}_o + (\vec{\omega}_o \cdot t) + \frac{1}{2} \cdot \vec{\alpha} \cdot t^2$ **UαM #3:** $\vec{\omega}^2 = \vec{\omega}_o^2 + 2 \cdot \vec{\alpha} \cdot (\vec{\theta} - \vec{\theta}_o)$	The UαM equations describe how objects move when subjected to a constant angular acceleration.
Newton's laws of motion	$\sum \vec{\tau}_{external} = I \cdot \vec{\alpha}$ $a_{radial} = R \cdot \omega^2$	The second law solves problems numerically with the help of Free Body Diagrams.

(Continued)

(Continued)

Name	Governing equations	Applications
Uniform Circular Motion		Using our linear-to-angular transformations, the radial acceleration of a point (P.) on a rotating body has a different form.
Law of Conservation of Energy (LCE):	$W_{other} + KE_{initial} + U_{g_{initial}} + U_{e_{initial}}$ $= KE_{final} + U_{g_{final}} + U_{e_{final}}$ with $KE = (\frac{1}{2} \cdot m \cdot v^2) + (\frac{1}{2} \cdot I \cdot \omega^2)$	The total energy of a system is 'conserved'—that is to say, fixed over time.
Law of Conservation of Linear Momentum (LCL):	If $\sum \vec{\tau}_{external} = 0$, then $\sum \vec{L}_{initial} = \sum \vec{L}_{final}$, where $\vec{L}$ is 'Linear Momentum,' define as $\vec{L} = I \cdot \vec{\omega}$.	In the absence of external forces, the total angular momentum of a system is 'conserved'—that is to say, fixed over time. This principle is most often used to analyze 'collisions' and 'explosions.'

3.3.13 Modeling

The linear-to-angular transformations have just *doubled* the number of problem-solving strategies we can use to analyze motion! Let's try some example problems using these transformations so see how our problem-solving strategies have increased in number. The key to solving each problem is to think about how we would solve the problem in the case of linear motion, then just transform the appropriate equations into the corresponding equations involving rotational motion.

❶ **EX 1:** A piece of gum is placed at the origin, on a bicycle wheel initially at rest. The wheel is accelerated with a constant angular acceleration of $\alpha = 0.5$ rad s^{-2} (or 28.6 degrees s^{-2}) in the Counter-Clockwise direction.

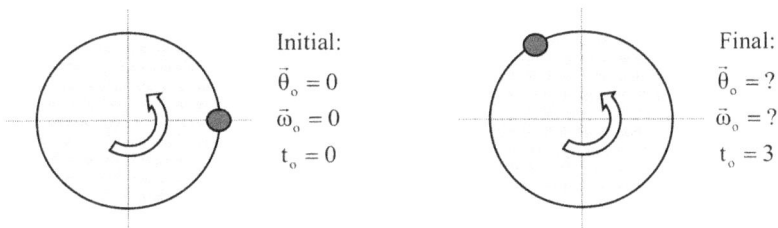

Initial:
$\vec{\theta}_o = 0$
$\vec{\omega}_o = 0$
$t_o = 0$

Final:
$\vec{\theta}_o = ?$
$\vec{\omega}_o = ?$
$t_o = 3$

a) What angle does the gum make at $t = 3$ s?
b) How fast will the gum be spinning at $t = 3$ s?
c) How fast will the gum be spinning at $\theta = 4.7$ rads (or 270°)?

☑ **ANSWER:** Because the angular acceleration is 'constant' at 0.5 rad s^{-2} ↺, this example should immediately remind you of a UAM problem. Because the angular acceleration is constant (as opposed to linear acceleration), we will tackle this with a '$U\alpha M$' problem-solving strategy as opposed to a 'UAM' problem-solving strategy. Previously, when confronted with a UAM problem, we label the 'initial' and 'final' situations; determine which parameters we have and which parameters we want; then solve with the appropriate equation. Let's take the same approach here.

a) The problem asks us to solve for the angular displacement θ, at $t = 3$ s. $U\alpha M$ #2 connects these two quantities:

$$\vec{\theta} = \vec{\theta}_o + (\vec{\omega}_o \cdot t) + \tfrac{1}{2} \cdot \vec{\alpha} \cdot t^2$$

$$\vec{\theta} = 0 + (0 \cdot 3) + \tfrac{1}{2} \cdot 0.5 \cdot 3^2$$

$$\vec{\theta} = 2.25 \ \text{↺ rad}$$

b) The problem now asks us to solve for the angular velocity ω, at $t = 3$ s. $U\alpha M$ #1 connects these two quantities:

$$\vec{\omega} = \vec{\omega}_o + \vec{\alpha} \cdot t$$

$$\vec{\omega} = 0 + (0.5 \cdot 3)$$

$$\vec{\omega} = 1.5 \ \tfrac{\text{rad}}{\text{sec}}$$

c) Last, the problem asks us to solve for the angular velocity ω, at $\theta = 4.7$ rad. $U\alpha M$ #3 connects these two quantities:

$$\vec{\omega}^2 = \vec{\omega}_o^2 + 2 \cdot \vec{\alpha} \cdot \left(\vec{\theta} - \vec{\theta}_o \right)$$

$$\vec{\omega}^2 = 0^2 + (2 \cdot 0.5 \cdot (4.7 - 0))$$

$$\vec{\omega}^2 = 4.7$$

$$\vec{\omega} = 2.17 \ \tfrac{\text{rad}}{\text{sec}} \text{or} 2.17 \ ↺ \ \tfrac{\text{rad}}{\text{sec}}$$

☺ **COMMENTS:** First, notice that our approach to solving the problem was to find an analogous problem involving linear motion. Once we realize that this problem is similar to a UAM problem, we can employ the problem-solving approaches we developed in an earlier chapter. You will notice that our approach to solving this problem was completely analogous to how we solved UAM problems in volume 1, chapter 3. Second, notice that we used $\alpha = +0.5$ rad s^{-2} (a positive value) since the problem states that the angular acceleration is Counter-Clockwise. Our convention for writing vectors that point 'out of' and 'into' the plane of this book was that positive vectors represented Counter-Clockwise motion and could be symbolized by '⊙,' '↺,' or 'CCW.' Last, the answer to part (c) may strike you as confusing;

however, the square-root operation returns both positive and negative roots (i.e., if $\omega^2 = 4.7$ then $\omega = \pm 2.17$). Since the disk is clearly moving in the Counter-Clockwise direction, you would select the positive root, $\overrightarrow{\omega} = 2.17$ ᶜrad/sec, as your answer.

❷ **EX 2:** A 1 kg block is attached, by a rope, to a **solid drum** of mass 3 kg and radius 0.5 m ($I_{\text{Solid Drum}} = 1/2\, M \times R^2$). The block is released. Solve for the linear acceleration of the block, the angular acceleration of the drum, and the tension in the rope.

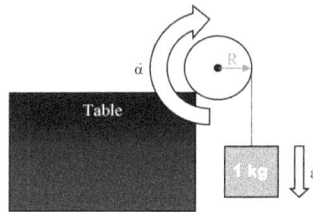

☑ **ANSWER:** Whenever you see ropes, pulleys and moving blocks, you should think of Newton's laws of motion. Anytime you use Newton's laws of motion, the first step is to draw Free Body Diagrams that display all of the forces acting on *each mass*. In previous problems, we ignored the pulleys because we said they were 'massless' and 'ideal' (Go back to all of the previous problems and you'll notice that all of the pulleys are labeled as 'massless'). Now that we know how to handle rotational motion, we can tackle more sophisticated motion by analyzing the translational movement of the block and the rotational movement of the spinning drum. Let's apply Newton's Second Law (for linear motion) to the falling block, then apply Newton's Second Law (for rotational motion) to the spinning drum:

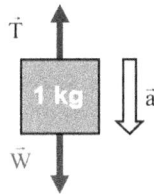

Free Body Diagram for Linear
Motion of Falling Block

Linear Motion:
$$\begin{cases} \sum \overrightarrow{F}_{\text{external}} = m \cdot \overrightarrow{a} \\ \overrightarrow{W}\text{eight} + \overrightarrow{T}\text{ension} = m \cdot \overrightarrow{a} \\ \begin{pmatrix} 0 \\ -1*9.8 \end{pmatrix} + \begin{pmatrix} 0 \\ +T \end{pmatrix} = 1 \cdot \begin{pmatrix} 0 \\ -a \end{pmatrix} \\ T = 9.8 - a \end{cases}$$

3-68

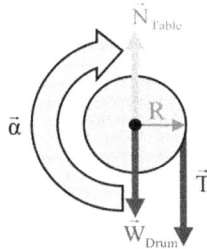

Free Body Diagram for
Rotational Motion of
Spinning Drum

Rotational Motion:

$$\sum \vec{\tau}_{external} = I \cdot \vec{\alpha}, \text{ where } I_{Solid\ Drum} = \frac{1}{2}M * R^2$$

$$\vec{\tau}_{\substack{Weight \\ of\ Drum}} + \vec{\tau}_{\substack{Normal \\ of\ Table}} + \vec{\tau}_{Tension} = I \cdot \vec{\alpha}$$

$$\begin{pmatrix} 0 \\ 0 \end{pmatrix} \times \begin{pmatrix} 0 \\ -3*9.8 \end{pmatrix} + \begin{pmatrix} 0 \\ 0 \end{pmatrix} \times \begin{pmatrix} 0 \\ N_{Table} \end{pmatrix} + \begin{pmatrix} 0.5 \\ 0 \end{pmatrix} \times \begin{pmatrix} 0 \\ -T \end{pmatrix} = \left(\frac{1}{2} \cdot 3 \cdot 0.5^2 \right) \cdot \vec{\alpha}$$

$$-0.5 \cdot T \circlearrowleft = -0.375 \cdot \alpha \circlearrowleft$$

$$T = 0.75 \cdot \alpha$$

Combine the two equations for tension and use the transformation, $a = R \cdot \alpha$, where R is the radius of the drum:

$a = 3.92\text{ms}^{-2}$, $\alpha = 7.84\text{rads}^{-2}$, and $T = 5.88$ Newtons.

☺ **COMMENTS:** Again, notice that the key to solving the problem is to find the analogous situation for linear motion. In this case, the problem is best tackled with Newton's second law of motion so we draw Free Body Diagrams; apply the second law; then combine results to solve numerically for the unknown quantities. Also, notice that in the case of the rotational motion, the weight of the drum and the normal force exerted by the table on the drum do not exert a torque on the drum. Neither force causes the drum to rotate since they both act on the drum *at its axis of rotation* (i.e., neither force is acting at a 'lever arm'). A force acting at the axis of rotation cannot cause an object to rotate. The only force acting on the drum at a 'lever arm' is the tension in the rope.

❸ **EX 3:** A ball is released, from rest, from the top of a frictionless 50 m high hill. How fast is the ball traveling at the 10 m valley if the ball is **a solid sphere** of mass 4 kg and radius 0.25 m? Note that $I_{Solid\ Sphere} = 2/5\ M \times R^2$.

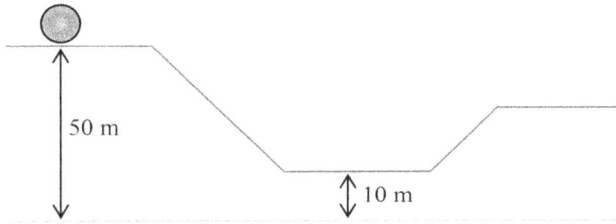

50 m

10 m

☑ **ANSWER:** This problem cries to be solved by LCE. Whenever we see an object that is moving and acted only upon by gravity, LCE is the problem-solving recipe of choice. First, let's solve the problem without considering the rotational motion of the solid sphere. In other words, when we calculate Kinetic Energy, we will only take into account the linear motion of the Center-of-Mass of the solid sphere, not the rotation of mass about its COM. Since no ropes, springs, or other 'non-conservative forces' are involved in the problem, the LCE equation simplifies to:

$$\cancel{W_{other}} + \cancel{KE_{initial}} + U_{g\,initial} + \cancel{U_{e\,initial}} = KE_{final} + U_{g\,final} + \cancel{U_{e\,final}}$$

$$m*g*y_{initial} = \frac{1}{2}*m*v_{final}^2 + m*g*y_{final}$$

$$4*9.8*(+50) = \frac{1}{2}*4*v^2 + 4*9.8*(+10)$$

$$1{,}960 = 2*v^2 + 392$$

$$1{,}568 = 2*v^2$$

$$v = 28.0 \ \frac{m}{s}$$

This is the result ($v = 28.0$ m s^{-1}) we would have obtained had we not incorporated rotational motion. However, we now know that energy is expended to rotate mass about the Center-of-Mass of the solid sphere. The Kinetic Energy of rotation is given by $\Delta KE = \Delta(1/2 \cdot I \cdot \omega^2)$:

$$\cancel{W_{other}} + \cancel{KE_{initial}^{COM\text{-}Linear}} + \cancel{KE_{initial}^{Rotational}} + U_{g\,initial} + \cancel{U_{e\,initial}} = KE_{final}^{COM\text{-}Linear} + KE_{final}^{Rotational} + U_{g\,final} + \cancel{U_{e\,final}}$$

$$4*9.8*(+50) = \frac{1}{2}*4*v^2 + \frac{1}{2}*\left(\frac{2}{5}*4*0.25^2\right)*\omega^2 + 4*9.8*(+10)$$

$$1{,}960 = 2*v^2 + 0.05*\omega^2 + 392$$

$$1{,}568 = 2*v^2 + 0.05*\omega^2$$

To solve for v, use the transformation, $v = R \cdot \omega$, where R is the radius of the solid sphere:

$$1568 = 2 \times v^2 + 0.05 \times \omega^2$$

$$1568 = 2 \times v^2 + 0.05 \times (\tfrac{v}{0.25})^2$$

$$1568 = 2 \times v^2 + 0.8 \times v^2$$

$$1568 = 2.8 \times v^2$$

$$v = 23.7 \text{ m s}^{-1}$$

☺ **COMMENTS:** Again, notice that the key to solving the problem is to find the analogous situation for linear motion. Also, the final speed dropped from 28 to 23.7 m s^{-1} when rotational motion was added to the analysis. This makes sense, right? In addition to moving the COM of the solid sphere, extra energy would be expended to rotate mass around the COM. The loss of energy needed to rotate the solid sphere implies less energy left over to move the sphere, thus the lower final speed of 23.7 m s^{-1}.

❹ **EX 4:** A thin, red pizza plate (i.e., **a thin disk** with $I_{\text{Thin Disk}} = 1/2\, M*R^2$) of mass 2 kg and radius 0.25 m rotates on a turntable with a Clockwise angular velocity of $\omega = 1$ rad s^{-1}. A second, translucent plate of mass 3 kg and radius 0.5 m is dropped on the original pizza plate while it is still rotating. What is the new angular velocity of both plates?

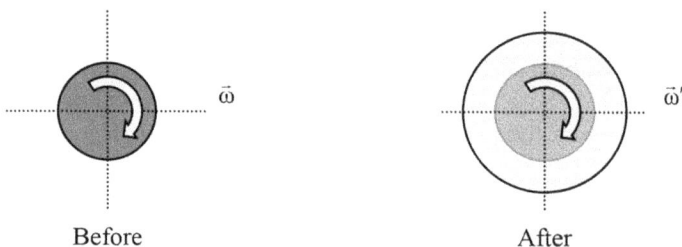

Before   After

☑ **ANSWER:** This is a tricky little problem! Looking at the situation, does it remind you of any problems we tackled during our discussions of linear motion? Two disks 'colliding' should tip you off that this is a 'Conservation of Momentum' problem. However, instead of conserving linear momentum $(\vec{p} = m \cdot \vec{v})$, you must now conserve **angular** momentum $(\vec{L} = I \cdot \vec{\omega})$. Linear momentum is only conserved if no external forces act on the system. Likewise, angular momentum is only conserved if no external torques act on the system. Notice that no one is pushing on the pizza plate nor the translucent plate; no springs exert a force on the plates to rotate them; and no ropes nor rods apply a force to them. Thus, no external torques are present. Therefore, the total angular momentum of the two plates, when taken together as a 'system,' is conserved. When writing our solution, remember that any motion occurring after a collision is usually represented by a 'prime' (i.e., the 'tick mark' or x'):

$$\sum \vec{L}_{\text{initial}} = \sum L'_{\text{final}}$$

$$\sum (I \cdot \vec{\omega})_{\text{initial}} = \sum (I \cdot \vec{\omega})'_{\text{final}}$$

$$\left[\begin{matrix} I_{\text{Plate 1}} * \omega_{\text{Plate 1}} \\ \text{Initial} \quad \text{Initial} \end{matrix} \right] + \left[\begin{matrix} I_{\text{Plate 2}} * \omega_{\text{Plate 2}} \\ \text{Initial} \quad \text{Initial} \end{matrix} \right] = \left[\begin{matrix} I_{\text{Plate 1}} * \omega_{\text{Plate 1}} \\ \text{Final} \quad \text{Final} \end{matrix} \right]' + \left[\begin{matrix} I_{\text{Plate 2}} * \omega_{\text{Plate 2}} \\ \text{Final} \quad \text{Final} \end{matrix} \right]'$$

$$\left[\left(\tfrac{1}{2}*2*0.25^2 \right)*1 \right] + \left[\left(\tfrac{1}{2}*3*0.5^2 \right)*0 \right] = \left[\left(\tfrac{1}{2}*2*0.25^2 \right)* \omega_{\substack{\text{Plate 1} \\ \text{Final}}} \right]' + \left[\left(\tfrac{1}{2}*3*0.5^2 \right)* \omega_{\substack{\text{Plate 2} \\ \text{Final}}} \right]'$$

$$0.0625*1 = 0.0625* \omega'_{\substack{\text{Plate 1} \\ \text{Final}}} + 0.375* \omega'_{\substack{\text{Plate 2} \\ \text{Final}}}$$

Realizing that after the collision, both plates rotate at the same angular velocity (i.e., $\omega'_{\text{Plate 1}} = \omega'_{\text{Plate 2}}$), we get a final answer of:
Final   Final

$$\omega'_{\text{Plate 1}} = \omega'_{\text{Plate 2}} = 0.143 \circlearrowleft \frac{\text{rad}}{\text{sec}}$$
Final   Final

☺ **COMMENTS:** Again, notice that the key to solving the problem is to find the analogous situation for linear motion. 'Collisions' and 'explosions' can happen with linear and rotational motion.

3.4 Keeping information

3.4.1 Closure

To close our discussion of rotational motion, let's re-visit the two problems in our Anticipatory Set.

We start with the amazing acrobatic professor of physics. The question posed in the Anticipatory Set was: 'How is the professor able to spin with increasing angular velocity, even though she is unable to grab something and exert a force to change her motion?' To tackle this question, let's assume that she is a solid cylinder of mass 80 kg (about 175 lbs) and radius 1 m (remember, her arms are extended horizontally when she begins to spin). Also, let's assume she starts her rotation at a relatively fast angular velocity of 1 revolution (2π rad) s^{-1} (i.e., $360°$ s^{-1}) in the Counter-Clockwise direction, as seen from above. Once she starts her rotation, she moves her arms inward so that she is now a solid cylinder of radius 0.5 m. Calculate what happens to her angular velocity.

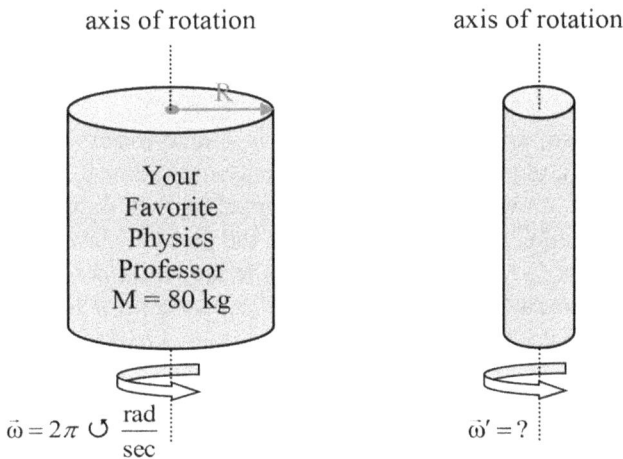

axis of rotation   axis of rotation

Your
Favorite
Physics
Professor
M = 80 kg

$\vec{\omega} = 2\pi \circlearrowleft \dfrac{\text{rad}}{\text{sec}}$ $\vec{\omega}' = ?$

Notice that no one is pushing on the professor; no springs exert a force on the professor to cause her to spin; and no ropes nor rods apply a force on her to cause her to spin. Therefore, no external torques are present. Therefore, the professor's total angular momentum is conserved.

$$\sum \overrightarrow{L}_{\text{initial}} = \sum L'_{\text{final}}$$

$$\sum (I \cdot \overrightarrow{\omega})_{\text{initial}} = \sum (I \cdot \overrightarrow{\omega})'_{\text{final}}$$

$$\left[\underset{\text{Initial}}{I_{\text{Prof}}} * \underset{\text{Initial}}{\omega_{\text{Prof}}} \right] = \left[\underset{\text{Final}}{I_{\text{Prof}}} * \underset{\text{Final}}{\omega_{\text{Prof}}} \right]'$$

$$\left[\left(\tfrac{1}{2}*80*1^2 \right)*2\pi \circlearrowleft \right] = \left[\left(\tfrac{1}{2}*80*0.5^2 \right)*\underset{\text{Final}}{\omega_{\text{Prof}}} \right]'$$

$$251.3 \circlearrowleft = 10*\underset{\text{Final}}{\omega'_{\text{Prof}}}$$

$$\underset{\text{Final}}{\omega'_{\text{Prof}}} = 2.51 \circlearrowleft \tfrac{\text{rad}}{\text{sec}}$$

So the professor's angular velocity increases to 2.51 rad s^{-1} (72° s^{-1}—Yikes!) merely by moving her arms inward and decreasing her moment of inertia. To answer the question in the Anticipatory Set—'*How is the professor able to spin with increasing angular velocity, even though she is unable to grab something and exert a force to change his motion?'*—The answer is simple: '*By conserving angular momentum!*' He decreased his moment of inertia by moving her arms closer to her body (which was acting as her axis of rotation). Since angular momentum was conserved, her body had to necessarily spin faster in response to this change in her moment of inertia. In summary, as the moment of inertia decreased, the angular velocity increased to compensate!

Now let's turn our attention to the clumsy cat. In the Anticipatory Set, the question was posed: 'How is the cat able to rotate his body and land on his feet, even though he is unable to grab something and exert a force to change his motion?' To tackle this question, assume the cat's body is upside down (feet pointing upward) and motionless. Also, assume the cat's body is a solid drum ($I_{\text{Solid Drum}} = 1/2\,M \times R^2$) of mass 4.5 kg (a reasonably-sized 10 lbs. cat) and radius 0.15 m. As soon as the cat is turned upside down and released, he senses he is in danger and instinctively begins twirling his tail Clockwise (as seen from the point of view facing the cat) at an angular velocity of $\pi \circlearrowright \text{rad s}^{-1}$. Last, assume the cat's tail is a rigid rod of mass 0.22 kg (0.5 lbs.) and length 0.31 m (12 inches). The Moment of Inertia of a rigid rod about an axis at the end of the rod is $I_{\text{End of Rod}} = 1/3\,M \times \ell^2$. Calculate what happens to the cat as he falls from the tree. Ignore the effects of the cat's feet or head on the rotational motion.

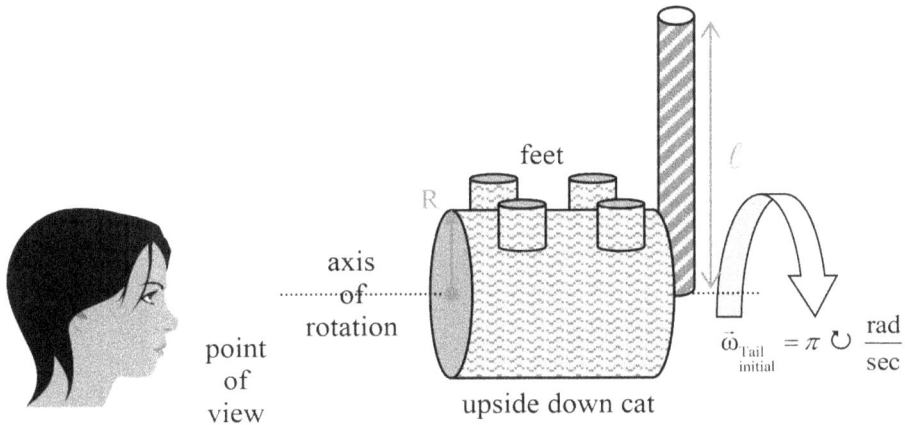

feet

R

axis
of
rotation

point
of
view

upside down cat

$\ddot{\omega}_{\text{Tail} \atop \text{initial}} = \pi \circlearrowleft \dfrac{\text{rad}}{\text{sec}}$

Profile of Girl Image Credit: Pixabay by OpenClipart-Vectors.

Like the previous example of the acrobatic professor, notice that no one is pushing on the cat; no springs exert a force on the cat to cause him to spin; and no ropes nor rods apply a force on the cat to cause him to spin. Therefore, no external torques are present. Therefore, the cat's total angular momentum is conserved.

$$\sum \overrightarrow{L}_{\text{initial}} = \sum L'_{\text{final}}$$

$$\sum (I \cdot \overrightarrow{\omega})_{\text{initial}} = \sum (I \cdot \overrightarrow{\omega})'_{\text{final}}$$

$$\left[\begin{matrix} I_{\text{Body} \atop \text{Initial}} * \omega_{\text{Body} \atop \text{Initial}} \end{matrix} \right] + \left[\begin{matrix} I_{\text{Tail} \atop \text{Initial}} * \omega_{\text{Tail} \atop \text{Initial}} \end{matrix} \right] = \left[\begin{matrix} I_{\text{Body} \atop \text{Final}} * \omega_{\text{Body} \atop \text{Final}} \end{matrix} \right]' + \left[\begin{matrix} I_{\text{Tail} \atop \text{Final}} * \omega_{\text{Tail} \atop \text{Final}} \end{matrix} \right]'$$

$$\left[\left(\tfrac{1}{2} * M * R^2\right) * \omega_{\text{Body} \atop \text{Initial}} \right] + \left[\left(\tfrac{1}{3} * M * \ell^2\right) * \omega_{\text{Tail} \atop \text{Initial}} \right] = \left[\left(\tfrac{1}{2} * M * R^2\right) * \omega_{\text{Body} \atop \text{Final}} \right]' + \left[\left(\tfrac{1}{3} * M * \ell^2\right) * \omega_{\text{Tail} \atop \text{Final}} \right]'$$

$$\left[\left(\tfrac{1}{2} * 4 * 0.15^2\right) * 0 \right] + \left[\left(\tfrac{1}{3} * 0.22 * 0.31^2\right) * 0 \right] = \left[\left(\tfrac{1}{2} * 4 * 0.15^2\right) * \omega_{\text{Body} \atop \text{Final}} \right]' + \left[\left(\tfrac{1}{3} * 0.22 * 0.31^2\right) * (-\pi) \right]'$$

$$0 = 0.045 * \omega'_{\text{Body} \atop \text{Final}} - 0.0221$$

$$\omega'_{\text{Body} \atop \text{Final}} = 2.03 \circlearrowleft \tfrac{\text{rad}}{\text{sec}}$$

This clumsy cat knows his physics! Once he is released, he instinctively rotates his tail in a Clockwise direction so that his body rotates in a Counter-Clockwise direction to conserve angular momentum. Whichever direction he rotates his tail, his body rotates in the opposite direction (in an amount proportional to the ratio of the Moments of Inertia of the tail-to-body) to conserve angular momentum. Cats possess a number of complex reflexes and abilities that allow them to usually 'land on all fours.' The reflex in cats to twirl their tails in an effort to maneuver their bodies into a position to land on all four feet is called the 'righting reflex' and is usually fully-developed in cats by 6–7 weeks of age. In reality, the righting reflex in cats is more than just twirling their tails. For example, cats use their eyesight and vestibular apparatuses to determine their orientation. Their small bodies are able to twist in the air. Their light bone structure and furry paws help soften their impact on the ground. Cats have even been seen billowing their bodies in flight to increase air resistance. Research has shown that if given as little as 12 inches of vertical distance, cats will have enough time to right themselves to land on all fours. Regardless of the complexities of the instincts of these amazing creatures, the basic principle of physics at work is conservation of angular momentum! To answer the question in the Anticipatory Set—*'How is the cat able to rotate his body and land on his feet, even though he is unable to grab something and exert a force to change his motion?* —The answer is simple: *'By conserving angular momentum!'*

3.4.2 Independent practice

3.4.2.1 The center-of-mass

1. Calculate the **Center-of-Mass** of the following four objects. Note that all point-masses are measured in kilograms while all rods are massless and have lengths measured in meters. The objects are not drawn to scale:

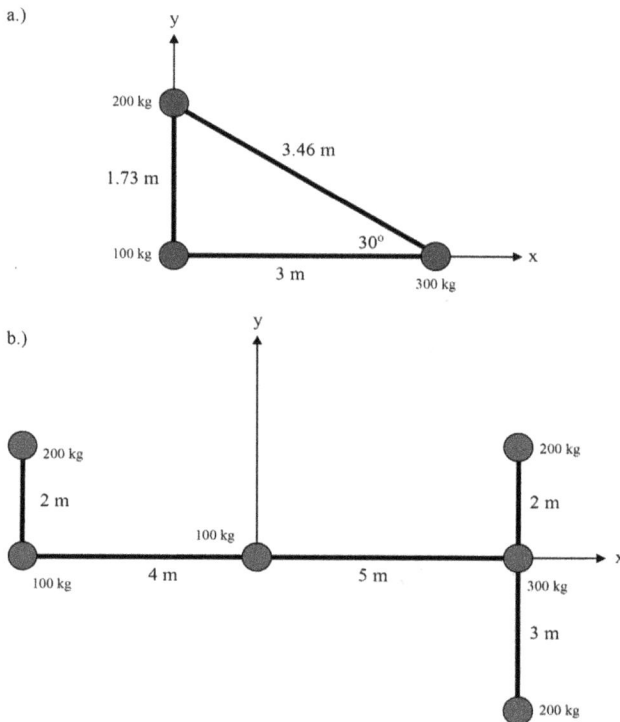

c.)

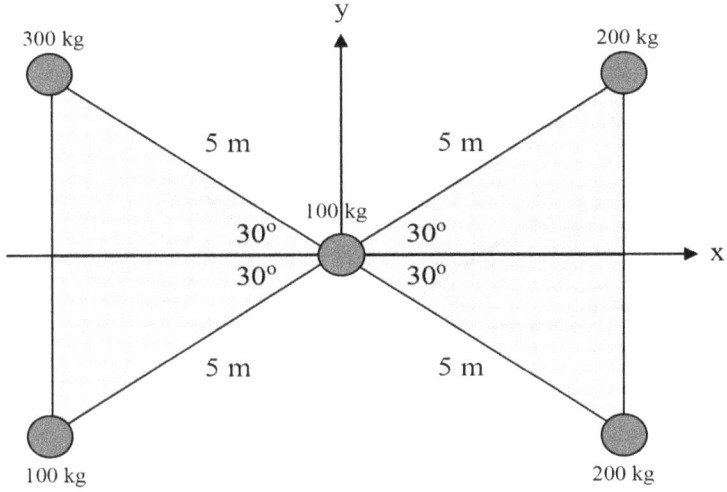

d.)

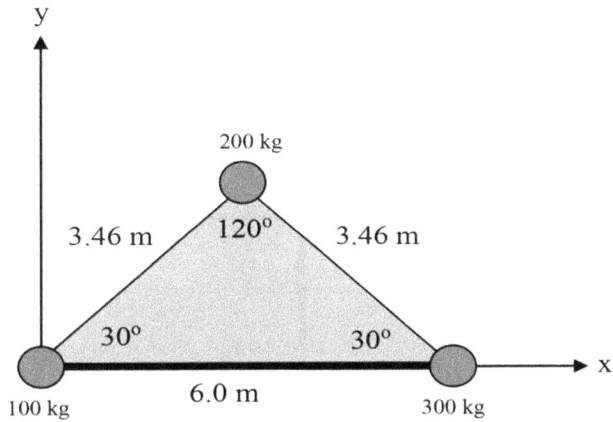

2. Calculate the **Center-of-Mass** of the following two objects—**THEY ARE NOT DRAWN TO SCALE**:

Every black ball has a mass of 10 kg.
Every gray ball has a mass of 20 kg.
Every white ball has a mass of 30 kg.

OBJECT #1

10 m

10 m

10 m

20 m   40°   50°   25 m

10 m

OBJECT #2

10 m

5 m

10 m   30°   10 m

40°

5 m

10 m

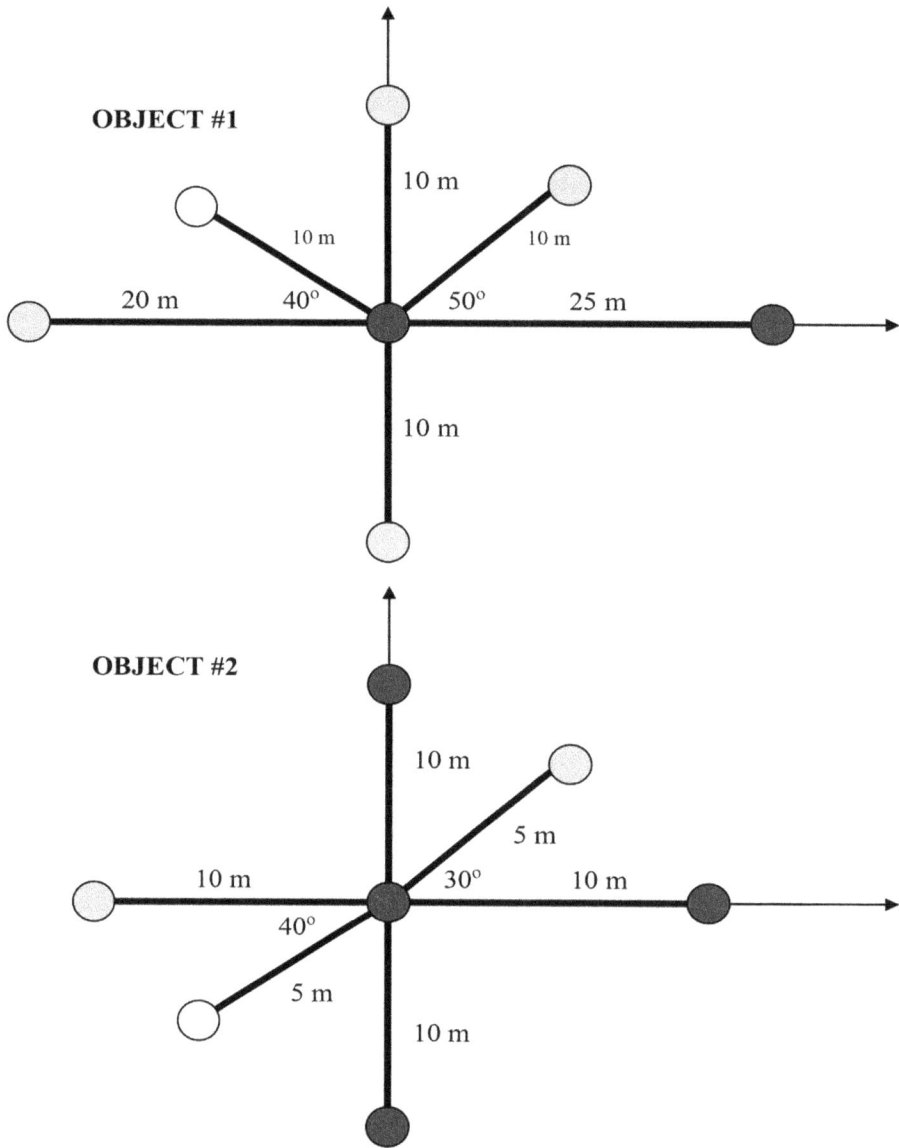

3.4.2.2 Moment of inertia

3. Calculate the **Center-of-Mass** and **Moment of Inertia** for each of the four objects shown below. When calculating the **Center-of-Mass**, use the given coordinate system. When calculating the **Moment of Inertia**, use the axis of rotation that is described. Note that all masses are in kilograms.

A.. An axis of rotation that is perpendicular to the paper, going through the origin of the given coordinate system

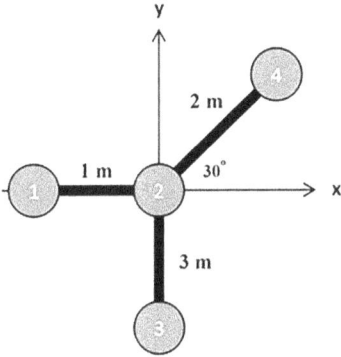

B.. An axis of rotation that lies in the plane of the paper (i.e., the xy-plane) and passes through the 1 kg and 4 kg masses. (Also, the distances on the object are symmetric)

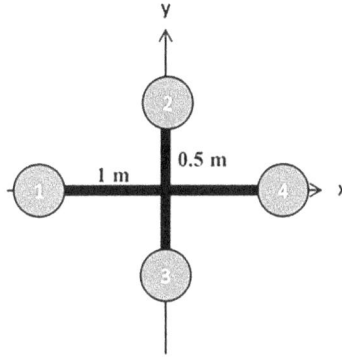

C.. An axis of rotation that is along the y-axis (Also, the distances on the object are symmetric)

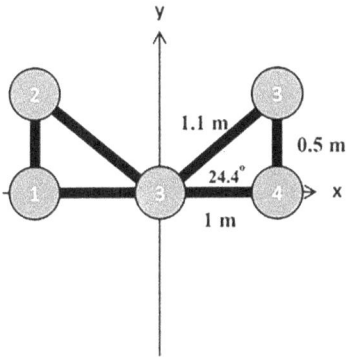

D.. An axis of rotation that is perpendicular to the paper, going through the 3 kg mass (Also, the distances on the object are symmetric)

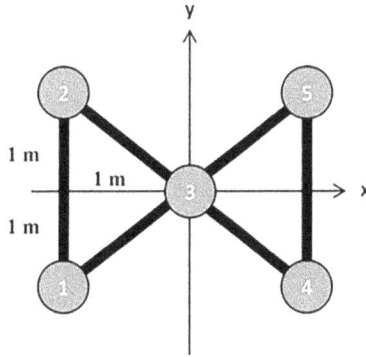

3.4.2.3 Torque

4. Calculate the **Torque** exerted by each force on the given object. The axis of rotation is always perpendicular to the paper and denoted by an ⊗.

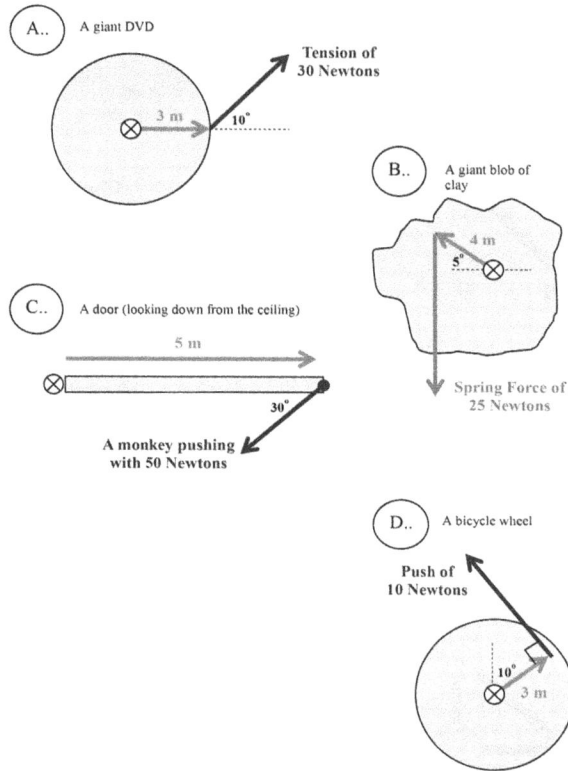

A.. A giant DVD

Tension of 30 Newtons

3 m   10°

B.. A giant blob of clay

4 m
5°

Spring Force of 25 Newtons

C.. A door (looking down from the ceiling)

5 m

30°

A monkey pushing with 50 Newtons

D.. A bicycle wheel

Push of 10 Newtons

10°   3 m

3.4.2.4 Rotational motion

5. A merry-go-round is spinning at an initial angular velocity of 1.5 rad s^{-1} in the CCW direction (as seen from above). A child is on the ride at an angle of $\pi/2$ radians (i.e., 90°). The ride begins to spin faster at a uniform acceleration of 0.25 rad s^{-2} in the CCW direction.
    a) Where is the child at $t = 4$ s later, and
    b) How fast is he spinning?

6. A 2 kg block is resting on a frictionless surface. The block is connected by a rope to a 4 kg block by passing-over an 8 kg pulley (that is actually a solid drum of radius 0.5 m). If the tension in the flat part of the rope is 7.84 Newtons (see diagram), find the tension in the vertical portion of the rope, the linear acceleration of the blocks, and the angular acceleration of the pulley. Note that $I_{\text{Solid Drum}} = 1/2\,M \times R^2$.

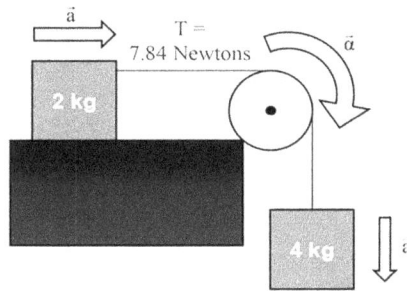

7. A solid ball, solid cylinder, and hoop have different masses and different radii. They are raced down a hill. They are simultaneously released, from rest, from the top of a frictionless hill that is 10 m high. Which object wins the race?

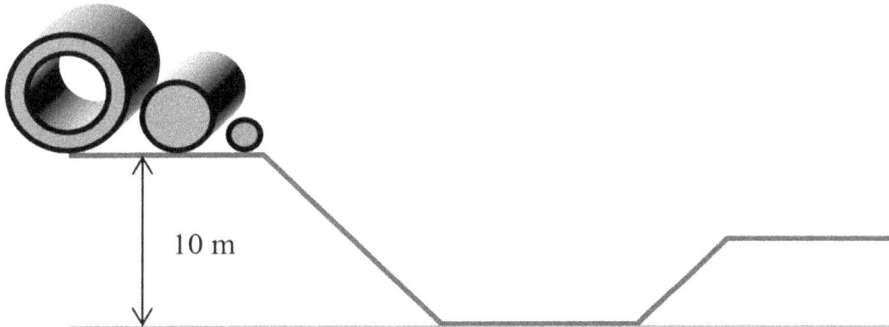

10 m

8. Let's pretend a cat's body is a solid cylinder of mass 20 kg, and radius 0.1 meters. Likewise, its tail is a rod (free to swing at its end) of mass 3 kg and length 0.25 meters. A cruel physics student holds a cat motionless, upside down, by its four legs and releases the cat hoping that the cat will land on its back. However, by sheer instinct, the cat begins to spin its tail at an angular speed of π rad/sec in the clockwise direction. What happens to the cat's body?

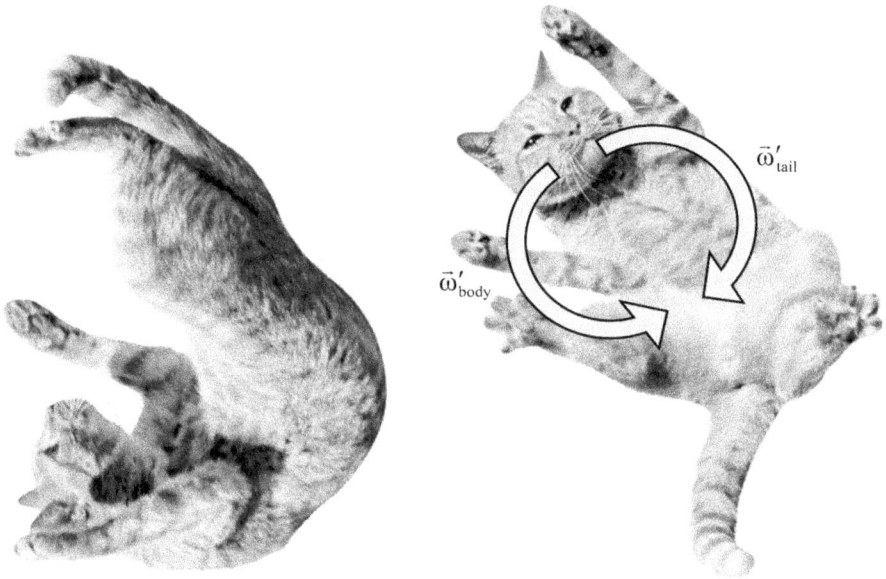

Before   After
Cat Image Credit: Vikafoto33/Shutterstock.com.

IOP Publishing

Simplified Classical Mechanics, Volume 2 (Second Edition)
Gravity and the conservation laws
Gregory A DiLisi

Chapter 4

Transition to classical electricity and magnetism

This chapter considers the question: *'Is gravity (and the interaction of masses) the only force we need worry about?'* Thus, the reader needs to consider a transition in thinking … a paradigm shift from the motion of objects with mass to those possessing electric charge. To help readers understand the paradigm shift from *Classical Mechanics* to *Classical Electricity and Magnetism*, parallels are drawn between the two fundamental forces. Similarities and differences are highlighted showing how the two forces are conceptually similar/different. If readers conceptually understand gravity, they better understand classical electricity and magnetism. The chapter concludes with some astonishing demonstrations of static electricity which the reader can do at home using household materials. The demonstrations help readers visualize concepts and retain the information presented throughout the chapter.

> *When placed in command, take charge.*
>
> —Norman Schwarzkopf

> *Being in charge sometimes means making people mad. Some days you have to overrule even the best advice, because you think it is not right.*
>
> —Colin Powell

> *If you put the federal government in charge of the Sahara Deserts, in 5 years there'd be a shortage of sand.*
>
> —Milton Friedman

> *Take charge of your life! The tides do not command the ship. The sailor does.*
>
> —Ogwo David Emenike

doi:10.1088/978-0-7503-6402-7ch4

Freedom is never an achieved state; like electricity, we've got to keep generating it or the lights go out.

—Wayne LaPierre

Talent is like electricity. We don't understand electricity. We use it.

—Maya Angelou

What is a soul? It's like electricity - we don't really know what it is, but it's a force that can light a room.

—Ray Charles

We will make electricity so cheap that only the rich will burn candles.

—Thomas A Edison

A smile costs less than electricity, but gives much light.

—Abbe Pierre

God is the electricity and we are the lamps.

—Marianne Williamson

The force of nonviolence is infinitely more wonderful and subtle than the material forces of nature, like electricity.

—Mahatma Gandhi

Without electricity, there can be no art.

—Nam June Paik

If your hate could be turned into electricity, it would light up the whole world.

—Nikola Tesla

Brunettes are full of electricity.

—Auguste de Villiers de l'Isle-Adam

Magnetism, as you recall from physics class, is a powerful force that causes certain items to be attracted to refrigerators.

—Dave Barry

I think I have kind of a natural magnetism.

—Wesley Snipes

The heat around young actors burns out. Natural ability and magnetism only get you so far. The rest is hard work.

—Ben Foster

Give a man a chance, and he'll be so casual that he won't bathe, shave, or comb his hair. He'll just depend on his animal magnetism to get him by.

—Elsa Schiaparelli

The journalists have obviously failed to capture my innate magnetism, humour and charisma, and they all need to be fired from their newspapers right away.

—Alexei Sayle

What's more unnerving than magnetism, ghosts, and unpurified water? Gadgetmongers who purport to protect us from metaphysical monsters that go bump in the New Age night.

—Chris Hardwick

The only thing that matters is to have charm and expression. Then comes that horrible gnawing doubt of our own magnetism. Is it possible that, though we are not lovely, we are not irresistible either? That we will have to go through life belonging neither to the triumphantly beautiful nor to the triumphantly ugly?

—Elizabeth Bibesco

4.1 Motivation

In volume 1, chapter 1, we developed some mathematical tools. The focus of that chapter was to develop our conventions for scalar notation and vector notation.

In volume 1, chapter 2, we defined four kinematic quantities (i.e., speed, velocity, acceleration, and jerk) that will be used throughout this series to describe how objects move.

In volume 1, chapter 3, we remained focused on describing **how** objects move and developing the UAM technique for solving problems. We saw that this technique involved three vector equations and is most applicable to 'ideal situations,' short-term accelerations, and free-falling objects (i.e., projectile motion). This was our first technique allowing for numerical analysis of motion.

In volume 1, chapters 4 and 5, we tackled the question of **why** objects move. Thanks to Isaac Newton and his three laws of motion, we saw that objects move because a net external force exists on them. This was our second technique allowing for numerical analysis of motion.

In volume 1, chapter 6, we adapted Newton's laws of motion to the special situation of 'Uniform Circular Motion' or 'UCM' In a sense, UCM is a sub-strategy of Newton's laws of motion. The focus of volume 1, chapters 4 and 5 was the general motion of objects when they are subjected to various common forces while volume 1, chapter 6 was a specific case of objects moving in a circle (or partial circle).

In volume 2, chapter 1, we digressed a little bit from our problem-solving strategies to examine more closely the origin of the gravitational field force that was introduced earlier in the series. We found that the formula for 'weight' that we had been using in the past was really a simplification of a much more elaborate law; namely, Newton's 'Universal Law of Gravitation' (abbreviated as the 'ULG'). The ULG determines the gravitational interaction of any two bodies—whether they are on the surface of the Earth or not.

In volume 2, chapter 2, we developed an entirely new approach to problem-solving. We defined two new physical quantities called 'work' (or 'energy') and 'linear momentum' and saw that these quantities were conserved; that is to say, they remained fixed over time. In certain situations, we were able to use principles of conservation to analyze a new set of everyday motions. For example, situations involving only kinetic, gravitational potential, and elastic potential energies could be solved using the Law of Conservation of Energy. On the other hand, collisions and explosions were particularly suited for analysis using the Law of Conservation of Momentum.

In volume 2, chapter 3, we quickly and efficiently doubled the number of techniques we had in our arsenal of problem-solving recipes. We took the same approach that paid us dividends in the past—we defined some new physical quantities and explored how these quantities allowed us to tackle different types of motion. In the process of defining these new physical quantities, we also developed one last new mathematical operation—the vector 'Cross'-product method for multiplying two vectors.

We conclude this series on *Simplified Classical Mechanics* with volume 2, chapter 4. This chapter allows us to transition from analyzing the motion of objects with mass to analyzing the motion of objects with electric charge, thus, entering the fascinating discipline of 'Classical Electricity and Magnetism.' This chapter is a conceptual one—we will look at the similarities and differences between the forces of gravity and electricity as well as examine some interesting demonstrations involving electric charge. The purpose of this chapter is to move us from the realm of 'Classical Mechanics' to that of 'Classical Electricity and Magnetism' and to prepare us for subsequent courses on electricity and magnetism.

4.2 Getting ready

4.2.1 Anticipatory set

4.2.1.1 Newton and the apple—there must be more to life than gravity
You have probably heard the story of how Isaac Newton discovered his Universal Law of Gravitation. The story is almost universally known and is described briefly in volume 1, chapter 4 of this series on *Simplified Classical Mechanics*. Let us now explore this story in greater detail. In 1666, the bubonic plague swept through Cambridge and the 23 year-old Newton was sent home to Woolsthorpe for his own safety. In the late summer, while sitting in his garden at Woolsthorpe Manor, Newton observed an apple falling from an apple tree.

Isaac Newton observes an apple fall from a tree in his garden—a chance encounter that changed our understanding of gravity, motion, and physics.

Image Credit: Author.

According to the University of York's School of Physics, Engineering, and Technology, Newton gave an account of his discovery to several acquaintances: the French philosopher, Voltaire … Newton's assistant at the Royal Mint, John Conduitt … Newton's niece, Catherine Barton … Newton's friend, William Stewkeley … and finally, a student at Cambridge named Chistopher Dawson. Furthermore, in 1726, the year of Newton's death, Conduitt published the first written notation of Newton's account of his discovery of the Universal Law of Gravitation. According to Conduitt, 'He (Newton) first thought of his system of gravitation which he hit upon by observing an apple fall from a tree.' Finally, according to Edmund Turnor's 'A History of the Town and Soak of Grantham,' published in 1806, the specific tree from which the apple fell had been identified because only one tree was growing in Newton's garden. Turnor states, 'The tree is still remaining and is showed to strangers.' Sadly, the tree was blown over in 1816 in a storm; however, portions of the tree were re-rooted at Woolsthorpe Manor and are still growing today.

We see from the historical record that that facts surrounding Newton's discovery of the Universal Law of Gravitation are well-established. Now let us enter the world of make-believe and imagine that Newton took his observations a few steps further.

First, let us imagine that Newton pulled out a knife and cut the apple in half; his purpose being to determine if gravity were the force responsible for holding the two halves of the apple together. In other words, let us imagine that Newton wanted to know if gravity was strong enough to hold the two halves of an apple together?

In our make-believe extension of Newton's encounter with the apple, he pulled out a knife and cut the apple in half. His goal was to determine if gravity was strong enough to hold the two halves of the apple together.

Image Credit: Author. Knife Image Credit: Pixabay by Jjuni.

We can imagine that Newton made a crude calculation as follows: A typical apple weighs about 1/3 of a pound, but for our calculations, let us assume that Newton's apple weighs exactly 2.2 pounds (you'll see why in a moment). If you think I'm exaggerating with the weight of Newton's apple, rest easy in knowing that the weight of the record-setting heaviest apple was 4 pounds, 1 ounce (grown by Chisato Iwasaki at his apple farm in Hirosaki City, Japan, in 2005), so a 2.2 pound apple is entirely possible. The reason for making Newton's apple weigh 2.2 pounds is that when cut in half, each half has a nice, easy-to-use mass of 1.1 pounds, or exactly half of a kilogram (i.e., 2.2 pounds = 1 kg, so when the apple is cut in half, each half has a mass of exactly 0.5 kg). If Newton then assumed that the centers of the two halves were approximately 1 cm (0.01 m) apart, his calculation for the force of gravity between the two halves would be, from volume 2, chapter 1:

$$\overrightarrow{F}_{g(\text{between two halves of apple})} = \frac{G \times M_{\text{half \#1 of the apple}} \times M_{\text{half \#2 of the apple}}}{(d_{\text{between halves}})^2}\hat{r}$$

$$= \frac{(6.673 \times 10^{-11}\frac{N \times m^2}{kg^2}) \times (0.5 \text{ kg}) \times (0.5 \text{ kg})}{(0.01m)^2}\hat{r}$$

$$= 1.67 \times 10^{-7}z, \text{ Newtons attractive}$$

Of course, Newton didn't have the value of G that we have today, nor did he have the m.k.s. system of units, but using the modernized formula for the Universal Law of Gravitation, he would have determined that the force of gravity between the two halves of the apple was smaller than the weight of a fruit fly (a typical fruit fly weighs about 2.5×10^{-6} Newtons). Therefore, if gravity were the *only* force that held an apple together, then every time a fruit fly landed on an apple, the apple would break in half … or fall to pieces. Clearly, gravity cannot be the force in nature that holds the two halves of the apple together. Newton could only draw one conclusion —*there must be another force in nature that is also attractive like gravity (i.e., so it can hold things together), but much stronger than gravity (i.e., so it is strong to hold the apple together)!*

Let us continue our make-believe extension of Newton's encounter with the apple. Imagine that after cutting the apple in half, Newton witnessed a worm crawl out from one of the halves.

In our make-believe extension of Newton's encounter with the apple, he witnessed a worm crawl out from one of the two halves of the apple.

Image Credit: Author.

Witnessing the worm crawl out from the apple would have had a profound impact on Newton. Not only would he no longer be interested in eating the apple, but he also would have realized that if gravity were the only force in nature, the worm would never be able to separate itself from the apple. Since gravity is always attractive, it can only hold objects together. If the worm were able to separate itself from the apple, Newton would have concluded that a *repulsive* force must be involved. Newton could only draw one conclusion—*there must be another force in nature that is repulsive (i.e., so a force exists that keeps objects apart)!*

Combining Newton's two observations from his make-believe extended encounter with the apple, we find:

1. Another force must exist in Nature that, like gravity, can be attractive, but can be much, much stronger than gravity (perhaps as much as 10^{36} as times as strong).
2. Another force must exist in Nature that is repulsive.

The force that incorporates both of these observations is called the 'electromagnetic' force and is the force responsible for both holding the apple together and allowing the worm to separate itself from the apple. Transitioning from the study of mass and gravity to the study of electric charge and the electromagnetic force is the focus of this chapter.

4.2.2 Objective

By the end of this chapter, you will be able to:
- Give examples of the 'fundamental properties of Nature' and their associated 'fundamental forces.'
- Define 'electric charge' and describe how this fundamental property enables objects to experience the electromagnetic fundamental force of Nature.
- Describe some characteristics associated with electric charge such as size and type.
- State the force law between two electric point-charges (i.e., 'Coulomb's Law').
- State the mechanism by which two electric charges interact even though they are not touching.

4.2.3 Purpose

This information is needed:
- Because mass is not the only fundamental property of Nature and therefore, gravity is not the only fundamental force of Nature. Looking back at volume 1 of this series (chapters 1–6) and the first three chapters of volume 2, you will notice that all the problems involve the motion of a *mass* subjected to the force of gravity. However, now a new fundamental property of Nature is introduced—electric charge. Since some objects possess another fundamental property of Nature called 'electric charge,' they experience another fundamental force of Nature, the electromagnetic force.
- Because analyzing the motion of objects with electric charge will greatly expand our arsenal of problem-solving techniques. Remember, our goal in this series on *Simplified Classical Mechanics* is to develop problem-solving techniques that handle different types of motion. Many new problem-solving techniques can be developed when analyzing the motion of objects with electric charge.

4.3 Giving information

4.3.1 Instructional input

4.3.1.1 Fundamental forces in Nature

Four fundamental fours exist in Nature. These forces affect us every day, even though we really only 'see' two of them affect our day-to-day lives. We will briefly describe each of the four forces below, starting with the weakest in relative strength and ending with the strongest in relative strength; however pay attention to the similarities and differences among the fundamental forces. Namely, notice that the forces all share conceptual properties like relative strength and effective range, but that the specific values associated with these quantities differ significantly. Also, notice that these forces play vastly different roles in our day-to-day lives.

❶ The first, and weakest, of the fundamental forces in Nature is gravity. Even though you feel it every day and it seems to exert such a 'heavy' (recall from volume 2, chapter 1 that the word 'gravity' comes from the Latin word 'gravis' that translates into 'heavy' or 'weighty') force on you, it is indeed the weakest of the four fundamental forces in Nature. Gravity is so weak that if we assign the strongest fundamental force in Nature a relative strength of 1, gravity would be 10^{-38} times as weak. In the classical model of physics, in order for an object to 'feel' the force of gravity, it must possess the fundamental property of mass. Herein lies a crucial concept—the fundamental property associated with gravity is mass—an object must possess mass in order to be subjected to the fundamental force of gravity. Despite its relatively weak strength, gravity's range is infinite, meaning that no matter how great the distance between two masses, an attractive force of gravity always exists between them (indeed we know from Newton's Universal Law of Gravitation that the force between to masses varies as $1/r^2$ ... so the force between two masses may be small, but it always exists because $1/r^2$ never reaches zero, no matter how large the distance r, is). We humans are quite familiar with the force of gravity. We feel it at all times and everywhere we go. Gravity is the force that exists between our bodies and the Earth so we have a very familiar sense of how this fundamental force works. Most of us also have an intuitive sense that gravity holds the planets together to form the solar system and on a greater scale, holds the stars together to form galaxies; however, most of us would be surprised to realize that gravity exists between us and our neighbor ... between us and the table ... between our neighbor and the table ... between *any* two objects with mass. We often think that gravity only exists between objects and the Earth because the interaction with the Earth is such a dominant interaction; rest assured gravity exists between you and your neighbor, however, that interaction is just overwhelmed by the interaction between you and the Earth so it appears to be negligible. We've devoted the entirety of volume 1, chapters 1–6 and volume 2, chapters 1–3 to developing problem-solving strategies for the motion of objects subjected to gravity. As I mentioned in the front matter of this series on *Simplified Classical Mechanics*, an introductory course in physics is typically divided into three courses, the first of which is *Classical Mechanics*. Such a course focuses entirely on the motion of objects subjected to the fundamental force of gravity.

❷ The second fundamental force in terms of relative strength is called the 'Weak' force or the 'Weak Nuclear' force. If we assign the strongest fundamental force in Nature a relative strength of 1, the weak force would be 10^{-13} times as weak. Compared to gravity, the weak force is 10^{25} times stronger; thus, it is often said that 'there is nothing weak about the weak force.' The weak force is the fundamental force responsible for changing one type of subatomic particle into another. The hallmark example of the weak force is beta decay in which a neutron transforms into a proton, emitting an electron and an antineutrino.

In this process, one type of subatomic particle is transformed into another type of subatomic particle and such a process is not possible unless a completely separate fundamental force in Nature exists. Unlike gravity, the effective range of the weak force is unimaginably small, 10^{-18} m, meaning a subatomic particle must stray to within less than the diameter of a proton before the weak force kicks in. In 1933, Italian physicist Enrico Fermi was the first to envision such a force so the weak force was originally named the 'Fermi interaction.' As you can imagine, we humans simply don't have too much intuition regarding the weak force. Afterall, our daily lives aren't often spent worrying about neutrinos straying too close to neutrons so that the neutron can transform into a proton while the neutrino becomes an electron. That's why most of us have never heard of the weak force. However, the weak force is crucial for the creation of elements within stars and radioactive materials. For instance, the weak force is important in the fusion of hydrogen into helium in stars. As a final example of the weak force in action, most of us have heard of the carbon-14 dating method that allows scientists to determine the age of artifacts. This technique relies on carbon-14 decaying into nitrogen-14 through the weak force.

❸ The third of the fundamental forces in Nature is the 'electromagnetic' (i.e., E&M) force. As the name suggests, the E&M force consists of two forces, the electric force and the magnetic force. Historically, these two forces were thought to be separate so that physicists believed there to be *five* fundamental forces in Nature. However, the electric and magnetic forces were unified into a single force by Michael Faraday and James Clerk Maxwell between 1820 and 1873. The focus of this chapter is to transition our mindset from mass and gravity to electric charge and the electric force. Unifying the electric force and magnetic force is something to be tackled in a subsequent series on *Simplified Classical Electricity and Magnetism*. The E&M force is so weak that if we assign the strongest fundamental force in Nature a relative strength of 1, the E&M force would be 100 times weaker. In the classical model of physics, in order for an object to 'feel' the E&M force, it must possess the fundamental property of electric charge or moving electric charge, also known as current. Herein lies a crucial concept—the fundamental property associated with electricity is electric charge and the fundamental property associated with magnetism is current—an object must possess electric charge in order to be subjected to the force of electricity and must possess electric current in order to be subjected to the force of magnetism. Despite its relatively weak strength, the E&M force's range is infinite, meaning that no matter how great the distance between two electric charges, an E&M force always exists between them. In fact, we will soon find that like gravity, the E&M force varies as $1/r^2$. Likewise, just like in the case of gravity, the force between two electric charges may be small, but it always exists because $1/r^2$ never reaches zero, no matter how large the distance r, is. We humans are quite familiar with the E&M force even though we might not be aware of its ubiquitous nature—it is responsible for friction, elasticity, the normal force, drag, and as mentioned in the anticipatory set of this chapter, for holding solids together, like an apple ... or keeping them apart, like a worm and an apple. As I mentioned in the front matter of this series on *Simplified Classical Mechanics*, an introductory course in physics is typically divided into three courses, the second of which is *Classical Electricity and Magnetism*. Such a course focuses entirely on the motion of objects subjected to the electro-magnetic fundamental force of Nature.

❹ The final fundamental force in terms of relative strength is called the 'Strong' force or the 'Strong Nuclear' force. As the name suggests, the strong force is the strongest of the fundamental forces in Nature. It is over a thousand trillion trillion trillion times stronger than gravity! The strong force is the fundamental force responsible for holding protons and neutrons together to form the nucleus of an atom. Since childhood, most of us have become familiar with the 'solar system' model of an atom in which a cluster of protons and neutrons forms the nucleus of the atom (representing the Sun) and is surrounded by orbiting electrons (representing the planets). However, this model should strike you as inherently unstable. Don't the protons at the center of the nucleus possess positive electric charge which would repel one another and cause the nucleus to blow itself apart? How can a cluster of positively-charged particles stay together? On the contrary, we find that the nucleus is physically stable. Before 1971, no fundamental force in Nature was known to exist that was 'stronger' than the electromagnetic repulsive force between positively-charged protons that would allow the nucleus to remain stable. A new fundamental force in Nature was needed to explain this phenomenon. Ultimately, thanks in large part to the work of physicists Murray Gell-Mann and George Zweig, the strong force was introduced to be the underlying force that holds the nucleus together. Like the weak force, the effective range if the strong force is unimaginably small, 0.5-to-3 femto-meters (1 femtometer is 10^{-15} meters). As you can imagine, we humans don't have too much intuition regarding the strong force so most of us have never heard of it.

A summary of the four fundamental forces is shown below:

Force	Relative strength	Range	Discoverers	Classical fundamental property	Examples
Gravity	10^{-38}	∞	Isaac Newton (1666)	Mass	Gravity is the force that holds us to the Earth, holds the planets together to form the solar system, and holds the stars together to form galaxies.

(Continued)

(Continued)

Force	Relative strength	Range	Discoverers	Classical fundamental property	Examples
Weak	10^{-13}	$<10^{-18}$ m	Enrico Fermi (1933)		The weak force is the force responsible for changing one type of subatomic particle into another.
Electro-magnetic	10^{-2}	∞	Michael Faraday and James Clerk Maxwell (1820–73)	Electric Charge or Current	The electromagnetic force is the force responsible for friction, elasticity, the normal force, drag, and for holding solids together … or keeping them apart.
Strong	1	$\sim 10^{-15}$ m	Murray Gell-Mann and George Zweig (1964)		The strong force is the force that holds the nucleus together.

4.3.1.2 Electric charge: units, sizes, types, and examples

Electricity, more formally known as the electric force, is just the force (i.e., the push or pull) an object experiences if it possesses *electric charge*. This is a crucial concept—the fundamental property that an object needs to possess in order to experience the electric force is electric charge. You cannot subject an object to electricity until you electrically charge that object. This notion of a fundamental property is a difficult one for students, and seasoned physicists, to grasp. The basic idea is that an object needs a certain 'ingredient' to experience a force, once it is placed in the field of another object. In other words, two objects interact with one another, through the interaction of their fields, provided they both possess a certain fundamental property.

Fundamental Property

A fundamental property is a physical property of matter that causes it to experience a force when placed in a field.

A fundamental property is not an object itself … it is a ***property*** of an object. We've already encountered a fundamental property, called mass, in volume 2, chapter 1 of this series when we discussed Newton's Universal Law of Gravitation. Our definition of a fundamental property certainly applies to *mass* because we know from volume 2, chapter 1, that mass, which we symbolized by m (or M) and measure in kilograms, was a physical property of matter that caused it to experience the force of gravity when placed in a *g*-field. The confusion surrounding fundamental properties is that we sometimes get too casual, or imprecise, with our language. For example, sometimes we say, 'There is a mass on the table ...' (implying that mass is an object), when we really should be saying, 'There is a rock, with mass, on the table ...' (more accurately conveying the notion that mass is a ***property*** of an object and not an object itself).

The same can now be said about electric charge. Electric charge is a fundamental property of an object, which we symbolize by q (or Q) and measure in Coulombs. Electric charge is a scalar physical quantity since it only has a magnitude and not a direction.

The fundamental property of electricity:
A fundamental property of electricity is electric charge.
    Symbol: q or Q
    Units: Coulombs
    Type of Quantity: Scalar

The unit of electric charge, the Coulomb, is named in honor of the French physicist, Charles-Augustin de Coulomb (1736–1806). Just like our use of the word 'mass,' we want to be precise in our use of the words 'electric charge' or just 'charge.' Confusion surrounding fundamental properties exists because we sometimes get too casual with our language. For example, sometimes we say, 'There is an electric charge on the table ...' (implying that electric charge is an object), when we really should be saying, 'There is a latex balloon, with electric charge, on the table ...' (more accurately conveying the notion that electric charge is a ***property*** of an object and not an object itself). To give you an idea of how big one Coulomb of electric charge is, look at some common values of electric charge on the chart below:

Object	Electric charge
Electric charge on an electron	$Q = 1.6 \times 10^{-19}$ C (negative)
Electric charge on a rubbed latex balloon	$Q \approx 10 \times 10^{-9}$ C to 100×10^{-9} C (negative)
Electric charge from walking across carpet	$Q \approx 1 \times 10^{-6}$ C
Electric charge from a lightning strike	$Q \approx 1$–15 C
Electric charge delivered by an AA battery over its lifetime	$Q \approx 5000$ C
	(Note that the comparison between the AA battery and the lightning strike can be misleading. The battery does indeed deliver 5000 C of electric charge over its lifetime, but these electrons have almost no energy while the small number of electrons delivered in a lightning strike carry huge amounts of energy.)

Biographical Information:

CHARLES-AUGUSTIN de COULOMB

Coulomb Image Credit: German Vizulis/Shutterstock.com

BORN: June 14, 1736 (Angoulême, Angoumois, France)
DIED: August 23, 1806 (Paris, France)

The Coulomb family was left penniless because of his father's poor financial investments and speculations. As a child, Coulomb argued with his mother about his career but eventually joined the French army. Coulomb obtained the rank of Lieutenant Colonel in the "Corp du Genie" before the French Revolution where, from 1764-1772, he commanded the corps that built Fort Bourbonin Martinique, in the West Indies. During the construction, Coulomb became sick and until his death, was chronically ill. He claimed: "Since my days at Fort Bourbon, I have never again been well."

Upon his return to Paris, Coulomb devoted much of his life to engineering and public service. He eventually became the supervisor of canals, harbors, and water systems in Paris. In November 1783, Coulomb was made the scapegoat of a harbor commission report and jailed for one week. Now with the rank of Captain, he discovered the inverse relationship of the force between electric charges and the square of the distance between them. In 1785, he published his findings—Coulomb's Law—in three papers on electricity and magnetism. Coulomb eventually retired from public service and devoted the remainder of his life to physics. In 1802, Coulomb married his mistress (Louise Francoise Le Proust Desormeaux) after she gave birth to his second child. After the French Revolution, his family was stripped of its money and Coulomb died in poverty. The unit of electric charge—**the COULOMB**—is named in his honor.

Busts of Charles-Augustin de Coulomb

Top image: Charles Coulomb Image Credit: German Vizulis/ Shutterstock.com. Bottom left image: Copyright: Thierry Blais, Angouleme Museum. Bottom right image: This Charles-Augustin de Coulomb - CNAM image has been obtained by the author from the Wikimedia website where it was made available by MOSSOT under a CC BY-SA 4.0 licence. It is included within this article on that basis. It is attributed to MOSSOT.

Since mass and electric charge are both fundamental properties which cause objects to experience forces, our natural tendency is to assume mass and electric charge share identical features or characteristics. Mass and electric charge may indeed be conceptually similar since they are both fundamental properties, but key distinctions distinguish them! The first important distinction between them is size.

Unlike mass, electric charge does not come in any size. For a moment, let us think back to our discussion of mass and gravity from volume 2, chapter 1. If you were to take a chunk of rock and start cutting it into smaller and smaller pieces, you would find that there is no limit to the size into which you could cut the rock. For instance, you could start with a huge boulder and cut it into a hand-sized stone ... then cut that into a tiny pebble ... then cut that into a speck of stone the size of a grain of salt ... and so on. Your initial boulder may have had a mass of $M_{boulder} = 180$ kg. However, once you cut it into a hand-sized stone, its new mass might have been $M_{stone} = 2.267\,96$ kg (notice that mass can assume any value). When you cut it into a tiny pebble, its new mass might have been $M_{pebble} = 0.004\,327$ kg. Finally, when you cut it into a speck of stone the size of a grain of salt, its new mass might have been $M_{salt-grain} = 0.000\,004\,123$ kg. The point here is that mass is a *continuous* variable, meaning it can assume any positive value. Mass is much like the positive numbers on a number line in that it can assume any value on the continuum. For instance, if I asked you to pick any positive number, you might select $N = 4$. However, you could just as easily have selected $N = 4.1$ ... or $N = 4.1234$ ... or $N = 4.123\,456\,78$. like positive numbers on the number line, mass can assume any value. When physicists say that a variable is 'continuous,' they mean that the variable can be assigned *any* value.

Continuous Variable:
A continuous variable is one that can be assigned any value. Mass is an example of a continuous variable.

Mass is a continuous variable and can assume any positive value, much like the positive numbers on a number line.

Image Credit: Author.

Electric charge is not a continuous variable—it behaves very differently than mass. If you were to take an electrically-charged particle and start cutting it into smaller and smaller pieces, you would find that *there is indeed a limit to the size into which you could cut the electric charge!* For instance, you could start by rubbing a hand-held plastic ball and giving it an initial negative electric charge.

Next, you could cut the ball into a quarter-sized shaving ... then cut that into a tiny sliver ... then cut that into a spec the size of a grain of salt ... then cut that down do a single atom of carbon ... then finally to a single electron. Your initial ball of plastic may have had an electric charge of $Q_{ball} = -1 \times 10^{-6}$ C. However, once you cut it into a quarter-sized shaving, its new electric charge might have been $Q_{shaving} = -2.5 \times 10^{-7}$ C. When you cut it into a tiny sliver, its new electric charge might have been $Q_{sliver} = -3.125 \times 10^{-8}$ C. Finally, when you cut it to a single electron, its final electric charge would have been $Q_{electron} = -1.6 \times 10^{-19}$ C. At this point, something interesting happens—even though the electric charge on the electron is an incredibly small amount of electric charge, you would find that you would not be able to divide the electric charge any further. In other words, a limit exists as to how much you can divide electric charge! The point here is that electric charge is a *discrete* or *'quantized'* variable, meaning it can assume only certain values. Recall from the front matter section of this series on *Classical Mechanics*, a 'quantum' refers to the smallest amount of a quantity that can exist.

Etymology:
The word 'quantum' comes from the Latin word 'quantum' (the plural is 'quanta') that translates into 'amount' or 'how much.' The implication here is that a 'quantum' represents the minimum amount of a quantity that can exist ... the smallest possible amount of something. Therefore, the word 'quantum' can be applied to many different quantities. For example, a quantum of light is called the 'photon,' a quantum of electricity is called the 'electron,' etc.

Since the smallest quantum of electric charge is the electric charge on the electron, the amount of electric charge any object can have must be an integer number of 1.6×10^{-19} C.

Electric charge can only assume certain values:
The electric charge on any object must be an integer multiple of the smallest quantum of electric charge, qe, the electric charge on the electron:

$$Q = n \times q_c = n \times (1.6 \times 10^{-19} \text{ C});$$

where n is an integer.

Electric charge is very similar to money, because money is also a discrete or quantized variable. Let us consider US currency. For instance, if I asked you to reach into your wallet and grab a $10 bill, you could easily divide that into two $5 bills. Likewise, you could divide one of the $5 bills into five $1 bills. Continuing, you could divide the $1 bill into four quarters. Finally, you could divide one of the four quarters into 25 pennies. However, just like electric charge, you would have noticed that with a penny, you have reached the limit of your ability to divide money. In other words, the penny is the 'quantum' of US currency—it is the smallest amount of allowable currency! In your wallet, you may possess $44 ... or $44.1 ... or $44.15 ... but you could never possess $44.156 since money does not come in quanta of $0.001 units. When physicists say that a variable is 'discrete' or 'quantized' they mean that the variable can be assigned *only certain* values.

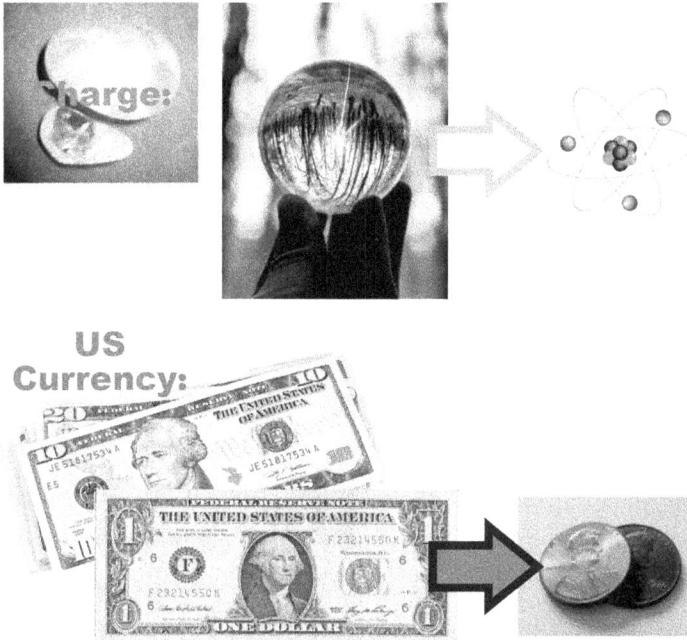

Electric charge is a discrete or quantized variable and can assume only certain values, much like US currency. In the case of electric charge, the electric charge on a glass ball cannot be reduced beyond the electric charge on an electron. In the case of US currency, the value of a $20 bill cannot be reduced beyond the value of a penny.

Amber Image Credit: Author. Atom Image Credit: Pixabay by geralt. Dollar Image Credit: Pixabay by HealthWyze. Glass Ball Image Credit: Pixabay by Moutasem. One Cent Image Credit: Pixabay by PublicDomainPictures. US Currency Image Credit: Pixabay by PublicDomainPictures.

Discrete or 'quantized' variable:
A discrete or 'quantized' variable is one that can be assigned only certain values. Electric charge is an example of a quantized variable

A nice pictorial representation of the difference between a continuous and discrete variable is shown below:

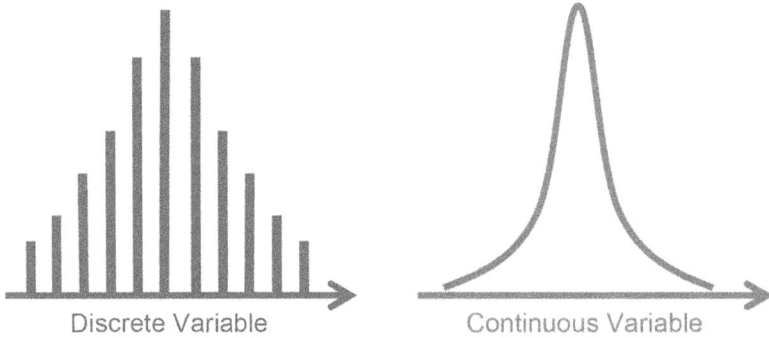

Discrete Variable   Continuous Variable

A discrete, or "quantized," variable can only assume certain values. In the case of electric charge, an object may only be electrically-charged to integer multiples of the electric charge on an electron, q_e (or $\pm 1.6 \times 10^{-19}$ C). Conversely, a continuous variable may assume any value on a continuum. In the case of mass, an object's mass may be any positive value.

Image Credit: Author.

Another distinction between mass and electric charge involves their types. Unlike mass, electric charge comes in two types and most of us know that two of the similar electric charges will exert a repulsive force on one another, while two of the unlike electric charges will exert an attractive force on one another. In other words, 'like charges repel one another, while unlike charges attract one another.' The American scientist, philosopher, and statesman, Ben Franklin adopted the conventional nomenclature used to label the two types of electric charge. He arbitrarily chose to label them 'positive' and 'negative.' If a *plastic* rod or ruler is rubbed with a cloth, it will accumulate an excess of negative electric charge. On the other hand, if a *glass* rod or ruler is rubbed with cloth, it will accumulate an excess of positive electric charge. You've probably experienced an 'electric shock' when brushing or combing your hair or when removing synthetic clothes from a dryer. Likewise, you've probably experienced a similar 'shock' when touching a metal knob after walking across a nylon carpet. These 'shocks' are excellent examples of objects becoming electrically-charged due to some type of rubbing process.

Biographical Information:
BENJAMIN FRANKLIN

BORN: January 17, 1706 (Boston, Massachusetts)
DIED: April 17, 1790 (Philadelphia, Pennsylvania)

Ben Franklin's father, Josiah Franklin, was a soaper and candlemaker who had two wives and 17 children. Josiah lived in Boston with his first wife, Anne Child, and had seven children. Following her death, Josiah married Abiah Folger in 1689. Josiah and Abiah had ten children, with Benjamin being their eighth (Ben was Josiah's 15th child overall, and 10th and final son). At age 10, Ben worked with his father as a soap boiler and by age 16, he was indentured to a printer. He was able to buy his freedom within a few years and moved to Philadelphia in 1730 where he began printing "The Pennsylvania Gazette" and in 1733, "Poor Richard: An Almanack."

Franklin began exploring the phenomenon of electricity in 1740. As a scientist, he formulated the "fluid theory" of electricity (which proved to be incorrect), first labeled electric charges as "positive" and "negative," successfully explained the Leiden Jar Effect, and first stated the conservation of electric charge. He was the first American to gain an international reputation as a scientist and was the first scientist to be recognized solely for his contributions to electricity. He is described as an expert machinist and gadgeteer. He invented the lightning rod, the rocking chair, bifocal lenses, the Pennsylvania and Franklin stoves, daylight savings, the flexible urinary catheter, and was the first scientist to correctly explain Gulfstream convection currents. Many pictures and paintings often depict Franklin standing in a storm, flying a kite, waiting for it to be struck by lightning, to demonstrate that lightning was electrical. In fact, Franklin only published a paper proposing such an experiment in his newspaper, "The Pennsylvania Gazette," on October 19, 1752 without mentioning that he had performed it. In 1767, Joseph Priestly read an account to the Royal Society mentioning that Franklin had indeed performed his kite-flying experiment, but had stood on an insulator and under a roof to avoid the danger of an electric shock. In appearance, he was described as "a natural man … simple in dress, quick with wisdom and wit, and gentle in manner." He was a womanizer and a tippler, meaning he especially liked to indulge in alcoholic beverages. He enjoyed skinny dipping in the English Channel and was once knocked unconscious while trying to kill a turkey with an electric shock. He suffered from obesity throughout his middle age and elder years which resulted in gout. He died from a pleuritic attack at his home in Philadelphia at aged 84.

Franklin's final resting place—The Christ Church Burial Ground in Philadelphia, Pennsylvania

Franklin appears on the US $100 bill.

Top image: Credit: National Portrait Gallery, Smithsonian Institute. Bottom left image: Image Credit: Author. Bottom right image: Currency Image Credit: Pixabay by benscripps.

The follow-up question that arises, of course, is '*Why* does the rubbing process charge objects electrically?' Electric charge cannot be created nor annihilated; it can only be exchanged or moved. The process of electrically charging two materials by rubbing them together depends on the electron configuration of the materials involved. Imagine two materials are brought near each other—the one on the left is labeled as an 'electron donor' because its outer electron shell is not nearly complete while the material on the right is labeled an 'electron acceptor' because its outer electron shell is nearly complete. When these two materials are brought into contact and rubbed together, some outer shell electrons may be captured by the material that is closer to filling its outer shell (i.e., the electron acceptor). This results in the atoms of one material having an excess of electrons and thus a negative (−) electric charge (i.e., the electron acceptor) and the other material missing electrons and having a positive (+) electrical charge (i.e., the electron donor).

(A.) Two neutral materials are brought near one another. The material on the left (Material #1) is an "electron donor" because its outer shell is not nearly complete while the material on the right (Material #2) is an "electron acceptor" because its outer shell is almost complete.	(B.) The two materials make contact. Outer shell electrons from material #1 transfer to material #2 ...	(C.)... to complete the outer shell of material #2.

(D.) As the electrons are transferred from material #1 to materials #2, material #1 becomes positively-charged (it has lost electrons) while material #2 becomes negatively-charged (it has gained electrons).	(E.) The two materials are now separated and they remained positively- and negatively-charged. The process of bringing the two neutral materials together and rubbing them together (i.e., friction) has resulted in an exchange of electric charge.

Styrofoam packing peanuts cling to the fur of a cat because of static electricity. The constant rubbing motion of the cat through the box of packing peanuts caused the exchange of electric charge between the cat's fur and the packing peanuts. The electric charge-exchange ultimately creates a force of attraction between the fur and the packing peanuts.

This Cat and styrofoam-electrostatic charge (235112299) image has been obtained by the author from the Wikimedia website where it was made available by Czar under a CC BY 2.0 licence. It is included within this article on that basis. It is attributed to Sean McGrath.

As we have discussed, electric charge cannot be created nor annihilated; yet can be exchanged or moved. The two types of electric charge exchange are called 'conduction' and 'induction.' Suppose a positively-charged rod **touches** a neutral rod. The free electrons in the neutral rod are attracted to the positive rod and pass over to it. Since the neutral rod is now missing some of its negative electrons, it will have a net positive electric charge. The process of electric charge exchange via **touching** is called 'conduction.'

CONDUCTION

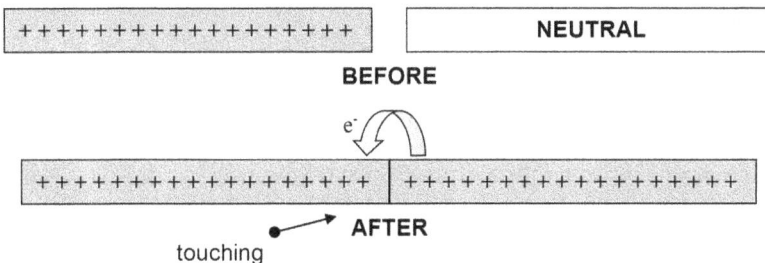

On the other hand, suppose the positive rod is brought near a neutral rod but does not actually touch it. The free electrons in the neutral rod are still attracted to the positive rod but do not pass over to it. Instead, one end of the neutral rod is left positive while the other end is left negative; the electric charge has been separated. If we now cut the rod in half, we would have two objects, one electrically-charged positively and one electrically-charged negatively. The process of electric charge exchange via *proximity* (but not touching) is called 'induction.'

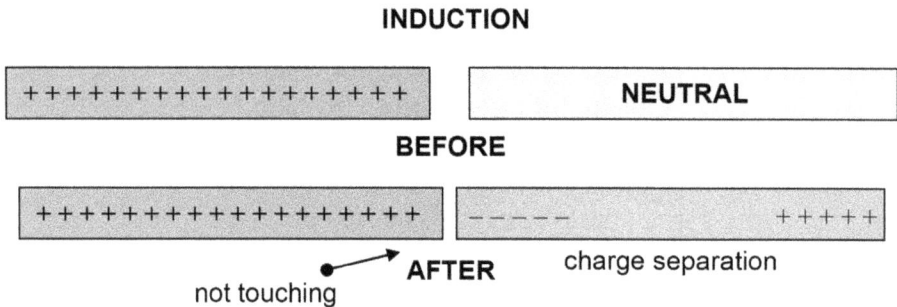

INDUCTION

+ + + + + + + + + + + + + + + + +	NEUTRAL

BEFORE

+ + + + + + + + + + + + + + + + + +	– – – – – + + + + +

not touching **AFTER**   charge separation

You may be wondering about everyday instances of electric charge. In other words, what are some examples of objects or situations which electric charge plays an important role? In the 'Closure' section of this chapter, we will perform some demonstrations, using simple household items, in which electric charge manifests itself in magnificent and striking ways; however for the present, let us just list a few quick examples. When you rub a latex balloon on your hair, the balloon becomes negatively-charged, leaving your hair with an excess of positive electric charge. As we mentioned before, the typical electric charge carried by a rubbed latex balloon is of the order of 10–100 nC (nanocoulombs). Likewise, a plastic comb, after moving through your hair, becomes negatively-charged, while your hair is left with a positive electric charge. When you walk across a carpet, friction between your shoes and the fabric of the carpet causes the transfer of electrons from the fabric to your body, leaving you negatively-charged. Then, when you touch a conductor, like a metal doorknob or metal kitchen appliance, the excess of electrons on your body quickly move to the conductor (via the process of conduction, explained above). This transfer of electrons is what you see and feel as an 'electric shock.' These everyday instances of small electric shocks are usually harmless. However, even small electric discharges can become extremely dangerous in certain situations; most notably, in the presence of flammable liquids or gases because only a small spark is needed to serve as an ignition source. Perhaps the most famous example of an electric discharge igniting a mixture of flammable gas and air was the *Hindenburg* disaster of May 6, 1937. Some tips to avoid static electric discharges include: simply wearing clothes made of natural fibers (which exchange electrons less easily), using humidifiers to increase moisture in the air (electric discharges occur less frequently during summer months when moisture is in the air), using anti-static products in your clothes dryer, or grounding yourself by touching a conductor before interacting with potentially dangerous materials.

Historical Information:

THE *HINDENBURG* DISASTER OF MAY 6, 1937

The *Hindenburg* disaster, May 6, 1937.

The final moments of the *Hindenburg*. As the bow angled upward, a blowtorch of fire erupted out of the nose.

On May 6, 1937, the German passenger zeppelin, *Hindenburg*, hovering three hundred feet in the air and held aloft by seven million cubic feet of hydrogen gas, burst into flames while preparing to dock at the Naval Air Station in Lakehurst, NJ. Amazingly, the ensuing fire consumed the massive airship in only 35 seconds!

Using eyewitness accounts to determine the origin of the fire proved confusing but the first sign of trouble appears to have been at the top, rear of the ship, just in front of the vertical fin. Both R. H. Ward (stationed with the port bow landing party) and R. W. Antrim (stationed atop the mooring mast) testified that they noticed a fluttering of the ship's outer cover at this location—suggesting hydrogen was leaking out of a rear interior bladder against the outer covering. Crewmen in the control stations of the lower fins testified hearing "muffled detonations" near the top of the ship. When they looked up, they saw bright red and yellow reflections of fire. By 7:25 pm, a yellow flame appeared on the outside of the ship at this spot. Within seconds, the tail section was engulfed in flames. The ship managed to stay afloat for a few seconds but eventually, the tail section sank, slamming crew and passengers 15-20 ft backwards into the rear walls of the control room, cabins, dining lounge, and promenade. As the *Hindenburg* tilted upward, the fire traveled inside the ship along the central axis until a blowtorch of fire erupted from the nose. Crewmen stationed in the bow were incinerated.

Subsequent investigations by the US and Germany were inconclusive in determining the cause of the fire. For years, scientists, politicians, and military personnel put forth several theories as to the underlying causes of the disaster. Was it sabotage? No evidence of sabotage was ever found. Was it a lightning strike? Unlikely—the outer covering of the ship had several burn-holes, some as large at five centimeters in diameter, proving the ship had survived in-flight lightning strikes during its first year of service. Today, a re-examination of the evidence strongly points to static electric discharge as the fire's source of ignition. As the *Hindenburg* passed through a storm off the New Jersey coast, friction between the air and the ship caused it to become electrically-charged. When the landing lines touched the ground prior to docking, they "earthed" the *Hindenburg's* steel frame, but not every panel of the ship's fabric covering. A spark between the electrically-charged panel of fabric and the grounded steel frame ignited some source of fuel.

So what was the source of fuel? The most likely explanation of is that the electrostatic discharge ignited leaking hydrogen gas. Experts agree that the ship was undoubtedly leaking hydrogen from the stern. What caused the leak? One theory suggests that just before docking, the ship executed an S-turn that was uncommonly tight and that one of the rudder's bracing cables may have been over-stressed to the point where it snapped and slashed through a gas cell. Or ... maybe something as simple as a sticky valve was at fault. Regardless of the cause of the leak, an explosive mixture of hydrogen gas and air floated above the ship's tail. In the aftermath, 35 of 97 people onboard died (13 passengers and 22 crewmen) plus 1 member of the ground crew.

Left image: Credit: National Archive. Right image: This Hindenburg disaster, 1937 image has been obtained by the author from the Wikimedia website, where it is stated to have been released into the public domain. It is included within this article on that basis.

4.3.1.3 Coulomb's force law

In volume 2, chapter 1, we examined Newton's discovery of the 'Universal Law of Gravitation,' which we abbreviated as the 'ULG.' The ULG determines the force of gravity between two masses, m_1 and m_2, separated by a distance r. As you may recall, unlike his laws of motion, Newton did not provide us with a simple, all-encompassing Latin phrase in the *Principia* to summarize his understanding of the ULG. Therefore, rather than presenting a Latin statement and providing a translation, we summarized the ULG in our own words:

Newton's Universal Law of Gravitation:
Every particle of matter in the universe attracts every other particle with a force that is directly proportional to the product of the masses of the particles and inversely proportional to the square of the distance between them.

We worked through this sentence to give a mathematical expression to the law:

$$\text{ULG: } \vec{F_g} = \frac{G \times m_1 \times m_2}{r^2}\hat{r} = \frac{(6.673 \times 10^{-11}) \times m_1 \times m_2}{r^2}\hat{r}$$

We now seek to solve the same problem, but between two stationary electric charges. In other words, we seek to find the electric force between two electric charges at rest, q_1 and q_2, separated by a distance r. Attempts to solve this problem date as far back to 600 BC when Thales of Miletus recorded that rubbing pieces of amber could lift small objects like animal fur or clippings of straw, thus giving us the etymology of the word 'electricity' (see the front matter of this series on *Classical Mechanics*). Scientists such as Allesandro Volta (inventor of the Voltaic pile ... the first electrical battery), Joseph Priestly (discoverer of oxygen), John Robinson (inventor of the siren and co-inventor of the steam car), and Henry Cavendish (who used the torsion balance to determine the universal constant of gravitation, G), all conducted experiments to determine the force of attraction/repulsion between two electrically-charged objects. Fortunately for us, French physicist Charles-Augustin de Coulomb published a series of papers in 1785 and settled the debate. Using a torsion balance, Coulomb determined (and we are using our own words here) that:

Coulomb's law:
The electric force between two electric point-charges is directly proportional to the product of the two electric charges and inversely proportional to square of the distance between them.

As we did with Newton's ULG, to understand Coulomb's law and to use it as a problem-solving instrument, we must translate Coulomb's discovery into an equation and carefully define what each word means. Since mathematics is just a concise symbolic language, we must ensure that the relations described by our syntax are embodied within our equation. Let us review our statement phrase-by-phrase so that Coulomb's discovery may guide us in the formation of our final equation. After every few words, we'll be able to add another symbol to our evolving equation:

We start with the phrase: 'The electric force between two electric point-charges ...'

- First, we need to define exactly what Coulomb's law gives us. In other words, 'What will Coulomb's law compute for us when we use it?' Coulomb's law determines **a force**. In volume 1, chapter 4, we saw that Newton devoted an entire law of motion, law #3, to precisely define what a force is: 'A force is any interaction between two bodies.' This is a crucial concept to understanding Coulomb's law; namely, it calculates the force between *only* two stationary objects with electric charge (i.e., two electric point-charges). If we want to calculate the interactions between more than two objects, we must use Coulomb's law between pairs of objects, then sum the results as vectors (i.e., we must use the principle of superposition).

- Also from volume 1, chapter 4, we know force is a vector and therefore two bits of numeric information are required to completely specify a force. We will write the force in vector notation, $\vec{F} = (F_x, F_y)$, and add or subtract our results according to our rules of vector addition and subtraction. Combining these first two interpretations we obtain:

$$\text{Coulomb's Law:} \quad \vec{F} = ?$$

Notice that we include the small arrow in our evolving equation to symbolize the vector nature of a force.

- We label this force as the 'Coulombic force' to honor Coulomb's work in the field of electrostatics. To update our evolving equation, let us add a subscript 'C' to our symbol for force to denote 'Coulombic':

$$\text{Coulomb's Law:} \quad \vec{F_C} = ?$$

- The Coulombic force can be attractive or repulsive, but is always directed along the line connecting the centers of the two objects. Masses pull each other together, they never push each other apart; however, as we all know, 'unlike charges attract one another (i.e., they exert an attractive force on one another) while like charges repel one another (i.e., they exert a repulsive force on one another).' Let us update our evolving equation by adding a symbol that reminds us that the Coulombic force is always directed along the line connecting the centers of the two electric point-charges. The symbol that physicists use to specify the direction is r and is spoken as '*r hat.*'

$$\text{Coulomb's Law:} \quad \vec{F_C} = ? \; \hat{r}$$

Notation:

We mentioned this previously in volume 2, chapter 1, but it is worth repeating. Scientists and engineers often use the '∧' (called a 'hat') to symbolize a pure direction. Traditionally, the '∧' is called a 'caret' or 'circumflex' and can be typed from the SHIFT character above the number-6 on a standard keyboard. In general, a hat only serves to specify the direction of a vector.

For example, $\vec{A} = 3\hat{x}$ miles (spoken as '3 x-hat miles'), translates into '3 miles along the x-axis' or, using our vector notation, $\vec{A} = (3, 0)$ miles.

As another example, $\vec{B} = 4\hat{x} + 5\hat{y}$ Newtons (spoken as '4 x-hat and 5 y-hat Newtons') translates into '4 Newtons along the x-axis and 5 Newtons along the y-axis,' or $\vec{B} = (4, 5)$ Newtons.

When you read our equation for Coulomb's law, the $\hat{r}$ simply translates to the following words: 'attractive or repulsive along the direction of the line joining the two electric point-charges.'

Let us continue to dissect our rendition of Coulomb's law that continues with the phrase: 'The electric force between two electric point-charges *... is directly proportional to the product of the two electric charges and ...*'

- Now Coulomb's law begins to take on some computational meat. The value or magnitude of the force, according to Coulomb's discovery, is proportional to the product of the two electric charges:

Coulomb's Law: $\vec{F_C} \propto (|q_1| \times |q_2|)\hat{r}$.

Notice that since electric charge can be positive or negative, we use absolute value signs in the formulation of Coulomb's law. This is done to avoid any confusion that negative signs associated with negative electric charges may introduce. Using absolute value signs in the formulation of Coulomb's law means that direction of force need only be calculated in the $\hat{r}$ portion of the formula. For example, if a positive electric charge and negative electric charge produce a Coulombic force directed into the first quadrant (requiring positive x- and y-components of force), the negative electric charge will not introduce an extra minus-sign into the calculation resulting in negative x- and y-components of force. At this point, you would verbalize this proportion as: 'The Coulombic force is proportional to the product of the absolute value of q_1 and the absolute value of q_2, and attractive or repulsive along the direction of the line joining the two electric point-charges.'

We conclude our rendition of Coulomb's law with the phrase: *'The electric force between two electric point-charges is directly proportional to the product of the two electric charges ... and inversely proportional to square of the distance between them.'*

- We continue the translation of Coulomb's law by including a symbol to represent the inverse square relationship of Coulomb's law with respect to separation distance. For instance, if you INCREASE the separation distance between two electric point-charges by a factor of 4, the electric force DECREASES by a factor of 4^2, or 16. The relationship is inverse (increasing distance, decreases force) and squared (changing distance by 4 changes force by 4^2 or 16)—amazingly, this is exactly as gravity behaves!

Coulomb's Law: $\vec{F}_C \propto \dfrac{(|q_1| \times |q_2|)}{r^2}\hat{r}.$

You would verbalize this proportion as: 'The Coulombic force is proportional to the product of the absolute value of q_1 and the absolute value of q_2, inversely proportional to the square of the distance between them, and directed along the line joining the two electric point-charges.'

Also at this time, we emphasize a critical notational convention used in Coulomb's law. The r appearing in the denominator of Coulomb's law denotes the distance between the **centers** of electric point-charges q_1 and q_2. We conceptualize any two electric charges as 'electric point-charges,' meaning that we can collapse their shapes to that of a point and treat the two electric charges as if all of their electric charge were concentrated at their centers. In effect, replace any object with a particle of equal electric charge located at its center. For example, if we separate two Styrofoam balls each of radius 0.5 m, by 4 m, the actual distance between their centers is: $4 + 0.5 + 0.5 = 5$ m. The situation is shown below:

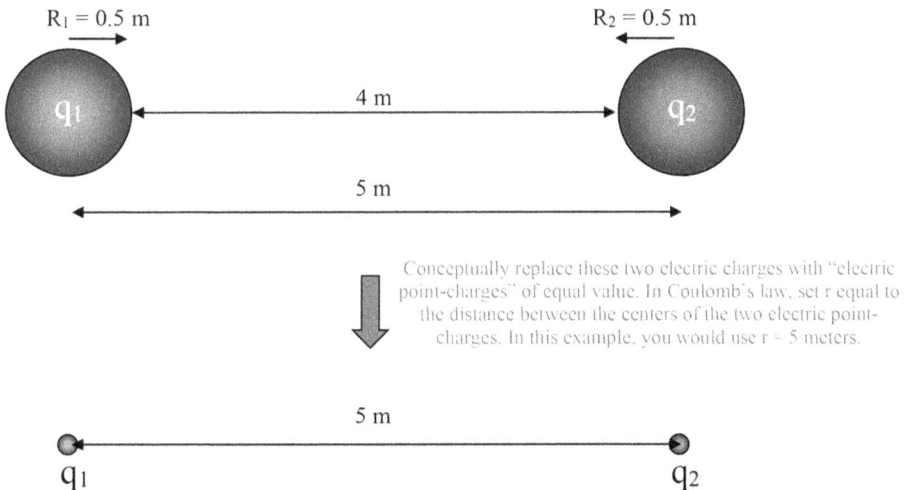

R$_1$ = 0.5 m   R$_2$ = 0.5 m

q$_1$ 4 m   q$_2$

5 m

Conceptually replace these two electric charges with "electric point-charges" of equal value. In Coulomb's law, set r equal to the distance between the centers of the two electric point-charges. In this example, you would use r = 5 meters.

5 m

q$_1$ q$_2$

- The final piece to writing Coulomb's law as an equation is to convert the proportionality into an equality. In general, any proportion can be converted into an equation by simply including a constant of proportionality—a fixed number that appropriately scales the proportionality to make the units work. In our case, the constant of proportionality is symbolized by 'k' and is called 'Coulomb's Constant.' The value of 'k' and how it was numerically determined is described in the next section.

Putting everything together, we get the final mathematical form of Coulomb's law ... and it will look striking similar to Newton's ULG. You would verbalize Coulomb's law as: 'The electric force equals some constant (i.e., k) times the product of the absolute value of q_1 and the absolute value of q_2, divided by the square of the distance between q_1 and q_2, and is directed along the line joining the two electric point-charges.'

$$\text{Coulomb's Law: } \vec{F}_C = \frac{k \times |q_1| \times |q_2|}{r^2} \hat{r}$$

4.3.1.4 Determining Coulomb's constant, k

In volume 2, chapter 1, we stated that:

Operational definition:
Newton's second law is chosen to be the operational definition of a force and the basic **m.k.s.** unit of force, the 'Newton.' Therefore, a Newton is **defined** to be the force of 1 kg accelerating at a rate of 1 m s^{-2}.

Thus, every subsequent equation (more will follow in later courses in physics) involving a force must include a factor, a constant of proportionality, to scale that equation to agree with the operational definition of force.

Therefore, like the ULG, we expect Coulomb's law to contain a factor—a constant of proportionality—to scale the equation to agree with our operational definition of the m.k.s. unit of a 'Newton.' Unlike the ULG, where G was a very, very, very small number (i.e., $G = 6.67 \times 10^{-11} \text{ N} \cdot \text{m}^2 \text{ kg}^{-2}$), in the case of Coulomb's law, we expect k to be a very, very, very large number. Recall when we first introduced the unit of 'Coulombs.' We said that 1 Coulomb was equivalent to the electric charge carried in a lightning strike—that's a huge amount of electric charge! Now imagine plugging unit values (i.e., simple values of 1) into every term of Coulomb's law. If we separated an object possessing 1 Coulomb of electric charge, from another object possessing 1 Coulomb of electric charge, by only 1 m, then we would expect a huge amount of Coulombic force to exist between the two objects. Thus, the scaling factor in Coulomb's law must be a very, very, very large number. In other words, since we used Newton's second law as our definition of force so that 1 Newton is equal to 1 kg of mass accelerated at 1 m s^{-1}-squared, then we must include an 'increasing scaling factor' in Coulomb's law that guarantees two 1 Coulomb electric point-charges separated by 1 m results in a very ... very ... very large force!

As discussed in volume 2, chapter 1, the value of G was experimentally determined using a device known as a 'torsion balance.' This device was also used by Coulomb to determine the value of k, Coulomb's constant. In Coulomb's version, he hung a bar (the 'inverted T') from a thin fiber which in was mounted near a curved scale. Two small metal-coated, spherical balls were attached to the ends of the inverted T.

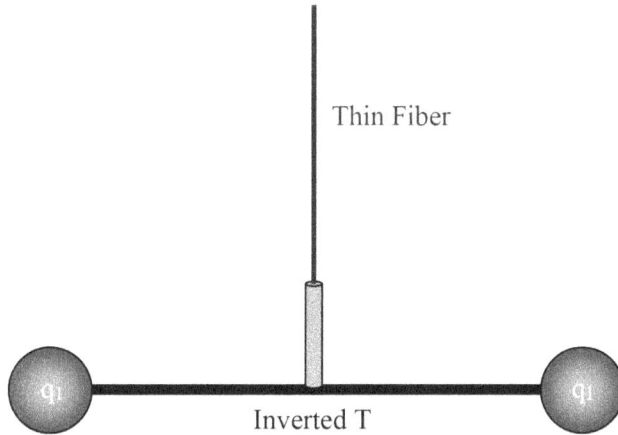

Thin Fiber

q_1 Inverted T q_1

To operate the torsion balance, Coulomb carefully electrically-charged the metal balls with a known value of static electric charge and then brought a second ball of the same polarity, near one of the metal balls. The two metal balls repelled one another and drove each other apart. He then measured the displacement of the inverted T along the scale. By knowing how much force was required to twist the thin fiber through a known angle (not an easy task), he was able to determine the force between the two electrically-charged metal balls and thus determine the constant of proportionality.

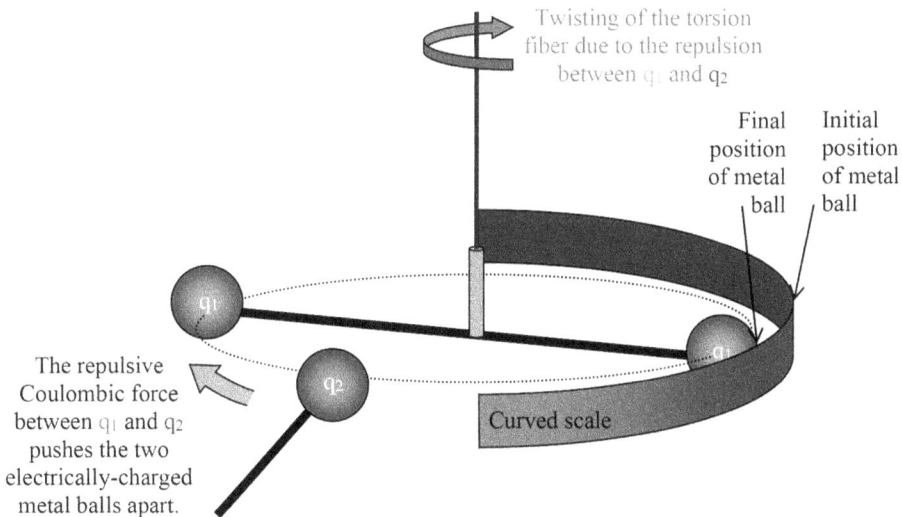

Twisting of the torsion fiber due to the repulsion between q_1 and q_2

Final position of metal ball

Initial position of metal ball

The repulsive Coulombic force between q_1 and q_2 pushes the two electrically-charged metal balls apart.

q_1

q_2

Curved scale

The currently accepted value for k is:

$$k = 8.987\ 551\ 7862 \times 10^{+9}\ \text{N} \cdot \text{m}^2\,\text{C}^{-2} \approx 9 \times 10^{+9}\ \text{N} \cdot \text{m}^2\,\text{C}^{-2}.$$

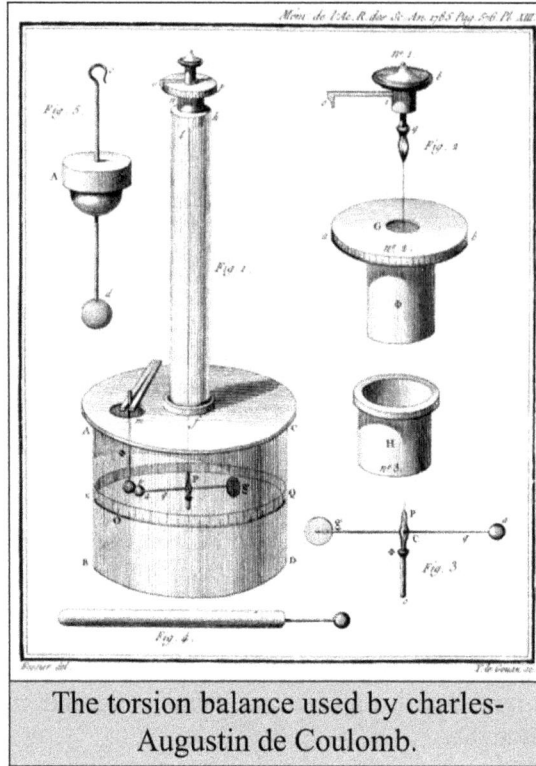

The torsion balance used by charles-Augustin de Coulomb.

This Coulomb image has been obtained by the author from the Wikimedia website, where it is stated to have been released into the public domain. It is included within this article on that basis.

We can now re-state Coulomb's law and translate it into a concise equation:

The electric force between two electric point-charges is directly proportional to the product of the two electric charges and inversely proportional to square of the distance between them.

$$\Downarrow$$

$$\textbf{Coulomb's Law: } \vec{F}_c = \frac{k*|q_1|*|q_2|}{r^2}\hat{r} = \frac{(9 \times 10^{+9})*|q_1|*|q_2|}{r^2}\hat{r}$$

Armed with a mathematical representation of Coulomb's law, we are now in a position to discuss the electric field and then add Coulomb's force law to our arsenal of problem-solving strategies.

4.3.1.5 The electric field

If you recall from volume 2, chapter 1, when we discussed Newton's Universal Law of Gravitation, we considered the question: 'How do two masses communicate with one another?' In other words, we can use the ULG to determine the gravitational force of attraction between two objects, but how does one object 'know' the other object exists? Do they 'see' each other? Do they somehow 'talk' to one another? Maybe one object transmits a signal that says, 'Calling all other objects ... I'm here!' What mechanism does gravity utilize so that masses can interact with one another? Ultimately, we found that masses exerted their force of gravity upon one another 'instantly' and 'over an infinite distance.' They do not need a medium, nor do they need to be in contact with each other, in order to exert the force of gravity upon one another. One way to summarize our findings is to state that masses interact via 'action at a distance'—it is the simple concept that an object with mass can be instantly affected by another object with mass even though those two objects are not in physical contact. This is indeed a simple concept to state in words ... but a difficult concept to grasp intellectually ... so difficult that Newton himself had difficulty wrapping *his* head around this idea! Can you imagine two objects being able to communicate with one another **instantly**, and **at infinite distances?** Therefore, we turned to British physicist Michael Faraday and his model of the 'gravitational field' in order to explain 'action at a distance.' This model of the 'field' is the classical interpretation of how physicists explain forces, such as gravity, which act as a distance with no contact between two objects. The idea suggests that a mass exists with a 'field' everywhere and at all times. This field fills space and interacts with other masses, causing them to experience forces even when they are not in direct contact with one another.

We now turn our attention to electric charges. How do *they* transmit the Coulombic force of attraction or repulsion between them? For purposes of this discussion, let us assume that we are dealing with two electric point-charges of opposite polarity (so that the electric charges attract one another). Therefore, how do two electric point-charges, q_1 and q_2, separated by a distance r, communicate with one another **instantly**, and **at infinite distances?**

We know from Newton's third law of motion that the Coulombic force will be equal in magnitude on both electric charges but opposite in direction—in the situation shown below, q_1 will be pulled to the right with a force of magnitude $(k \times |q_1| \times |q_2|)/r^2$ while q_2 will be pulled to the left with a force of the same magnitude:

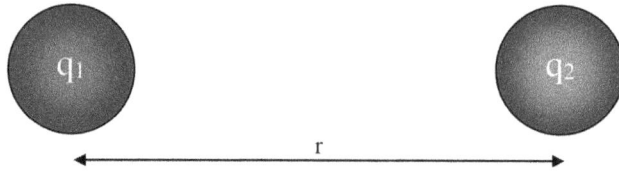

The question we now consider is how the Coulombic force (as given by Coulomb's law) is communicated between these two masses. How does q_1 'know' q_2 exists and likewise, how does q_2 'know' that q_1 exists? Let us consider several possible explanations:

1. One possible explanation is that the electrostatic force is a 'contact force,' such as the normal force or the frictional force. However, our everyday experience tells us this is not true. For instance, rub a latex balloon against your sweater and hold the balloon a few inches **above** some pieces of paper confetti lying on your kitchen table. Without touching the confetti, you'll see the balloon lifting the confetti off the table. Two electric charges need not be in contact in order for the Coulombic force to exist between them. Therefore, we conclude the electrostatic force is not a contact force. Let us try something else.

2. How about action through a medium? Perhaps a 'medium' must exist between two electric charges in order for them to be able to communicate. In other words a medium, such as air ... or water ... or a gas ... or a special liquid, must exist between the two electric charges in order for the electrostatic force to exist. Using the example described in the previous paragraph, would an electrically-charged latex balloon be able to pick up confetti from the surface of the Moon, where no atmosphere exists? Trust me when I say that electric charges do not need a medium to communicate. The Coulombic force exists between two electric point-charges in a total vacuum. Let us try something else.

3. Having excluded the possibilities of a 'contact force' and 'action through a medium,' we are left with our good ole friend—the concept of 'action at a distance.' The idea suggests that an electric charge exists with an 'electric field' everywhere and at all times. This electric field fills space and interacts with other electric charges, causing them to experience Coulombic forces even when they are not in direct contact with one another.

Just like with the gravitational field, the electric field explains the Coulombic attraction or repulsion between two electric charges by invoking a two stage process: (i) Each electric charge (every electric charge in the universe—for our purposes, let us call this electric charge q_1) carries with it an invisible *'electric field'* that exists everywhere and at all times. The field did not 'grow' from q_1. The field exists as part of q_1. In other words, electric charges carry fields with them. If q_1 moves, its field moves instantly everywhere with it. (ii) Another electric charge q_2 located at point (P.) a distance r from q_1, interacts with the electric field from q_1 and therefore experiences the Coulombic force given by Coulomb's law:

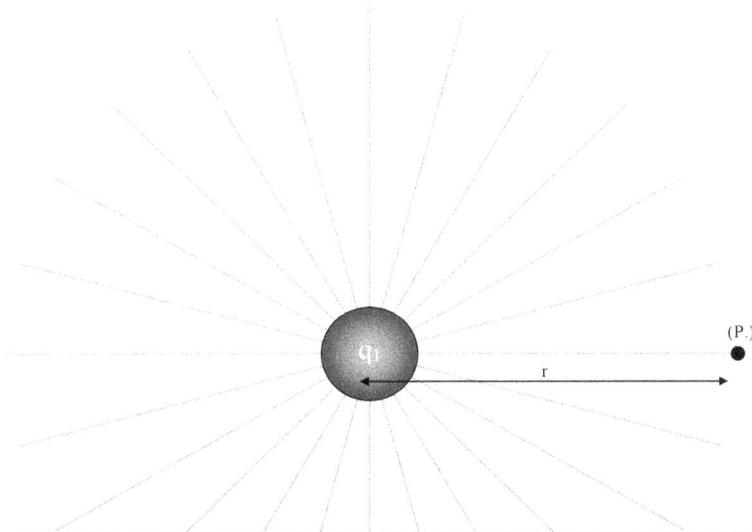

STEP #1:

An electric charge q_1 (q_1 could be any electric charge in the universe-an electron, a latex balloon, a packing peanut, etc.) carries an invisible "electric field," which, even though it may actually be invisible, we depict with the green dashed lines. This field exists "everywhere," meaning the green dashed lines should extend well off the paper, above the paper, below the paper, etc. We can't possibly draw all of the field lines so we only draw a representative few. Also, the field exists "at all times," meaning the green dashed lines were always present and always will be present-they are part of q_1.

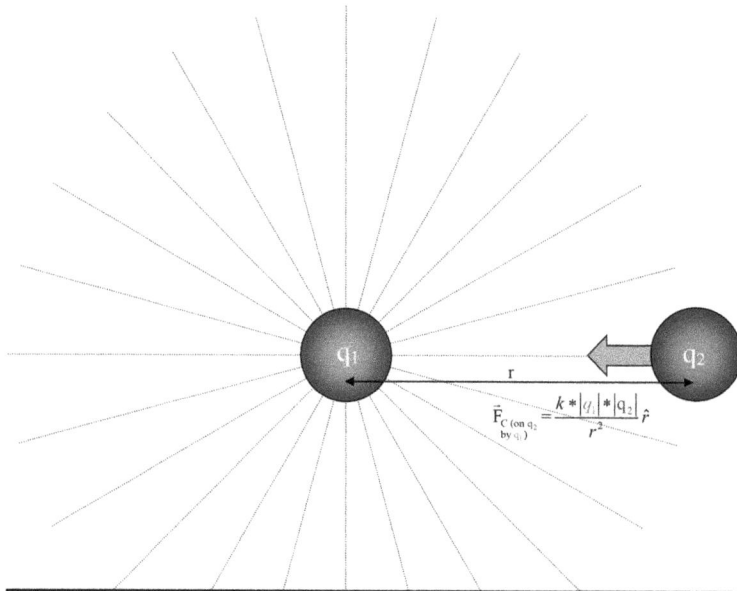

$$\vec{F}_{C \text{ (on } q_2 \text{ by } q_1)} = \frac{k * |q_1| * |q_2|}{r^2} \hat{r}$$

STEP #2:
Another electric charge q_2 (q_2 could also be any electric charge in the universe-an electron, a latex balloon, a packing peanut, etc.) now is located at point (P.) which we said is a distance r from q_1. This second electric charge q_2, is "caught" by q_1's electric field and therefore experiences the Coulombic force given by Coulomb's law.

Recall from volume 2, chapter 1 the analogy of the fly being captured by the spider. That analogy also holds for electric charges. Imagine a fly and a spider are separated by some distance r—in this analogy, the spider represents q_1 and the fly represents q_2. The question we wish to answer is 'How is the spider able to capture the fly and exert a force on it?' Very similar to the concept of an electric field, the spider invokes a two stage process: (i) the spider carries with it an invisible 'web' that exists everywhere and at all times, and (ii) the fly interacts or 'gets caught' by this web and is pulled by the spider.

4.3.1.6 Calculating the electric field
The recipe that we use to calculate the electric field is almost identical to that used to calculate the gravitational field; there is only one small, but important difference. As we develop this two-step recipe, keep in mind that the electric field, just like the gravitational field, is vector quantity so we have to be careful and use vector notation as well as the rules of vector addition and subtraction. The general recipe is outlined below:

Problem: Consider an electric charge Q located at the origin of some coordinate system. A point (P.) is located a distance r, from the electric charge. Calculate the electric field of the electric charge Q at the point (P.)

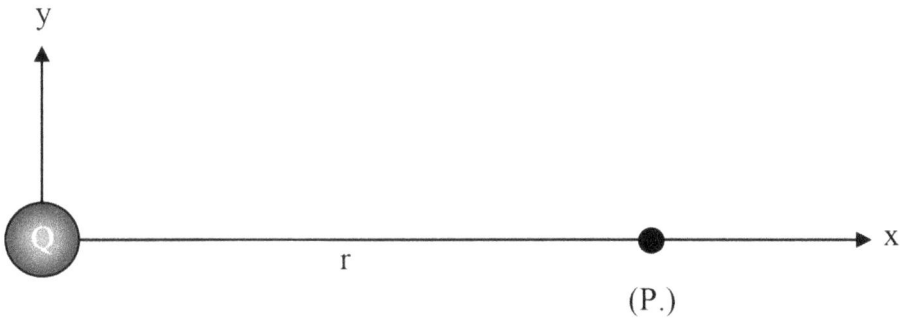

Recipe:

i. Drop off a small *positive* electric 'test-charge' at the point (P.) and calculate the Coulombic force exerted on that electric test-charge, by Q, using Coulomb's law. The positive electric test-charge can have any magnitude, but it must be positive! Here lies a crucial difference between calculating the electrical field versus calculating the gravitational field. In the case of calculating the gravitational field, our first step was to place a test-mass at point (P.). Since mass only comes in one type (i.e., we don't have such things as 'positive mass' or 'negative mass' ... we just have 'mass'), we didn't have to worry about whether our test-mass was a 'positive' or 'negative' type of mass. In the case of calculating the electric field, our first step is to place an electric test-charge at point (P.). However, electric charge does indeed come in two types—positive and negative—so we have to agree on a convention to always drop off a positive electric test-charge to sample the electric field at our point of interest. We like to think of this step as placing a small electric point-charge at point (P.) then calculating the Coulombic force on that small electric point-charge ... but always remember that the electric field of a charge Q *is determined for its interaction with a positive electric test-charge.*

$$\vec{F}_{C \text{ (on } q_{\text{test}} \text{ from } Q)} = \frac{k * |Q| * |q^+_{\text{test}}|}{r^2} \hat{r} \text{ Newtons}$$

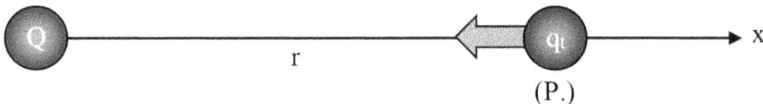

ii. Now remove the positive electric test-charge by dividing your answer by the value of the positive electric test-charge. As you perform this division, your answer will change from a 'force' (symbolized by $\vec{F}$) to a 'field' (symbolized by $\vec{E}$) and the units will change from 'Newtons,' to 'Newtons/C.'

$$\vec{E} = \frac{\vec{F}_{C \ (\text{on } q^-_{test} \text{ from } Q)}}{q^+_{test}} = \frac{k * |Q|}{r^2} \hat{r} \ \frac{\text{Newtons}}{C}$$

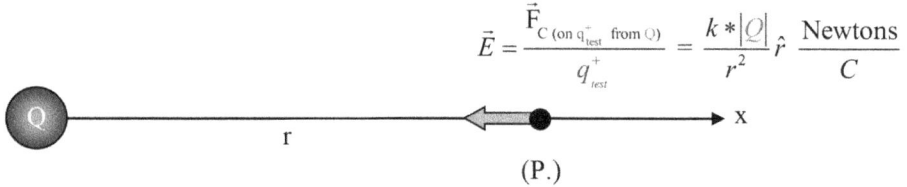

(P.)

4.3.2 Modeling

❶ **EX 1:** A point-sized positive electric charge, $q_1 = +3 \times 10^{-4}$ C is placed at the origin. A second point-sized negative electric charge, $q_2 = -4 \times 10^{-4}$ C is placed at the location (8,0) m. No other electric charges are near these two electric charges. Calculate the Coulombic force of q_2 on q_1 and the Coulombic force of q_1 on q_2.

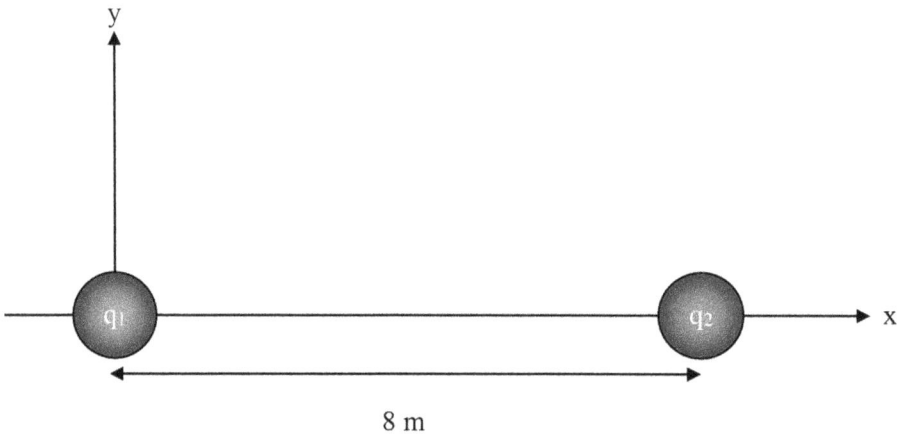

8 m

☑ **ANSWER:** Coulomb's law computes a force. Since all electric charges are given in Coulombs and all distances in meters, we can simply plug the numbers into Coulomb's law and interpret the results:

$$\vec{F}_C = \frac{k * |q_1| * |q_2|}{r^2} \hat{r}$$

$$= \frac{\left(9 \times 10^{+9}\right) * 3 \times 10^{-4} * 4 \times 10^{-4}}{8^2} \hat{r}$$

$$= 16.9 \ \hat{r} \ \text{Newtons}.$$

Plug in the numbers but keep $\hat{r}$ separate— $\hat{r}$ does not enter into any of the computations; instead, it is only a reminder that the Coulombic force is a vector so our final answer must be written in vector notation.

However, we've been a bit cavalier about interpreting the 'directional reminder' $\hat{r}$ and writing our answer as a vector. Recall as we translated our rendition of Coulomb's law into equation form, we included the $\hat{r}$ to remind us that the Coulombic force can be *attractive or repulsive along the direction of the line joining the two particles.* In this example, the x-axis is the line that joins the two particles. Since we know two charges of opposite polarity **always attract one another,** q_1 is *attracted toward* q_2 (i.e., along the **positive** x-axis) while q_2 is *attracted toward* q_1 (i.e., along the **negative** x-axis)—this is an action/reaction pair according to Newton's third law of motion. Therefore, we must write the force on q_1 as 16.9 Newtons **to the right** and the force on q_2 as 16.9 Newtons **to the left.** Thankfully, we have vector notation that allows us to mathematically keep track of directions.

$$\vec{F}_{C\,(\text{on } q_1 \text{ from } q_2)} = (16.9, 0) \text{ Newtons} \qquad \vec{F}_{C\,(\text{on } q_2 \text{ from } q_1)} = (-16.9, 0) \text{ Newtons}$$

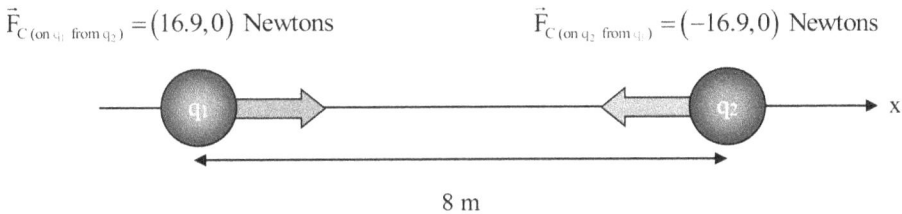

8 m

☺ **COMMENTS:** Notice that our answer for force is a vector and therefore must have two bits of numeric information:

$\vec{F}_{C\,(\text{on } q_1 \text{ from } q_2)} = (16.9, 0) \text{ Newtons}$ ⎧ This answer contains 2 pieces of numeric information: (i.) a force directed to the right, and (ii.) no force directed up or down.

$\vec{F}_{C\,(\text{on } q_2 \text{ from } q_1)} = (-16.9, 0) \text{ Newtons}$ ⎧ This answer contains 2 pieces of numeric information: (i.) a force directed to the right, and (ii.) no force directed up or down.

Also, notice the function of the directional reminder $\hat{r}$. At no point in our **numeric** calculation does $\hat{r}$ play a role. To compute the numeric value of the gravitational force, we only need to employ the formula: $F_C = (k \times |q_1| \times |q_2|)/r^2$. However, $\hat{r}$ **symbolically** reminds us that we must write the Coulombic force as a vector—and the direction of this vector is up for debate! Sometimes the direction is attractive along the line joining the two masses (when the two electric charges under consideration are of different polarity) and sometimes the direction is repulsive along the line joining the two masses (when the two electric charges under consideration are of the same polarity). Essentially, $\hat{r}$ is just a short-handed symbol that constantly screams at us: 'Hey, don't forget to write your answer in vector notation!'

❷ **EX 2:** Let's revisit the previous problem but solve it using a different approach. A point-sized positive electric charge, $q_1 = +3 \times 10^{-4}$ C is placed at the origin.

a. Calculate the electric field from q_1 at the location (8,0) m.
b. Armed with the value of the electric field at the location (8,0) m, drop off a second point-sized negative electric charge, $q_2 = -4 \times 10^{-4}$ C at the location (8,0) m. Calculate the Coulombic force of q_1 on q_2.

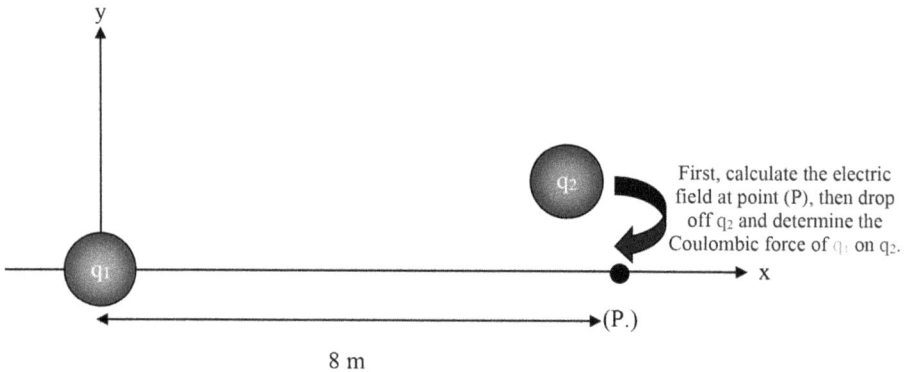

First, calculate the electric field at point (P.), then drop off q_2 and determine the Coulombic force of q_1 on q_2.

8 m

☑ **ANSWER:** Let us place a positive electric test-charge at point (P.), determine the net electric force on that electric test-charge based on our formula for Coulomb's law, then divide that force by the value of the electric test-charge we placed at point (P.). We're free to place **any** electric test-charge at point (P.) as long as we divide it out as the last step in our recipe. The easiest choice is to put a +1 Coulomb electric test-charge at point (P.), even though that is indeed a huge amount of electric charge:

a. Start by place +1 C at point (P.) and calculating the Coulombic force on that charge from q_1:

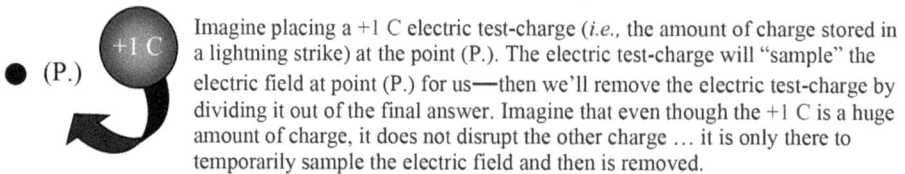

Imagine placing a +1 C electric test-charge (*i.e.*, the amount of charge stored in a lightning strike) at the point (P.). The electric test-charge will "sample" the electric field at point (P.) for us—then we'll remove the electric test-charge by dividing it out of the final answer. Imagine that even though the +1 C is a huge amount of charge, it does not disrupt the other charge ... it is only there to temporarily sample the electric field and then is removed.

$$\vec{F}_{C \text{ (on +1 C from } q_1)} = \frac{\left(9 \times 10^{+9}\right) * 3 \times 10^{-4} * 1}{8^2} \hat{r}$$

$$= 4.22 \times 10^{+4} \, \hat{r} \text{ Newtons}$$

$$= \left(4.22 \times 10^{+4}, 0\right) \text{ Newtons}$$

+1 C is repelled away from q_1 along the positive x-axis. Thus, in vector notation, the Coulombic force is direction along the positive axis.

Now divide the net force by 1 Coulomb, the value of our electric test-charge. *You see why we chose to use +1 C as our electric test charge ··· because our numeric value does not change—only the units of our answer! Instead of a force, measured in 'Newtons,' we now have an electric field, 'measured in 'Newtons per Coulomb.'*

$$\vec{E} = \frac{\vec{F}_{C(\text{net on } +1\,C)}}{+1\ C} = \frac{(4.22 \times 10^{+4},\ 0)\ \text{Newtons}}{+1\ C} = (4.22 \times 10^{+4},\ 0)\ \text{Newtons}\ C^{-1}.$$

b. Now armed with the electric field, we place $q_2 = -4 \times 10^{-4}$ C at the location (8,0) m. Calculate the Coulombic force of q_1 on q_2.

$$\vec{F}_{C(\text{on } q2 \text{ from } q_1)} = q_2 \times \vec{E}_{(\text{at (P.) from } q_1)}$$
$$= (-4 \times 10^{-4}\ C) \times (4.22 \times 10^{+4},\ 0)\ \text{Newtons}\ C^{-1}$$
$$= (-16.9,\ 0)\ \text{Newtons}.$$

☺ **COMMENTS:** Notice that the value of the electric test-charge is unimportant to the calculation; you can make it any value that you like, as long as you make it positive and divide your final answer by that same number. Also, notice that the units of the electric field are 'Newtons/Coulomb' and suggest interpreting the electric field as a 'force per unit of electric charge.' In the previous example, we determined that the electric field at the point (8,0) m was $\vec{E} = (4.22 \times 10^{+4},\ 0)$ Newtons C^{-1}. This value suggests that every Coulomb of electric charge located at the point (8,0) m, experiences a pull of $4.22 \times 10^{+4}$ Newtons to the right. We can now ask: 'What would the force be on 2 Coulombs of electric charge located at that point?' Obviously, 2 Coulombs of electric charge would experience $2 \times 4.22 \times 10^{+4}$ Newtons or 84 400 Newtons to the right. Likewise, 3 Coulombs of electric charge would experience a force of $3 \times 4.22 \times 10^{+4}$ Newtons or 126 600 Newtons to the right. Once the electric field is determined at a point (P.), we can easily multiply our answer by an electric charge to determine the force acting on that electric charge. Think of the electric field as a nifty vector quantity that tells us the **force per unit of electric charge** at a point (P.)—we can then simply multiply the electric field by an electric charge to determine the actual force acting on that electric charge.

Concept Map:

$$\vec{E} \xrightarrow{\;*\;q\;} \vec{F}_C$$

❸ **EX 3:** Let's add to the complexity of our charge configuration by adding an additional charge to the set-up. A point-sized positive electric charge, $q_1 = +3 \times 10^{-4}$ C is placed at the origin. A second point-sized negative electric charge, $q_2 = -4 \times 10^{-4}$ C is placed at the location (8,0) m. Finally, a third charge, $q_3 = +5 \times 10^{-4}$ C is placed at the location (0,6) m. No other electric charges are near these three electric charges.

    a. Calculate the net Coulombic force on q_1.

    b. Calculate the net Coulombic force on q_2.

    c. Calculate the net Coulombic force on q_3.

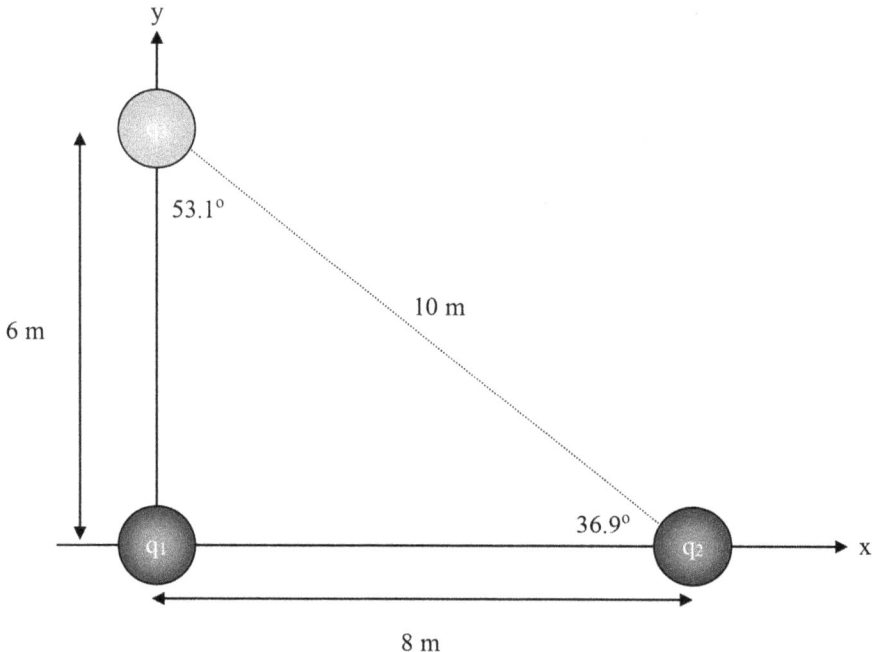

☑ **ANSWER:** Let us plug the numbers into Coulomb's law, $F_C = (k \times |q_1| \times |q_2|)/r^2$, then assign directions based on our knowledge of vector notation.

a. Calculate the net Coulombic force on q_1.

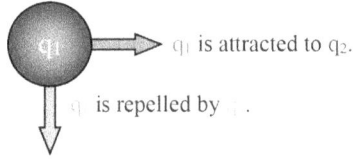

q1   →   q₁ is attracted to q₂.

↓   q₁ is repelled by q₂.

$$\vec{F}_{C\,(on\ q_1\ from\ q_2)} = \frac{\left(9 \times 10^{+9}\right)*3\times10^{-4}*4\times10^{-4}}{8^2}\,\hat{r}$$

$$= 16.9\,\hat{r}\ \text{Newtons}$$

$$= \left(16.9, 0\right)\ \text{Newtons}$$

q₁ is attracted to q₂ along the positive x-axis.

$$\vec{F}_{C\,(on\ q_1\ from\ q_3)} = \frac{\left(9 \times 10^{+9}\right)*3\times10^{-4}*5\times10^{-4}}{6^2}\,\hat{r}$$

$$= 37.5\,\hat{r}\ \text{Newtons}$$

$$= \left(0, -37.5\right)\ \text{Newtons}$$

q₁ is repelled by q₃ along the negative y-axis.

Therefore, the net force is the vector sum: $\vec{F}_{C(total\ on\ q_1)} = (16.9, -37.5)$ Newtons.

b. Calculate the net Coulombic force on q_2.

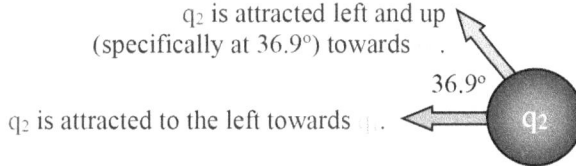

q₂ is attracted left and up
(specifically at 36.9°) towards q₃.

36.9°

q₂ is attracted to the left towards q₁. ←   q2

$$\vec{F}_{C\,(on\ q_2\ from\ q_1)} = \frac{\left(9 \times 10^{+9}\right)*4\times10^{-4}*3\times10^{-4}}{8^2}\,\hat{r}$$

$$= 16.9\,\hat{r}\ \text{Newtons}$$

$$= \left(-16.9, 0\right)\ \text{Newtons}$$

q2 is attracted to q₁ along the negative x-axis.

$$\vec{F}_{C\,(on\ q_2\ from\ q_3)} = \frac{\left(9 \times 10^{+9}\right)*4\times10^{-4}*5\times10^{-4}}{10^2}\,\hat{r}$$

$$= 18\,\hat{r}\ \text{Newtons}$$

$$= \left(-18*\cos\left(36.9°\right), 18*\sin\left(36.9°\right)\right)\ \text{Newtons}$$

$$= \left(-14.39, 10.81\right)\ \text{Newtons}$$

q2 is attracted to q₃ along the line at 36.9° up from the negative x-axis.

Therefore, the net force is the vector sum: $\vec{F}_{C(total\ on\ q_2)} = (-31.29, 10.81)$ Newtons.

c. Calculate the net Coulombic force on $q3$.

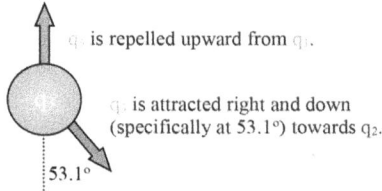

q_3 is repelled upward from q_1.

q_3 is attracted right and down (specifically at 53.1°) towards q_2.

53.1°

$$\vec{F}_{C \, (on \; q_3 \; from \, q_1)} = \frac{\left(9 \times 10^{+9}\right) * 5 \times 10^{-4} * 3 \times 10^{-4}}{6^2} \hat{r}$$

$= 37.5 \, \hat{r}$ Newtons

$= (0, 37.5)$ Newtons

q_3 is repelled away from q_1 along the positive y-axis.

$$\vec{F}_{C \, (on \; q_3 \; from \, q_2)} = \frac{\left(9 \times 10^{+9}\right) * 5 \times 10^{-4} * 4 \times 10^{-4}}{10^2} \hat{r}$$

$= 18 \, \hat{r}$ Newtons

$= \left(18 * \cos\left(36.9°\right), -18 * \sin\left(36.9°\right)\right)$ Newtons

$= (14.39, -10.81)$ Newtons

q_3 is attracted to q_2 along the line at 36.9° down from the positive x-axis.

Therefore, the net force is the vector sum: $\overrightarrow{F}_{C(total \, on \; q_3)} = (14.39, 26.69)$ Newtons.

☺ **COMMENTS:** We now see the full complexity of the symbolic directional reminder $\hat{r}$. We must use Coulomb's law $F_C = (k \times |q_1| \times |q_2|)/r^2$ to compute *only* the magnitude of the electrostatic force between two electric charges. After the magnitude is computed, $\hat{r}$ reminds us to write this Coulombic force along a direction that is attractive or repulsive along the line connecting the two electric charges. Often, writing the force as a vector requires us to calculate angles and decompose vectors into their x- and y-components.

EX 4: A point-sized electric charge $q_2 = -4 \times 10^{-4}$ C is placed at the location (8,0) m. Another point-sized electric charge $q_3 = +5 \times 10^{-4}$ C is placed at the location (0,6) m. Finally, another point-sized electric charge $q_4 = -6 \times 10^{-4}$ C is placed at the location (5,7) m. No other electric are near these three electric charges.

a. Calculate the electric field at the origin.
b. A point-sized electric charge $q_1 = +3 \times 10^{-4}$ C is placed at the origin. What is the Coulombic force on q_1 from q_2, q_3, and q_4?

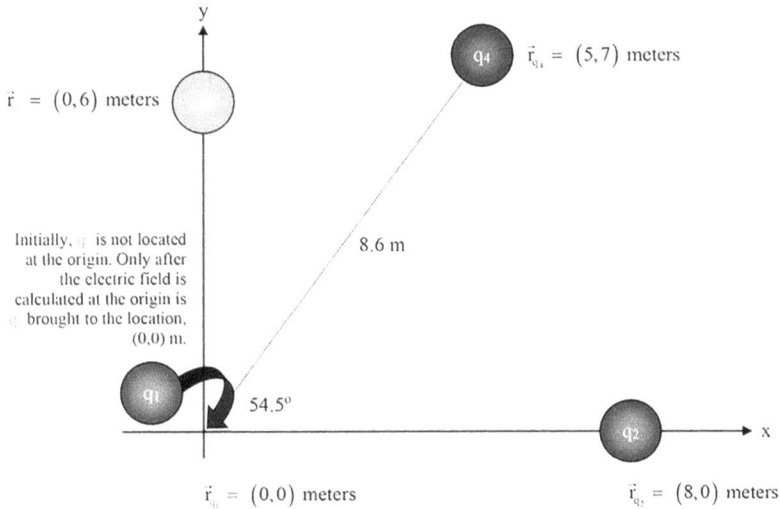

$\vec{r} = (0,6)$ meters

q_4 $\vec{r}_{q_4} = (5,7)$ meters

Initially, q_1 is not located at the origin. Only after the electric field is calculated at the origin is q_1 brought to the location, $(0,0)$ m.

8.6 m

q_1 54.5°

q_2

$\vec{r}_{q_1} = (0,0)$ meters

$\vec{r}_{q_2} = (8,0)$ meters

☑ **ANSWER:**

a. Our procedure for calculating the electric field is to drop off a positive electric test-charge at our point of interest and calculate the Coulombic force on that electric test-charge. As mentioned in example #2, the easiest choice to make, from a computational standpoint, is to drop off a +1 Coulomb electric test-charge at our point of interest. Therefore, let us drop off a +1 Coulomb electric test-charge at the origin and calculate the Coulombic force on it from charges q_2, q_3, and q_4 (remember q_1 has not been placed at the origin yet). Note that a +1 C electric test-charge is attracted to electric point-charges q_2 and q_4, while repelled from electric point-charge q_3.

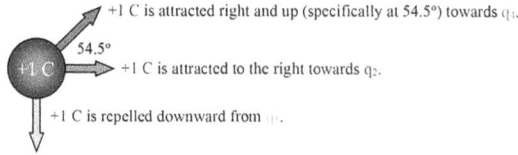

+1 C is attracted right and up (specifically at 54.5°) towards q_1.

54.5°

+1 C   +1 C is attracted to the right towards q_2.

+1 C is repelled downward from q_3.

$$\vec{F}_{C\,(on\,+1\,C\,from\,q_2)} = \frac{\left(9 \times 10^{19}\right)*4\times10^{-4}*1}{8^2}\,\hat{r}$$

$$= 5.625 \times 10^{14}\,\hat{r}\text{ Newtons}$$

$$= \left(5.625 \times 10^{14}, 0\right)\text{ Newtons}$$

The +1 C electric test-charge is attracted to q_2 along the positive x-axis.

$$\vec{F}_{C\,(on\,+1\,C\,from\,q_3)} = \frac{\left(9 \times 10^{19}\right)*5\times10^{-4}*1}{6^2}\,\hat{r}$$

$$= 1.25 \times 10^{15}\,\hat{r}\text{ Newtons}$$

$$= \left(0, -1.25 \times 10^{15}\right)\text{ Newtons}$$

The +1 C electric test-charge is repelled from q_3 along the negative y-axis.

$$\vec{F}_{C\,(on\,+1\,C\,from\,q_4)} = \frac{\left(9 \times 10^{19}\right)*6\times10^{-4}*1}{(8.6)^2}\,\hat{r}$$

$$= 7.3 \times 10^{14}\,\hat{r}\text{ Newtons}$$

The +1 C electric test-charge is attracted to q_1 at 54.5° up from the positive x-axis.

$$= \left(7.3 \times 10^{14}*\cos\left(54.5°\right), 7.3 \times 10^{14}*\sin\left(54.5°\right)\right)\text{ Newtons}$$

$$= \left(4.24 \times 10^{14}, 5.94 \times 10^{14}\right)\text{ Newtons}$$

Therefore, the net force is the vector sum: $\vec{F}_{C(\text{total on}\,+\,1\,C)} =$ $(9.865 \times 10^{+4}, -6.56 \times 10^{+4})$ Newtons.

To determine the electric field, divide the net force by 1 Coulomb, the value of our electric test-charge.

$$\vec{E} = \frac{\vec{F}_{C(\text{net on}\,+\,1\,C)}}{+1\,C}$$

$$= \frac{(9.865 \times 10^{+4}, -6.56 \times 10^{+4})\text{ Newtons}}{+1\,C}$$

$$= (9.865 \times 10^{+4}, -6.56 \times 10^{+4})\text{ Newtons C}^{-1}.$$

b. Now armed with the electric field, we place a point-sized electric charge $q_1 = +3 \times 10^{-4}$ C at the origin. Therefore, we can calculate the Coulombic force on q_1.

$$\vec{F}_{C(\text{on}\,q_1\,\text{from}\,q_2,\,\ldots\,,\,\text{and}\,q_4)} = q_1 \times \vec{E}_{(\text{at (P.) from}\,q_2,\,\ldots\,,\,\text{and}\,q_4)}$$

$$= (3 \times 10^{-4}\,C) \times (9.865 \times 10^{+4}, -6.56 \times 10^{+4})\text{ Newtons C}^{-1}$$

$$= (29.6, -19.7)\text{ Newtons}.$$

☺ **COMMENTS:** Again, we see the beauty of the electric field. The field gives us the force-per-unit-charge ... in other words, the force for every unit of electric charge at a given point. One only needs to multiply the electric field by the value of the electric charge that is placed at a particular point and the product is the Coulombic force on that electric charge!

4.3.3 Checking for understanding

To check for understanding, let us perform one last calculation in which we solve for the Coulombic force among a series of electric point-charges by first determining the electric field at a point of interest, then calculating the Coulombic force by placing various electric point-charges at the location where the electric field has been computed. Consider three electric point-charges placed on the vertices of a 6 m^2. $q_1 = -7 \times 10^{-4}$ C is placed at the origin. Another electric point-charge $q_2 = 8 \times 10^{-4}$ C, is placed at the location (6,0) m. Finally, an electric point-charge $q_3 = -9 \times 10^{-4}$ C, is placed at the location (0,6) m.

  a. Calculate the net electric field at point (P.), located at (6,6) m.
  b. Calculate the net Coulombic force on a -3×10^{-4} C—if it were placed at point (P.).
  c. Calculate the net Coulombic force on a $+4 \times 10^{-4}$ C—if it were placed at point (P.).

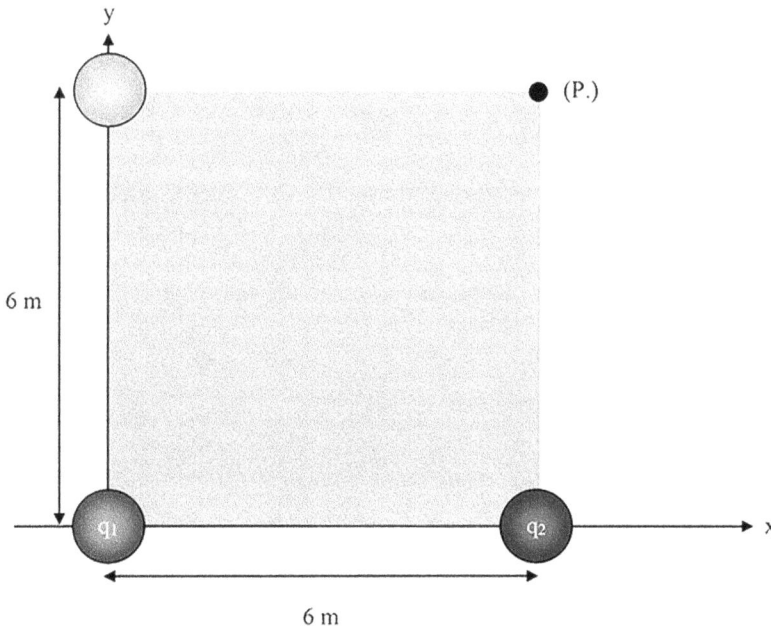

☑ **ANSWER:**

  a. To calculate the electric field, we place a $+1$ C electric test-charge at point (P.) then determine the net Coulombic force on that electric test-charge based on Coulomb's law. We are free to place **any** electric test-charge at point (P.) as long as we divide it out as the last step in our recipe; however, we have seen that the easiest computational choice is to use a $+1$ C electric test-charge since the value of $+1$ C does not numerically affect any of the calculations we perform.

● (P.) $+1$ C

Imagine placing a $+1$ C electric test-charge at the point (P.). The electric test-charge will "sample" the electric field at point (P.) for us—then we'll remove the electric test-charge by dividing it out of the final answer.

$$\vec{F}_{C\,(on\,+1\,C\,from\,q_1)} = \frac{\left(9 \times 10^{19}\right)*7 \times 10^{-4} *1}{\left(\sqrt{6^2 + 6^2}\right)^2}\hat{r}$$

The $+1$ C electric test-charge is attracted to q_1 at 45° into the third quadrant (i.e., along the negative x-axis and along the negative y-axis).

$$= 8.75 \times 10^{14}\,\hat{r} \text{ Newtons}$$

$$= \left(-8.75 \times 10^{14}*\cos\left(45°\right), -8.75 \times 10^{14}*\sin\left(45°\right)\right) \text{ Newtons}$$

$$= \left(-6.187 \times 10^{14}, -6.187 \times 10^{14}\right) \text{ Newtons}$$

$$\vec{F}_{C\,(on\,+1\,C\,from\,q_2)} = \frac{\left(9 \times 10^{19}\right)*8 \times 10^{-4} *1}{\left(6\right)^2}\hat{r}$$

$$= 2.0 \times 10^{15}\,\hat{r} \text{ Newtons}$$

The $+1$ C electric test-charge is repelled from q_2 along the positive y-axis.

$$= \left(0, 2.0 \times 10^{15}\right) \text{ Newtons}$$

$$\vec{F}_{C\,(on\,+1\,C\,from\,q_3)} = \frac{\left(9 \times 10^{19}\right)*9 \times 10^{-4} *1}{\left(6\right)^2}\hat{r}$$

$$= 2.25 \times 10^{15}\,\hat{r} \text{ Newtons}$$

The $+1$ C electric test-charge is attracted to q_3 along the negative x-axis.

$$= \left(-2.25 \times 10^{15}, 0\right) \text{ Newtons}$$

Therefore, the net force on the $+1$ C electric test-charge is the vector sum: $F_{C(total\,on\,+1\,C)} = (-2.869 \times 10^{+5}, 1.38 \times 10^{+5})$ Newtons.

Now dividing the net force by $+1$ C, the value of our electric test-charge, we get a value for the net electric field at point (P.):

$$\vec{E} = \frac{\vec{F}_{C(net\,on\,+1\,C)}}{+1\,C}$$

$$= \frac{(-2.869 \times 10^{+5}, 1.38 \times 10^{+5}) \text{ Newtons}}{+1\,C}$$

$$= (-2.869 \times 10^{+5},\, 1.38 \times 10^{+5}) \text{ Newtons } C^{-1}.$$

b. To calculate the net Coulombic force on a -3×10^{-4} C—if it were placed at point (P.) —simply multiply our answer in part (a) by -3×10^{-4} C:

$$\vec{F}_{C(net\,on\,-3 \times 10^{-4}\,C)} = (-3 \times 10^{-4}\,C) \times \vec{E}$$

$$= (-3 \times 10^{-4}\,C) \times (-2.869 \times 10^{+5}, 1.38 \times 10^{+5}) \text{ Newtons } C^{-1}$$

$$= (86, -41.4) \text{ Newtons}.$$

c. To calculate the net Coulombic force on a $+4 \times 10^{-4}$ C—if it were placed at point (P.) —simply multiply our answer in part (a) by $+4 \times 10^{-4}$ C:

$$\vec{F}_{\text{C(neton}+4\times10^{-4}\text{ C)}} = (+4 \times 10^{-4} \text{ C}) \times \vec{E}$$
$$= (+4 \times 10^{-4} \text{ C}) \times (-2.869 \times 10^{+5}, 1.38 \times 10^{+5}) \text{ Newtons C}^{-1}$$
$$= (-114.8, 55.2) \text{ Newtons.}$$

4.4 Keeping information

4.4.1 Closure

In the 'Instructional Input' section of this chapter, we discussed some everyday instances in which the electric force manifests itself in ways which we can readily see. Some examples included rubbing a latex balloon, combing your hair, and walking across a carpet. However, we promised to close this chapter by performing some demonstrations which showcase the electric force manifesting itself in very magnificent and striking ways! As we noted at the beginning of this chapter, the most common force in Nature with which we humans interact on a daily basis is gravity. We simply cannot escape the gravitational pull of the Earth on us—it is always there and we have just come to take this example of the force of gravity for granted. The Earth's pull of gravity on us and everyday objects is so commonplace that like it or not, we have come to appreciate gravity on some level. On the other hand, we rarely get the chance to see the electric force in action. Our lack of interaction with the electric force deprives us of the chance to truly appreciate this force of Nature. The 'Closure' section of this chapter provides you with the opportunity to safely experiment with electricity and to develop a true appreciation and awe for this force of Nature. These demonstrations are simply magical! You will see the electric force in action and may even find yourself spellbound by what you see. Most of these demonstrations are simple experiments you can perform easily in your home—you only need a few household items and the open-mindedness to watch the magic unfold before your eyes!

4.4.1.1 Electricity in action: soda cans

In the first example, take a *plastic* rod or ruler and electrically charge it by rubbing it with a cloth or animal fur. Recall when we first introduced the different types of electric charge (i.e., positive and negative), we noted that a *plastic* rod or ruler will accumulate an excess of *negative* electric charge when rubbed with a cloth (see step A below). Next, lie a neutral aluminum soda can on its side on a smooth surface and slowly bring the negatively-charged rod near the soda can (see step B below). The key to understanding the demonstration is to realize that an object labeled as 'neutral' possesses as much positive electric charge as negative electric charge. Students have the misconception that a neutral object is simply void of electric charge. That may indeed be true, but an object may also be neutral because it possesses a balance of positive and negative electric charges, thus rendering the object with a net neutral charge. In our demonstration, consider the soda can not to be lacking electric charge, but instead to possess a balance of positive and negative electric charge. As the negatively-charged plastic rod is brought near the soda can, the negative electric charges in the aluminum soda can are repelled from the rod and move away from it, while the positive electric charges in the aluminum soda can are attracted to the rod and move toward it. This movement of electric charges creates an imbalance of electric charges in the aluminum can. The net effect is that the Coulombic forces of the electric charges in the aluminum can move it toward the plastic rod (see step C below).

Note that this demonstration (and all of the subsequent demonstrations) works equally well by rubbing a *glass* **rod or stick with a cloth or animal fur and using the glass rod to move the soda can. We know from our previous discussions that a glass rod will accumulate an excess of** *positive* **electric charge when rubbed with a cloth. In other words, you can interchangeably use positive or negative electric charges to move the soda can.** *We arbitrarily chose to use the plastic rod in this chapter.*

(Step A.) By rubbing the plastic rod with a cloth or animal fur, electrons are transferred from the animal fur to the plastic rod, giving the plastic rod an excess of negative electric charge.

Image Credit: Author.

(Step B.) As the negatively-charged plastic rod is brought near the neutral soda can, the separation of electric charges occurs within the soda can. The positive electric charges are attracted toward the plastic rod via the Coulombic force while the negative electric charges are repelled away from the plastic rod.

Image Credit: Author.

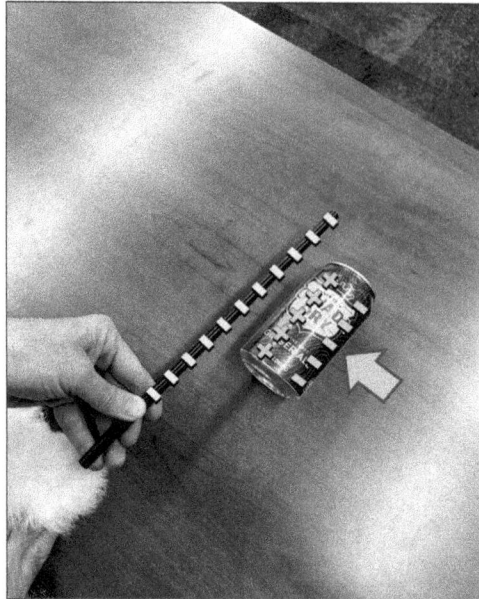

(Step C.) The net effect is that the plastic rod pulls on the aluminum soda can via the Coulombic force. The soda will literally roll on the smooth tabletop and follow the electrically-charged plastic rod.

Image Credit: Author.

4.4.1.2 Electricity in action: wooden sticks

Our next example is very similar to the first example; however, rather than using the Coulombic force to move an aluminum soda can, we will use it to move a wooden stick. The reason we focus on moving a wooden stick is because when we used the Coulombic force to move the aluminum soda can, students often develop a misconception that the conducting nature of the soda can's aluminum metal had something to do with the dramatic effect seen in the first example. In other words, students often mistakenly think that perhaps we can only create the charge separation in conductors (like aluminum or other metals) when moving objects with the Coulombic force. To dispel this misconception, let us try moving an insulator, like a wooden stick!

First, balance a wooden stick on a fulcrum. For our demonstration, we use a painter's stirring stick as our wooden stick, but a wooden ruler or dry twig will work just as well.

For the fulcrum, the top of a capped water bottle works nicely. The point here is that the wooden stick has to be able to easily swivel 360° on the fulcrum. For our demonstration, we use a lens placed on the bottom of a soda can—the purpose of the lens is only to give the wooden stirring stick an easy means of swiveling about the soda can (see step A below). Next, take a plastic rod or ruler and electrically charge it by rubbing it with a cloth or animal fur. Like the previous example, we note that a plastic rod or ruler will accumulate an excess of negative electric charge when rubbed with a cloth (see step B below). Just like in the example above, slowly bring the negatively-charged rod near the wooden stick (see step C below). Again, consider the wooden stick to be neutral—not because it lacks electric charge but because it possesses a balance of positive electric charge and negative electric charge. As the negatively-charged plastic rod is brought near the wooden stick, the negative electric charges in the wooden stick are repelled from the rod and move away from it, while the positive electric charges in the stick are attracted to the rod and move toward it. This movement of electric charges creates an imbalance of electric charges in the stick. The net effect is that the Coulombic forces of the electric charges in the stick swivel it toward the plastic rod (see step D below).

(Step A.) A lens is placed on the bottom of a soda can and is used only to act as a fulcrum so that the wooden stick may easily swivel 360° about the soda can.

Image Credit: Author.

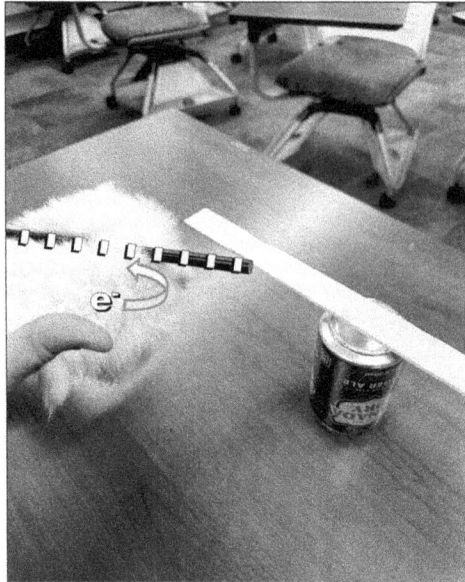

(Step B.) A wooden stirring stick is placed atop the fulcrum. Next, by rubbing the plastic rod with a cloth or animal fur, electrons are transferred from the animal fur to the plastic rod, giving the plastic rod an excess of negative electric charge.

Image Credit: Author.

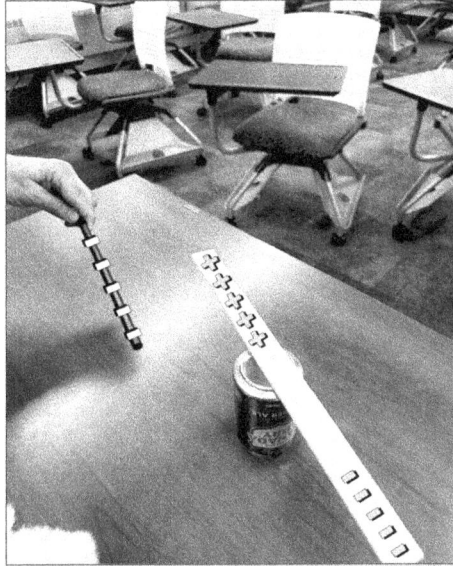

(Step C.) As the negatively-charged plastic rod is brought near the wooden stirring stick, the separation of electric charges occurs within the wooden stick. The positive electric charges are attracted toward the plastic rod via the Coulombic force while the negative electric charges are repelled away from the plastic rod.

Image Credit: Author.

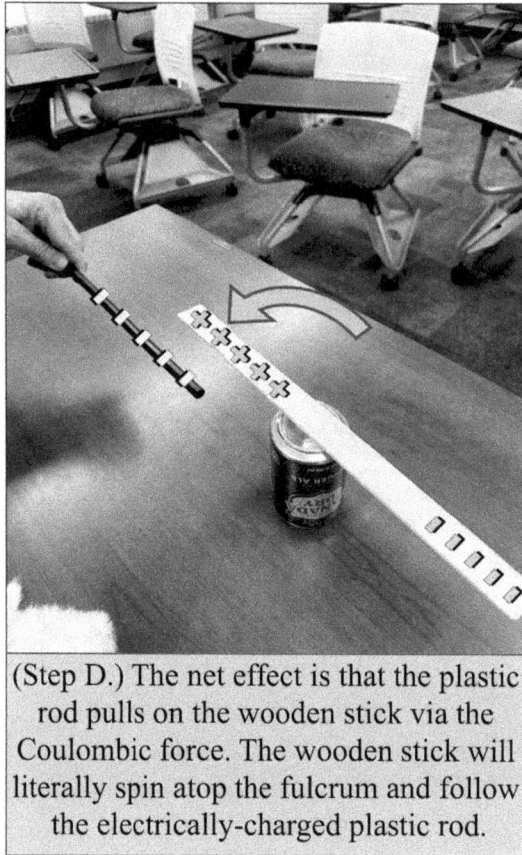

(Step D.) The net effect is that the plastic rod pulls on the wooden stick via the Coulombic force. The wooden stick will literally spin atop the fulcrum and follow the electrically-charged plastic rod.

Image Credit: Author.

4.4.1.3 Electricity in action: polar molecules

Water is a polar molecule, meaning that electric charge is not evenly-distributed across the molecule because of an uneven distribution of electrons among different atoms in the molecule. The oxygen atom in the molecule has a negative electric charge while the hydrogen atoms have a positive electric charge, resulting in an uneven distribution of electric charge across the molecule (see the depiction of the water molecule in step A below).

Again, take a plastic rod or ruler and charge it electrically by rubbing it with a cloth or animal fur. Like the previous examples, we note that a plastic rod or ruler will accumulate an excess of negative electric charge when rubbed with a cloth. Next, we need to generate a thin (i.e., pencil-thin) stream of water. We can achieve this in a number of ways. First, we can simply use a faucet but usually a stream of water inside a faucet will be difficult for a group of people to see. On the other hand, if you wish to perform this demonstration in front of a group of friends, fill a plastic 2-L soda bottle with water and poke a tiny pin-sized hole near the base of the bottle.

When you loosen the bottle's cap, a nifty pencil-thin stream of water will emerge from the bottle (see steps A and B below). Simply tighten the cap to stop the stream of water. Just like in the examples above, slowly bring the negatively-charged rod near the thin stream of water (see step C below). As the negatively-charged plastic rod is brought near the polar water molecules, the uneven distributions of electric charge in the water molecules create an imbalance of Columbic force on the thin stream of water. The net effect is that the Coulombic force on the steam of water deflects it from its normal trajectory and draws it toward the plastic rod (see step D below).

(Step A.) The water molecule is "polar," meaning it has an uneven distribute of electric charge across the molecule.

Image Credit: Author.

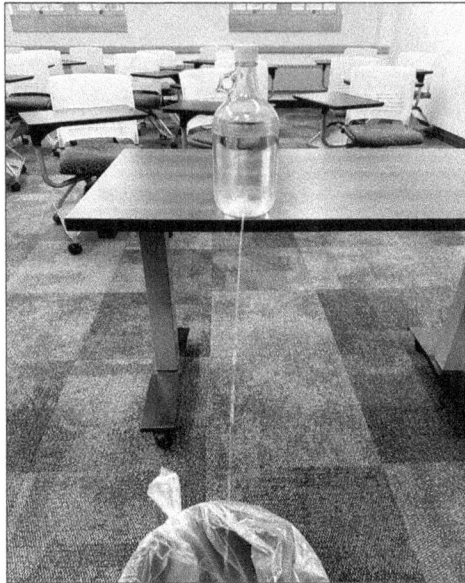

(Step B.) You can generate a pencil -thin steam of water by filling a plastic 2-liter soda bottle with water then poking a pin-sized hole near the base of the bottle. By simply loosening the cap, a stream of water will emerge from the hole. Tightening the cap will stop the steam of water.

Image Credit: Author.

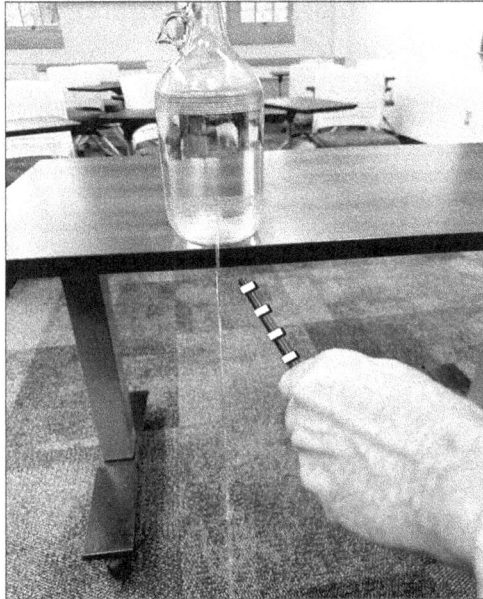

(Step C.) As the negatively-charged plastic rod is brought near the stream of water, the uneven distribution of electric charge across the water molecule creates an imblance of force across the steam of water.

Image Credit: Author.

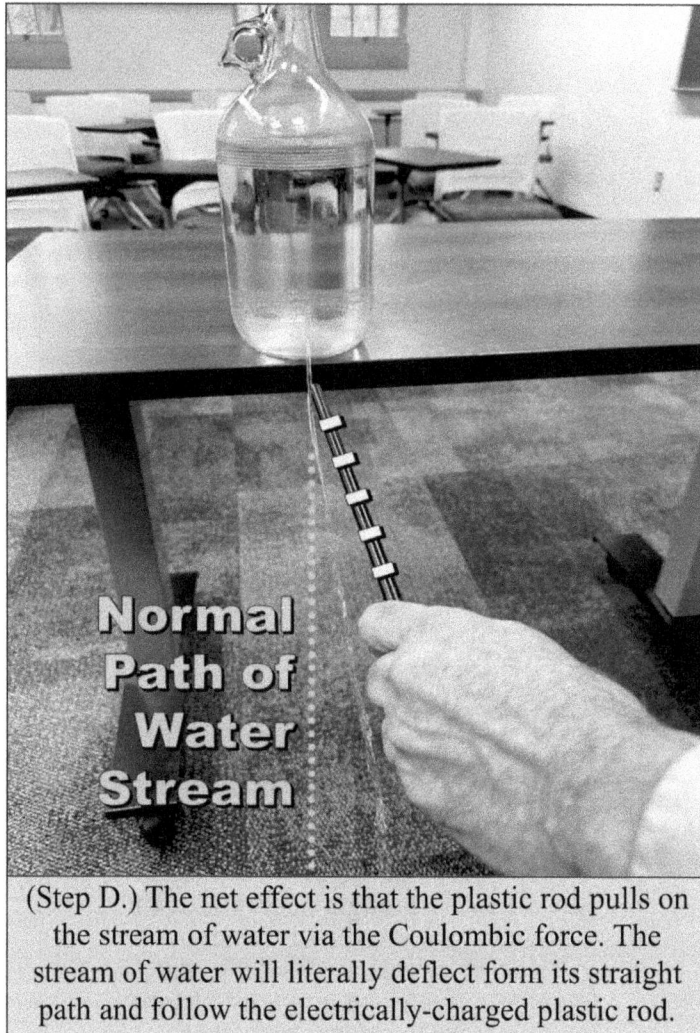

Normal
Path of
Water
Stream

(Step D.) The net effect is that the plastic rod pulls on the stream of water via the Coulombic force. The stream of water will literally deflect form its straight path and follow the electrically-charged plastic rod.

Image Credit: Author.

4.4.1.4 Electricity in action: the Van De Graaff generator

The Van de Graaff generator is the central piece in the widely-seen demonstration whereby a volunteer places his or her hand on a large metallic sphere and in a matter of seconds, the audience witnesses the volunteer's hair standing on end. Although you will not be able to perform this demonstration in your home (unless you own a Van de Graaff generator), you have probably seen a Van de Graaff generator at a local Science Center or Discovery Center.

In 1931, American physicist Robert J Van de Graaff (1901–67) invented the particle accelerator that now bears his name. The main component of the accelerator is the Van de Graaff generator that serves to illustrate several aspects of electric charge and the electric field. To construct a Van de Graaff generator, a large, hollow conducting sphere is supported by an insulating, vacuum-filled column above the base. The sphere can be given an excess of positive electric charge as follows:

i. A motor drives a rubber belt around two pulleys. The lower pulley is surrounded by a cloth or animal fur. The rubbing or the rubber belt against the cloth causes an excess of positive electric charge to be deposited on the belt. In addition, a voltage of typically 50 000 V is applied to a conductor (A) which pulls electrons off the rubber belt onto the electrode at (A). Hence the rubber belt becomes highly positively-charged (see step A below).

ii. The motor continues to drive the belt which then carries the positive electric charges up the column into the interior of the sphere.

iii. The positive electric charges are then removed from the belt by a conducting 'comb' at point (B) (see step B below).

iv. The positive electric charges then spread out over the exterior of the metal sphere.

v. As more and more charges are brought up the column, the sphere becomes more highly positively-charged; thus creating a huge electric field as the sphere reaches higher and higher voltages. The process of building up charge requires energy which is supplied by the motor driving the belt. Typically, Van de Graaff generators can accelerate particles with 30 MeV of kinetic energy.

So … how does the Van de Graaff generator make someone's hair stand on end? First, the volunteer needs to stand on an insulating platform or chair so that any build-up of electric charges is not neutralized by the ground which has an infinite supply of positive and negative electric charges. With the volunteer's hand placed upon the Van de Graaf's metal sphere, the generator is turned on in order to build up positive electric charge on the metal sphere. As the electric charges move onto the metal sphere, the volunteer acts as a conductor; thus causing the electric charges to accumulate on the outside of the volunteer's head. The key to understanding the demonstration is to realize that because of friction, the volunteer's hair has been rubbing against fabric (perhaps a hat or sweater) throughout the day and has thus accumulated an excess of positive electric charge. As more positive electric charges accumulate on the volunteer's head from the Van de Graaf generator, the volunteer's hair will stand on end because of Coulombic repulsion (i.e., the volunteer's positively-charged hair will be repelled by his or her positively-charged head). The volunteer's head has literally become the conducting sphere of the generator.

A nifty way to demonstrate the same effect as the volunteer's hair standing on end is to use Styrofoam packing peanuts or tinsel. Styrofoam packing peanuts and tinsel carry an excess of positive electric charge. For example, place a small bundle of tinsel atop the Van de Graff generator (see step C below). When the Van de Graaff generator is turned on, an excess of positive electric charge will build up on the metal sphere; hence scattering the tinsel into the air, because of Coulombic repulsion (see step D below). You can imagine that the tinsel is mapping out the electric field in such a way that you can 'see' the invisible field by which electric charges exert the Coulombic force on one another.

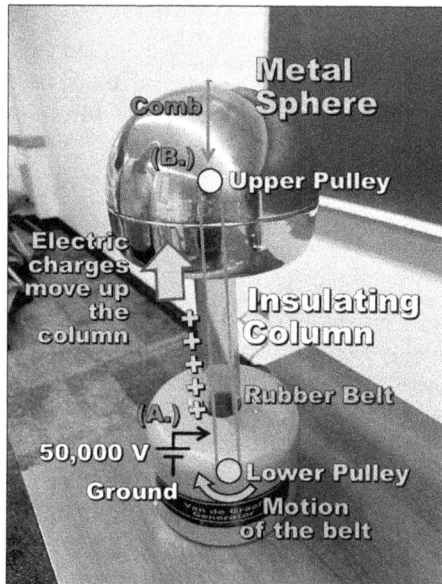

(Step A.) The components of a Van de Graaff generator. The motion of a belt around two pulleys moves positive electric charges up the column to the location of point (B.). At point (B.), a metallic "comb" provides the electric charges a path from the belt to the metallic sphere.

Image Credit: Author.

(Step B.) A look at the upper section of the Van de Graaff generator, with the metal sphere removed. The fingers of the metallic "comb" rest just above the moving belt, thus providing a path for positive electric charges to move from the belt to the metal sphere.

Image Credit: Author.

(Step C.) Light-weight and continuously subjected to friction, tinsel carries an excess of positive electric charge. With the Van de Graaff generator turned off, tinsel is subjected only to the force of gravity and falls toward the Earth.

Image Credit: Author.

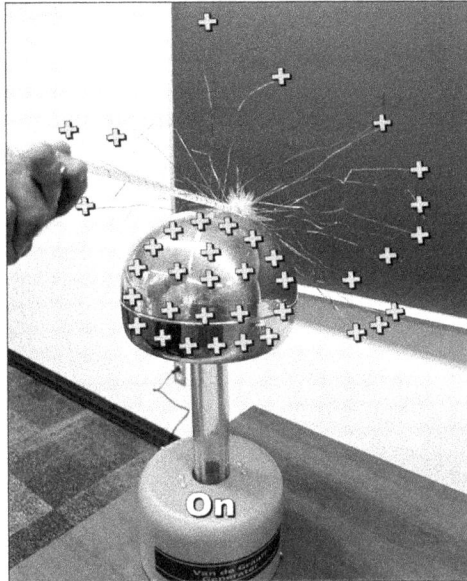

(Step D.) With the Van de Graaff generator turned on, tinsel is subjected to both gravity and the electric force. We see here that the Coulombic repulsive force, directed away from the Van de Graaff's metal sphere, is greater than the force of gravity and the tinsel stands on end!

Image Credit: Author.

4.4.1.5 Electricity in action: styrofoam packing peanuts, cereal, and tinsel

As usual, start off by electrically charging a plastic rod or ruler with an excess of negative electric charge by rubbing it with a cloth or animal fur (see step A below). Grab a handful of light-weight Styrofoam packing peanuts (or cereal … or tinsel) and wave the negatively-charged rod near the packing peanuts. In our demonstration, the packing peanuts are kept in a small, clear plastic tube so that the packing peanuts do not scatter all over the room or continually stick to our clothes. When waving the plastic rod near the packing peanuts, try moving it up-and-down, side-to-side, and over-and-around the packing peanuts … in every direction you can imagine (see steps B and C below). You will see the packing peanuts move, dance, and jump to match the motion of the plastic rod. The key to understanding the demonstration is two-fold. First, Styrofoam packing peanuts, cereal, and tinsel typically carry an excess of positive electric charge; therefore, as you move the negatively-charged plastic rod near them, each individual packing peanut will be attracted to the rod, while repulsed from other packing peanuts, because of the Coulomb force. Next, Styrofoam packing peanuts, cereal, and tinsel are light-weight; therefore, the force of gravity on these objects does not overwhelm the electric force. The electrostatic repulsion among the packing peanuts and the attraction to the charged rod is strong enough to overcome gravity so the packing peanuts can fly through the air! The net result of these two effects is a very chaotic motion for the packing peanuts.

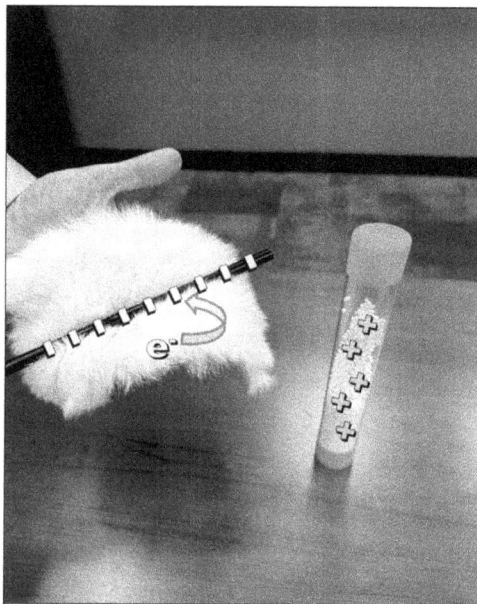

(Step A.) As we have come to expect, by rubbing the plastic rod with a cloth or animal fur, electrons are transferred from the animal fur to the plastic rod, giving the plastic rod an excess of negative electric charge

Image Credit: Author.

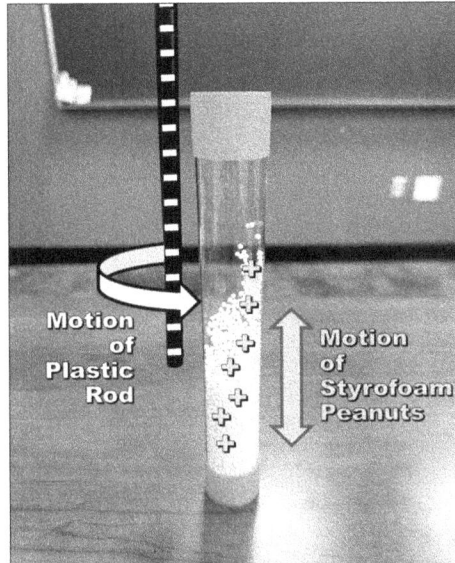

(Step B.) Moving the electrically -charged rod near a handful of Styrofoam packing peanuts results in a very chaotic and exciting motion of the packing peanuts! The packing peanuts are kept in a clear, plastic tube only to prevent them from scattering about the room or sticking to our clothes.

Image Credit: Author.

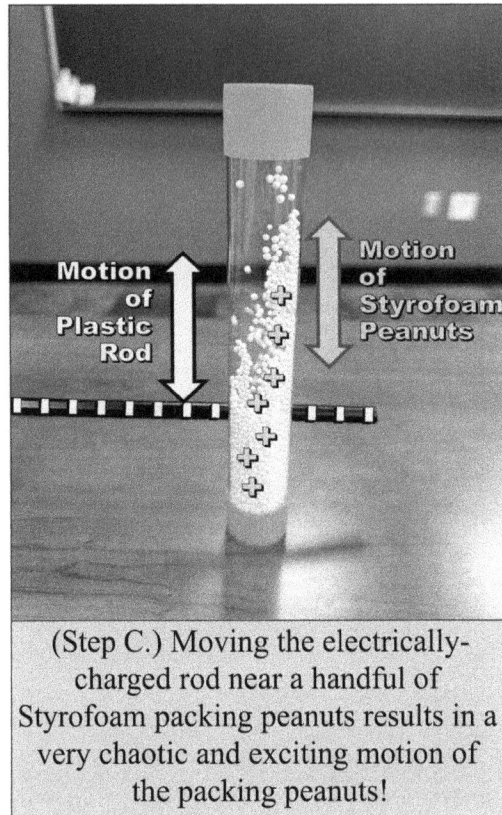

(Step C.) Moving the electrically-charged rod near a handful of Styrofoam packing peanuts results in a very chaotic and exciting motion of the packing peanuts!

Image Credit: Author.

4.4.1.6 Electricity in action: picking-up confetti

In a shocking (pun intended) turn of events, we will not start off by rubbing a plastic rod or ruler with a cloth or animal fur. Instead, rub a latex balloon with a cloth or animal fur. Recall when we first introduced the different types of electric charge (i.e., positive and negative), we noted that a *latex balloon* will accumulate an excess of *negative* electric charge when rubbed with a cloth because of the transfer of electrons from the cloth to the balloon (see step A below). Next, scatter a handful of confetti on a table. The confetti can be made of tiny squares of paper or tissue paper. The key here is that the pieces of confetti must be light-weight (see step B below). Wave the negatively-charged balloon over the confetti and watch the confetti fly off the table and stick to the balloon (see step C below). Here we literally see the electric force overcome the force of gravity so that the pieces of confetti lift off the table and stick to the balloon! Once again, the key to understanding the demonstration is to appreciate that the confetti is light-weight; therefore, the force of gravity on these pieces of paper (or tissue) does not overwhelm the electric force. The electrostatic attraction between the balloon and paper is strong enough to overcome gravity and the pieces of paper fly through the air!

(Step A.) By rubbing the latex balloon with a cloth or animal fur, electrons are transferred from the animal fur to the balloon, giving it an excess of negative electric charge.

Image Credit: Author.

(Step B.) Tissue paper is cut into tiny square pieces of confetti. Typically, these pieces of tissue paper carry an excess of positive electric charge.

Image Credit: Author.

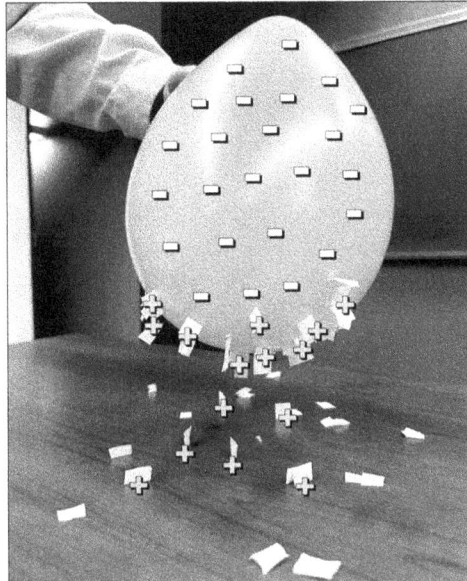

(Step C.) Moving the negatively electrically-charged latex balloon over the positively electrically-charged pieces of confetti lifts the confetti off of the table. Notice the pieces of confetti on the table which are standing, waiting to jump from the table to the balloon!

Image Credit: Author.

4.4.1.7 Electricity in action: the electroscope

The electroscope is a scientific instrument first developed in the 1730s to detect the presence of electric charge. The electroscope's typical configuration was perfected in 1787 by British physicist Abraham Bennet. The instrument is made by hanging a metal plate from the end of a metal rod. Also attached to the end of the metal rod is a thin metal foil (often made of gold, but for our purposes, we will settle for Mylar). To protect the foil from air drafts, the metal plate and hanging foil are enclosed in a case (often made of glass or plastic). The metal rod is insulated from the case by a rubber O-ring so that electric charge may be applied to the metal rod, plate, and foil without being distributed to the case (ssee step A below). To conduct this demonstration, rub our ole friend, the plastic rod or ruler, with a cloth or animal fur. As we know, the plastic rod will accumulate an excess of negative electric charge when rubbed with a cloth because of the transfer of electrons from the cloth to the rod (see step A below). Next, touch the plastic rod to the top of the metal rod of the electroscope. Via conduction, negative electric charge will distribute across the metal rod, metal plate, and metal foil. The metal rod and plate are rigid and remain in place. However, the metal foil is loose and free to move. Because of the Coulombic repulsion between the negative electric charges on the rigid metal plate and the free-hanging metal foil, the metal foil will swing away from the metal plate (see step B below). Since most of us don't own an electroscope, a homemade version is easy to construct—simply hang two latex balloons (or pieces of cereal) from the ceiling using fishing line. Rub each balloon with cloth or animal fur. Each balloon obtains an excess of negative electric charge and hence repel one another as indicated by the fixed separation between them (see step C below). This is the same effect seen by the hanging foil inside the electroscope.

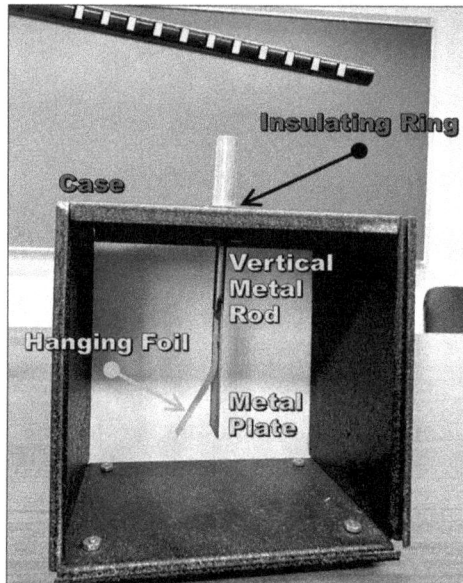

(Step A.) The components of the electroscope. A metal rod hangs inside a case. The metal rod is insulated from the case by a rubber O-ring. At the end of the metal rod, a rigid metal plate is attached. Also at the end of the metal rod, a free-hanging metal foil is attached.

Image Credit: Author.

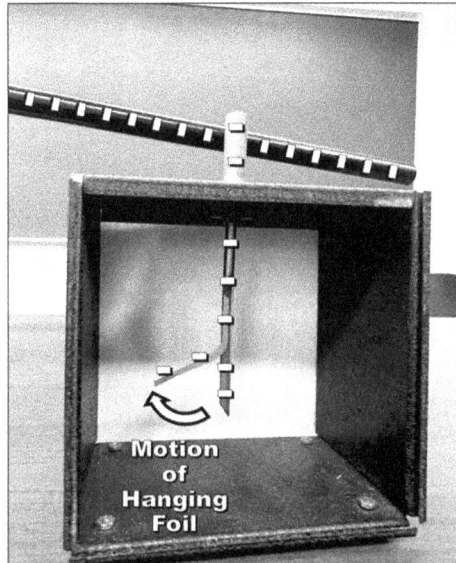

(Step B.) When the negatively electrically-charged plastic rod touches the top of the metal rod, negative electric charges rush across the metal rod, metal plate, and metal foil. The electrostatic repulsion between the free-hanging foil and the rigid metal plate causes the foil to swing outward.

Image Credit: Author.

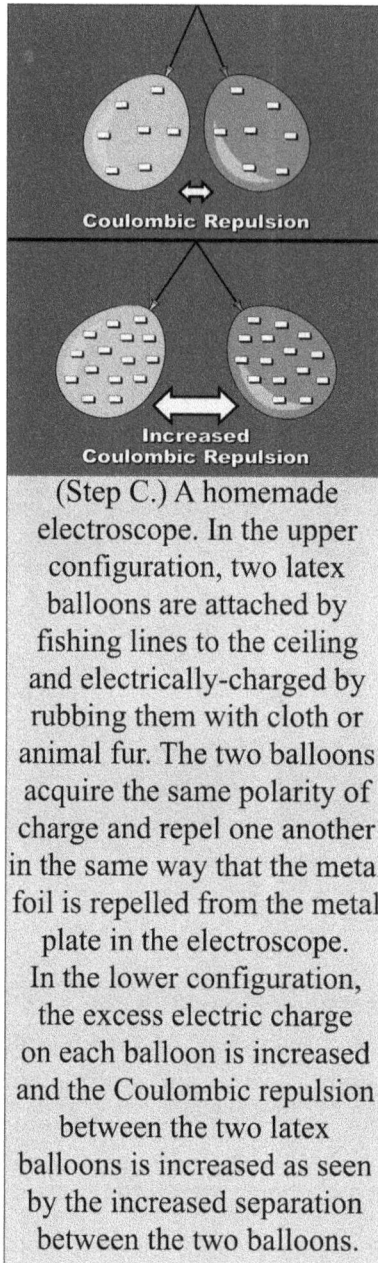

(Step C.) A homemade electroscope. In the upper configuration, two latex balloons are attached by fishing lines to the ceiling and electrically-charged by rubbing them with cloth or animal fur. The two balloons acquire the same polarity of charge and repel one another in the same way that the metal foil is repelled from the metal plate in the electroscope. In the lower configuration, the excess electric charge on each balloon is increased and the Coulombic repulsion between the two latex balloons is increased as seen by the increased separation between the two balloons.

Blue and Green Baloon Images Credit:
Pixabay by Clker-Free-Vector-Images. Image Credit: Author.

4.4.1.8 Electricity in action: the funny fly stick

The 'Funny Fly Stick' is a children's toy you can purchase that generates electric charge using a motor that is powered by two AA-batteries. The Funny Fly Stick then transfers that charge to Mylar shapes (constructed to look like butterflies, hoops, and dumbbells), causing them to levitate in the air. The underlying principle of the Funny Fly Stick is a miniaturized version of the previously-mentioned Van de Graaff generator. Inside the stick, which is actually a cardboard tube, a rubber belt runs around two pulleys. While the pulley closest to the batteries deposits electric charge on the rubber belt via friction, the pulley closest to the tip of the fly stick extracts the electric charges with a small wire comb. As the rubber belt continues to run, more and more electric charges build up on the tip of the fly stick. When the tip of the Funny Fly Stick is touched to Mylar shape, the Mylar shape is electrically-charged to the same polarity as the tip of the Funny Fly Stick. The Mylar shape is then instantly repulsed from the tip of the Funny Fly stick and 'flies' into the air (see below). Again, since Mylar is light-weight, the electric force is not overwhelmed by the force of gravity and the Mylar shape is able to levitate in the air.

A student uses the Funny Fly Stick to levitate a Mylar atomic ring-shape. In the same way the Van de Graaff generator electrically-charged the tinsel in a previous demonstration and levitated the tinsel, the electric charge on the tip of the Funny Fly Stick repels the electric charge on the Mylar shape and levitates the shape in the air.

Image Credit: Author.

4.4.1.9 Electricity in action: the dancing ghost
Here's a great demonstration you can do with kids near the holidays. Rub a latex balloon with a cloth or animal fur. We know that a latex balloon will accumulate an excess of electric charge when rubbed with a cloth because of the transfer of electrons from the cloth to the balloon (see step A below). Next, cut a holiday-themed shape out of tissue paper and tape the base of the shape to a table. For example, for Halloween, you can cut out the shape of a ghost; for Christmas, you can cut out the shape of a Christmas tree; for Easter, you can cut out the shape of a bunny; etc The key here is that the holiday-themed shape must be light-weight so be sure to use single-ply tissue. Wave the negatively-charged balloon over the holiday-themed shape and watch the shape 'dance' as it is attracted to the latex balloon (see step B below). In step B below, we literally see the electric force overcome the force of gravity so that the 'dancing ghost' levitates off the table and 'dances' with the balloon! Once again, the key to understanding the demonstration is to appreciate that the ghost is light-weight; therefore, the force of gravity on this tissue does not overwhelm the electric force. The electrostatic attraction between the balloon and ghost is strong enough to overcome gravity and the ghost is able to dance!

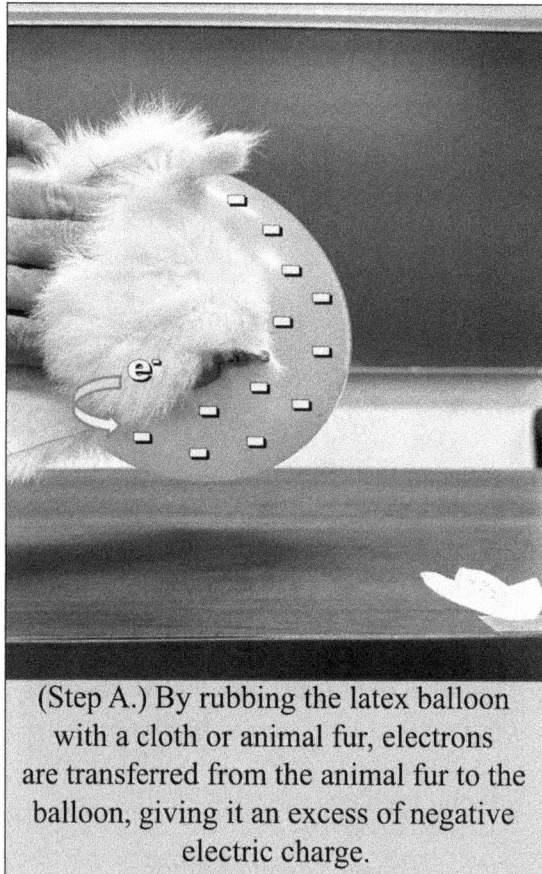

(Step A.) By rubbing the latex balloon with a cloth or animal fur, electrons are transferred from the animal fur to the balloon, giving it an excess of negative electric charge.

Image Credit: Author.

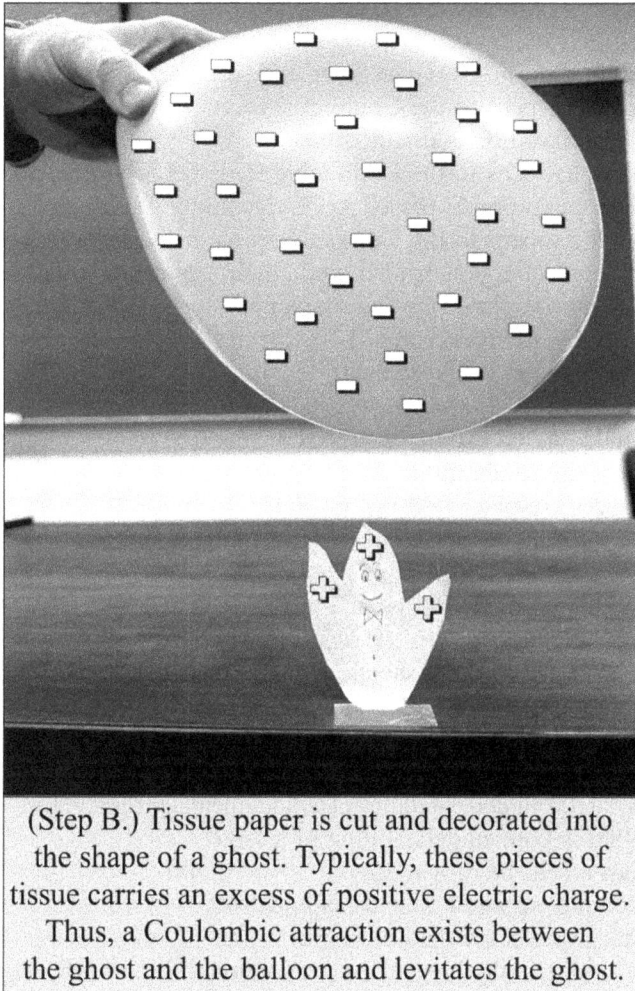

(Step B.) Tissue paper is cut and decorated into the shape of a ghost. Typically, these pieces of tissue carries an excess of positive electric charge. Thus, a Coulombic attraction exists between the ghost and the balloon and levitates the ghost.

Image Credit: Author.

4.4.2 Independent practice

1. Two electric charges q_1 and q_2 experience a Coulombic force between them. What happens (numerically) to the Coulombic force if:
    a) Their separation distance is increased by a factor of 8?
    b) Their separation distance is decreased by a factor of 4?
    c) The magnitude of q_1 is doubled?
    d) Their separation distance is reduced by a factor of 3?
    e) The magnitude of q_1 is increased by a factor if 2 while the magnitude of q_2 is reduced by a factor of 3?
    f) q_1 keeps its magnitude but changes its sign?
    g) Their separation distance is increased by a factor of 4 while each charge's magnitude is increased by a factor of 4?

h) Their separation distance is decreased by a factor of 4 while each charge's magnitude is increased by a factor of 4?

2. Give simple quantitative answers to the situations below:
   a) What happens to the Coulombic force between two electric charges if their separation is doubled?
   b) What happens to the Coulombic force between two electric charges if their separation is tripled while the value of each charge is halved?
   c) What happens to the Coulombic force between two electric charges if only one charge is tripled while their separation is halved?

3. Four charges are placed on the xy-plane:
   - $q_1 = +30$ μC is placed at $(-0.4, 0)$ m.
   - $q_2 = -40$ μC is placed at $(0, -0.3)$ m.
   - $q_3 = -50$ μC is placed at $(+0.4, 0)$ m.
   - Finally, $q_4 = +10$ μC is placed at $(0, +0.3)$ m.

   What is the resultant Coulombic force on q_4?

4. Two charges are placed on the xy-plane:
   - $q_1 = +30$ μC is placed at $(-0.4, 0)$ m.
   - $q_2 = -40$ μC is placed at $(0, -0.3)$ m.
   - Finally, point (P.) is located at $(0, +0.3)$ m.
      a) What is the electric field at point (P.)?
      b) Make a rough sketch of the electric field in the region surrounding q_1 and q_2.
      c) What is the Coulombic force on a $+3$ μC placed at point (P.)?

5. In a few brief sentences, answer the following:
   a) What is your understanding of a fundamental property? List some examples.
   b) What is your understanding of a field?
   c) Can a field be measured?
   d) How do we know a field really exists?
   e) Can two electric field lines ever cross?
   f) What is usually bigger, the gravitational force or the Coulombic force between two objects?

6. Two charges are placed on the x-axis:
   - $q_1 = +5$ C is placed at $(-2, 0)$ m.
   - q_2 has an unknown positive magnitude but is placed at $(+5, 0)$ m.

   What magnitude must q_2 have so that the electric field at the origin is zero?

7. Three charges are placed on the xy-plane:
   - $q_1 = +6$ μC is placed at $(0, +2)$ m.
   - $q_2 = +5$ μC is placed at $(-2, 0)$ m.
   - $q_3 = -4$ μC is placed at $(0, -2)$ m.
   - Finally, point (P.) is located at $(+2, 0)$ m.
      a) What is the electric field at point (P.)?
      b) What is the Coulombic force on a -3 μC placed at point (P.)?

IOP Publishing

Simplified Classical Mechanics, Volume 2 (Second Edition)
Gravity and the conservation laws
Gregory A DiLisi

Appendix

Solutions to independent practice problems

Chapter 1: The universal law of gravitation

f.) $W_{ħ} = \dfrac{G M_{ħ} M}{R_{ħ}^2} = \dfrac{6.67\times10^{-11} \cdot 5.68\times10^{26} \cdot M}{(5.85\times10^7)^2} = 11.1 M$

g.) $W_{\odot} = \dfrac{G M_{\odot} M}{R_{\odot}^2} = \dfrac{6.67\times10^{-11} \cdot 2\times10^{30} \cdot M}{(7\times10^8)^2} = 272 M$

5.

$F_{g \text{ on perry from Fred}} = \dfrac{G \cdot 60 \cdot M}{7^2} \hat{r} \implies = \left(\dfrac{G \cdot 60 \cdot M}{7^2}, 0\right) \approx (0,0)$

$F_{g \text{ on perry from Tim}} = \dfrac{G \cdot 65 \cdot M}{0.5^2} \hat{r} \implies = \left(-\dfrac{G \cdot 65 \cdot M}{0.5^2}, 0\right) \approx (0,0)$

$F_{g \text{ on perry from North Star}} = \dfrac{G \cdot 1\times10^{30} \cdot M}{(1\times10^{17})^2} \hat{r} = \left(0, \dfrac{G \cdot 1\times10^{30} \cdot M}{(1\times10^{17})^2}\right) \approx (0,0)$

$F_{g \text{ on perry from EARTH}} = \dfrac{G \cdot 5.98\times10^{24} \cdot M}{(6.38\times10^6)^2} \hat{r} = (0, -9.8 M)$

$\vec{F}_{TOTAL} \approx (0, -9.8 M)$ ALL forces are negligible except for the force of gravity from the EARTH

6 NO — IT VARIES AS $\dfrac{1}{r^2}$

7 a.) $\vec{F_g} = \dfrac{G \cdot M_{\oplus} \cdot 50}{R_{\oplus}^2} = (0, -9.8 * 50) = (0, -490)$ Newtons

b.) $-9.8\ m/s^2$ or $9.8\ m/s^2$ towards center of the Earth

c.) $\vec{F_g} = \dfrac{G \cdot M_{\oplus} \cdot 50}{(R_{\oplus} - 4m)^2} = \dfrac{(6.67\times10^{-11})(5.98\times10^{24}) \cdot 50}{(6.38\times10^6 - 6437)^2} = 9.82 * 50 = 491$ Newtons

or $(0, -491)$ Newtons

d.) $-9.82\ m/s^2$ or $9.82\ m/s^2$ towards center of the Earth

8. a) $F_g = \dfrac{G M_\odot \cdot 40}{R_\odot{}^2} = \dfrac{6.67 \times 10^{-11} \cdot 2 \times 10^{30} \cdot 40}{(7 \times 10^8)^2} = 272 \cdot 40 = 10{,}889$ Newtons

b) $272 \, ^m/_{s^2}$ towards the center of the Sun

9. a) $F_g = \dfrac{G \cdot M_\oplus \cdot M_{SAT}}{(R_\oplus + 3.2 \times 10^5)^2} = \dfrac{(6.67 \times 10^{-11})(5.98 \times 10^{24})(2500)}{(6.38 \times 10^6 + 3.2 \times 10^5)^2} = 22{,}213.5$ New

b) $g = \dfrac{F_g}{2500 \, Kg} = 8.89 \, ^m/_{s^2}$ towards the center of the Earth

R (Distance) M-km	T (period) yr
57.9	0.241
108.2	0.615
149.6	1
227.9	1.88
778.3	11.86
1,427	29.5
2,870	84
4,497	165
5,900	248

Period vs. Distance

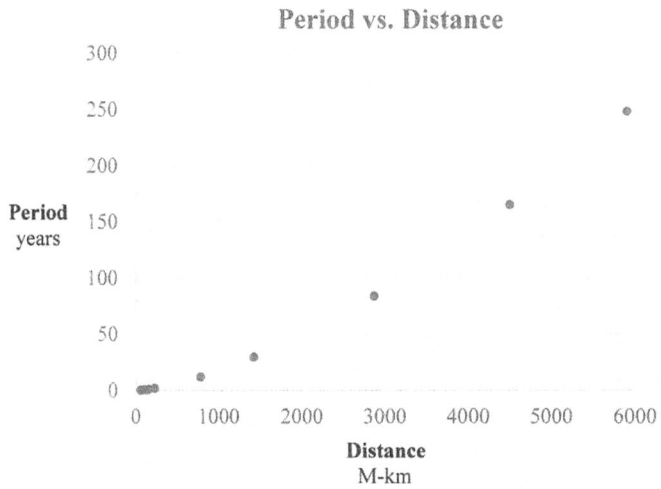

R (Distance) M-km	R^3/2 (Distance)^3/2 M-km^3/2	T (period) yr
57.9	441	0.241
108.2	1125	0.615
149.6	1830	1
227.9	3440	1.88
778.3	21713	11.86
1,427	53906	29.5
2,870	153753	84
4,497	301567	165
5,900	453188	248

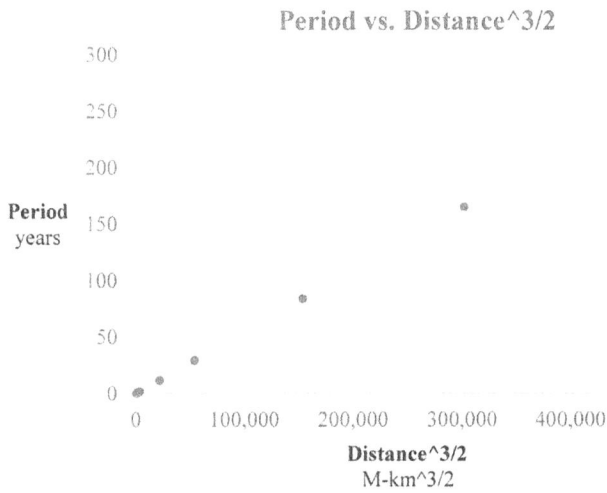

Period vs. Distance^3/2

Period years

300

250

200

150

100

50

0

0   100,000   200,000   300,000   400,000

Distance^3/2
M-km^3/2

Chapter 2: The conservation laws

INDEPENDENT PRACTICE - CH 2 (LCE)

1. a.) $W_{NORMAL} = \binom{0}{N} \cdot \binom{2}{0} = 0$; $W_{GRAV} = \binom{0}{-mg} \cdot \binom{2}{0} = 0$

$W_{FRIC} = \binom{-0.4\,mg}{0} \cdot \binom{2}{0} = -0.4 * m * 9.8 * 2$ (IN Joules)

b.) $W_{FRIC} + \Delta \Sigma = 0 : -0.4(9.8)(2) = -\frac{1}{2} \not{m} v^2 \Rightarrow V = \sqrt{0.4 \cdot 9.8 \cdot 2 \cdot 2} = 3.96$ m/s

2.

a.) $18.5\,m/s$ $37°$ $27.5\,m$ USE UAM #3 $\binom{v_x^2}{v_y^2} = \binom{(18.5\cos35°)^2}{(18.5\sin35°)^2} + 2\binom{0}{-9.8}\binom{(0-0)}{(0-27.5)}$

$\binom{v_x}{v_y} = \binom{14.77}{-25.7}$ m/s

These are THE SAME ANSWER

$V = \sqrt{14.77^2 + (-25.7)^2}$

$= 29.7$ m/s

USE LCE: $\frac{1}{2} m v_0^2 + mgh = \frac{1}{2} m v^2$

$\frac{1}{2} \not{m} (18.5)^2 + \not{m} (9.8)(27.5) = \frac{1}{2} \not{m} v^2 \Rightarrow V = 29.7$ m/s

* NOTICE THAT LCE & UAM #3 GIVE THE SAME ANSWER, BUT

IN DIFFERENT FORM

b.) SAME AS PART (A.)

3. a.) $W_F = \vec{f} \cdot \vec{d} = \binom{-0.4 \cdot 6 \cdot 9.8 \sin 36.9°}{0} \cdot \binom{4}{0} = -56.5$ Joules $\vec{f} = \mu N$

$= \mu mg \sin 36.9°$

b.) $W_g = \vec{W} \cdot \vec{d} = \binom{-6.9.8 \cos 36.9°}{-6.9.8 \sin 36.9°} \cdot \binom{4}{0} = 188.1$ Joules

c.) $W_N = \vec{N} \cdot \vec{d} = \binom{0}{N} \cdot \binom{4}{0} = 0$

d.) $W_{TOT} = -56.1 + 188.1 = 131.6$ Joules

$36.9°$ $53.1°$ $\vec{d}$ $\vec{W}$

4. a.) LCE: $20 \cdot 9.8 \cdot 2 = 20 \cdot 9.8 \cdot (1 + \cos 30°) + \frac{1}{2} 20 v^2 \Rightarrow V = 1.62$ m/s

b.) UCM: $\binom{20 \cdot 9.8 \cos 60°}{-20 \cdot 9.8 \sin 60°} + \binom{0}{N} = \binom{20a}{-20 \cdot 1.62^2}$ From PART (a) & use $a = \frac{v^2}{R}$ $\Rightarrow N = 117$ Newtons

c.) LCE $20 \cdot 9.8 \cdot 2 = \frac{1}{2} 20 v^2 \Rightarrow V = 6.2$ m/s

5. $0.1 \cdot 9.8 \cdot 0.5 + W_f = \frac{1}{2} \cdot 0.1 \cdot 1.8^2 \Rightarrow W_f = -0.328$ Joules

6. a) $80 \cdot 9.8 \cdot 75 = \frac{1}{2} \cdot 80 \cdot V^2 \Rightarrow V = 38.3 \text{ m/s}$

b)
$$\frac{1}{2} \cdot 80 \cdot 38.3^2 + \begin{pmatrix} -0.2 \cdot 80 \cdot 9.8 \\ 0 \end{pmatrix} \cdot \begin{pmatrix} 225 \\ 0 \end{pmatrix} = \frac{1}{2} \cdot 80 \cdot V^2 \Rightarrow V = 24.1 \text{ m/s}$$

c) $\vec{P} \cdot \vec{d} + \frac{1}{2} \cdot 80 \cdot 24.1^2 = 0$

$$\begin{pmatrix} -P \\ 0 \end{pmatrix} \cdot \begin{pmatrix} 2.5 \\ 0 \end{pmatrix} + \frac{1}{2} \cdot 80 (24.1)^2 = 0 \Rightarrow \vec{P} = \begin{pmatrix} -9292 \\ 0 \end{pmatrix} \text{ New}$$

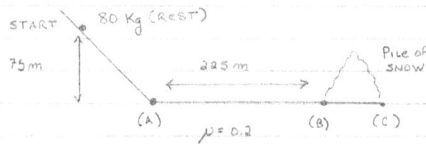

Picture below

START • 80 Kg (REST)

75 m

(A) $\mu = 0.2$ (B) 225 m (C) Pile of SNOW

7. a) $2 \cdot 9.8 \cdot 4 \sin 53.1° - 0.2 \cdot 2 \cdot 9.8 \sin 36.9° \cdot 4 = \frac{1}{2} \cdot 2 \cdot V^2 \Rightarrow V = 7.29 \text{ m/s}$

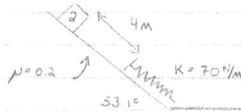

2 Kg 4 M K = 70 N/m $\mu = 0.2$ 53.1°

b) $\frac{1}{2}(2)(7.29)^2 - 0.2 \cdot 2 \cdot 9.8 \sin 36.9° \cdot d = \frac{1}{2} \cdot 70 \cdot d^2 - 2 \cdot 9.8 \cdot d \sin 36.9°$

(QUAD FORMULA WITH $A = \frac{1}{2} \cdot 70$

$B = 0.2 \cdot 2 \cdot 9.8 \sin 36.9° - 2 \cdot 9.8 \sin 36.9°$

$C = -\frac{1}{2}(2)(7.29)^2$

$d = 1.43$ meters

c) $\frac{1}{2}(2)(7.29)^2 - 0.2 \cdot 2 \cdot 9.8 \sin 36.9° \cdot d' = -2 \cdot 9.8 \sin 36.9° \cdot d'$

$d' = 2.42$ meters (above the spring)

8. a) $\frac{1}{2} 150 x^2 = 0.05 \cdot 9.8 \cdot 1.8 \sin 40° \Rightarrow x = 0.087$ m

   b) $\frac{1}{2} 150 (0.087)^2 = KE + 0.05 \cdot 9.8 \cdot 0.8 \sin 40° \Rightarrow KE = 0.315$ Joules

9. a) $\frac{1}{2} 2.5 \cdot 3^2 + 2.5 \cdot 4 = \frac{1}{2} \cdot 2.5 v^2 \Rightarrow v = 4.12$ m/s

   b) $a = 2.5 N / 2.5 kg = 1$ m/s^2 OR UAM #3 $4.12^2 = 3^2 + 2a(4) \Rightarrow a = 1$ m/s^2

   c) See part (b)

10. $0 = \frac{1}{2} 20 x^2 - 4 \cdot 9.8 x \sin 30° - 0.25 \cdot 4 \cdot 9.8 \sin 60° x + 30 \cdot x$

    $\qquad\quad U_{spring} \qquad\quad U_g \qquad\qquad \vec{f} \cdot \vec{d} \qquad\qquad \vec{P} \cdot \vec{d}$

    $x = 4.11$ meters

11. $W_g = \begin{pmatrix} 0 \\ -70 \cdot 9.8 \end{pmatrix} \cdot \begin{pmatrix} 0 \\ 0.5 \end{pmatrix} = -343 J \Rightarrow \therefore W_{human} = +343$ Joules $\begin{pmatrix} \text{opposite of} \\ \text{WHAT gravity} \\ \text{DOES!} \end{pmatrix}$

12. yes - LCE does not depend on direction (it is a scalar equation)

13. NO - NOT WITHOUT THE addition of some other force

14. $\frac{1}{2} m v_0^2 = mgh \Rightarrow h = \frac{v_0^2}{2g} = 1.6$ m

15. a) $mg \cdot 0.17 \sin 35° = \frac{1}{2} m v^2 \Rightarrow v = 1.38$ m/s

    b) $mg \cdot 0.17 \sin 35° + \frac{1}{2} m 2^2 = \frac{1}{2} m v^2 \Rightarrow v = 2.43$ m/s

16. $\frac{1}{2} m v_0^2 = mgh + \frac{1}{2} m v^2 \Rightarrow \frac{1}{2} v_0^2 = g \cdot 1.1 + \frac{1}{2} (6.5)^2 \Rightarrow v_0 = 8 \frac{m}{s}$

INDEPENDENT PRACTICE - CH 2 (LCP)

1.   "Before"   "After"

1kg   3kg v_1' 1kg   3kg   0.5 m/s

LCE: $\frac{1}{2}(10)x^2 = \frac{1}{2} 1 v_1'^2 + \frac{1}{2}(3)(0.5)^2$

LCP: $1\binom{0}{0} + 3\binom{0}{0} = 1\binom{-v_1'}{0} + 3\binom{0.5}{0} \Rightarrow v_1' = 1.5$ m/s $x = 0.55$ m

2.

5 m/s

2 m/s

0.045 kg   0.145 kg

$\vec{P}_{golf} = (0, 0.225)$ kg m/s

$\vec{P}_{base} = (-0.29, 0)$ kg m/s

$\vec{P}_{TOTAL} = (-0.29, 0.225)$ kg m/s

3.

"Before"   "After"

77 kg   92 kg   77 kg   92 kg

13 m/s 5 m/s 2.5 m/s v'

LCP: $77\binom{13}{0} + 92\binom{-5}{0} = 77\binom{2.5}{0} + 92\binom{v'}{0} \Rightarrow v' = 3.78$ m/s

4.

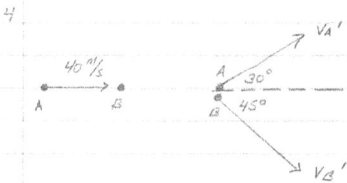

40 m/s

A   B

$M\binom{40}{0} + M\binom{0}{0} = M\binom{v_A'\cos 30°}{v_A'\sin 30°} + M\binom{v_B'\cos 45°}{-v_B'\sin 45°}$

Solve $v_A' = 29.3$ m/s & $v_B' = 20.6$ m/s

Rest

300 kg

5.

50 m

Frictionless

(6°)

70Kg 80Kg

(A)   FRICTIONLESS   (B)

40m

Solve speed at (A):

LCE: $U_g = KE \Rightarrow mg(50\sin 6°) = \frac{1}{2}m V_A^2 \Rightarrow V_A = \sqrt{2 \cdot 9.8 \cdot 50 \sin 6°} = 10.12 \text{ m/s}$

Solve speed at (B) - After Lone Ranger & Tonto jump in.

LCP: $300 \binom{10.12}{0} = 450 \binom{V_B}{0} \Rightarrow V_B = 6.75 \text{ m/s}$

Solve for time to get to canyon:

$t = d/V_B = 40/6.75 = 5.9 \text{ sec} \Leftarrow$ Since the heroes only need 5 sec,
They will make it.

6.   "Before"   "After"

0.3Kg 0.6 m/s   1.5 m/s 0.2 Kg   0.3 Kg V_1' 0.2 Kg V_2'

LCP: $0.3\binom{0.6}{0} + 1.5\binom{0.2}{0} = 0.3\binom{V_1'}{0} + 0.2\binom{V_2'}{0}$ Solve:

LCE: $\frac{1}{2}0.3(0.6)^2 + \frac{1}{2}1.5(0.2)^2 = \frac{1}{2}0.3 V_1'^2 + \frac{1}{2}0.2 V_2'^2$ $V_1' = -1.08 \text{ m/s}$

$\Big\lfloor$ KE is conserved only for an "Elastic" collision $V_2' = +1.02 \text{ m/s}$

7   200Kg 5 m/s ──────→ x "Before"

a) "After"   180 Kg $\vec{V_F}'$ 20 Kg 5 m/s $200\binom{5}{0} = 180\binom{V_F'}{0} + 20\binom{5}{-2}$

   ↓ 2 m/s $V_F' = 5 \text{ m/s}$

b) "After"   20 Kg   180Kg $\vec{V_F}'$ $200\binom{5}{0} = 180\binom{V_F'}{0}$

                                          $V_F' = 5.55 \text{ m/s}$

c) "After"   200 Kg 5 m/s   6 m/s 20 Kg $200\binom{5}{0} + 20\binom{-6}{0} = 220\binom{V_F'}{0}$

                                          $V_F' = 4 \text{ m/s}$

8 LCP $M_{man} \vec{V}_{man} + M_{shark} \vec{V}_{shark} = M_{man} \vec{V}_{man}' + M_{shark} \vec{V}_{shark}'$ NOTE: IF THE MAN

$12000 \binom{0}{0} + M_s \binom{0}{0} = 12000 \binom{150}{0} + M_s \binom{-850}{0}$ moved 150 ft in t seconds, then the

↳ Across X $\quad 0 = 12000 \cdot 150 - M_s \cdot 850$ shark moved -850 ft in t secs.

$M_{shark} = 2,117.6$ lbs

9

"Before" $\qquad$ "After"

REST

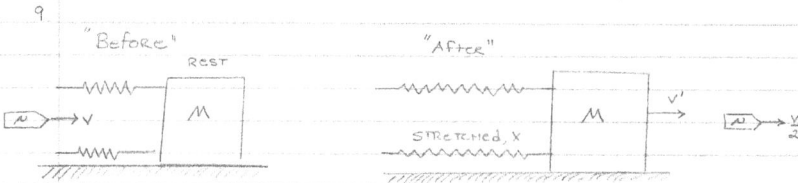

LCP: $N \binom{V}{0} = N \binom{V/2}{0} + M \binom{V'}{0} \Rightarrow NV = \frac{NV}{2} + MV' \Rightarrow V' = \frac{NV}{2M}$

LCE $\frac{1}{2} N V^2 = \frac{1}{2} N \left(\frac{V}{2}\right)^2 + \frac{1}{2} M V'^2 + \frac{1}{2} K x^2 + \frac{1}{2} K x^2$ } SOLVE X

$\frac{3}{8} N V^2 - \frac{1}{2} M V'^2 = K x^2$ } USE $V' = \frac{NV}{2M}$ from LCP

$x = \sqrt{\frac{N V^2}{8K} \left(3 - \frac{N}{M}\right)}$

Chapter 3: Rotational motion

INDEPENDENT PRACTICE - CH 3

1 a.) $\vec{r}_{com} = \dfrac{100\binom{0}{0} + 300\binom{3}{0} + 200\binom{0}{1.73}}{600} = \binom{1.5}{0.576}\,m$

b.) $\vec{r}_{com} = \dfrac{100\binom{0}{0} + 300\binom{5}{0} + 200\binom{5}{2} + 200\binom{5}{-2} + 100\binom{-4}{0} + 200\binom{-4}{+2}}{1100} = \binom{2.09}{0.182}\,m$

c.) $\vec{r}_{com} = \dfrac{200\binom{5\cos 30^{\circ}}{5\sin 30^{\circ}} + 200\binom{5\cos 30^{\circ}}{-5\sin 30^{\circ}} + 100\binom{0}{0} + 300\binom{-5\cos 30^{\circ}}{+5\sin 30^{\circ}} + 100\binom{-5\cos 30^{\circ}}{-5\sin 30^{\circ}}}{900}$

$= \binom{0}{0.555}\,m$

d.) $\vec{r}_{com} = \dfrac{100\binom{0}{0} + 300\binom{6.0}{0} + 200\binom{3.46\cos 30^{\circ}}{3.46\sin 30^{\circ}}}{600} = \binom{4.00}{0.576}\,m$

2.

OBJECT # 1:

$\vec{r}_{com} = \dfrac{10\binom{0}{0} + 10\binom{25}{0} + 20\binom{0}{-10} + 20\binom{0}{+10} + 20\binom{-20}{0} + 20\binom{10\cos 50^{\circ}}{10\sin 50^{\circ}} + 30\binom{-10\cos 40^{\circ}}{-10\sin 40^{\circ}}}{130}$

$= \binom{-1.93}{-0.305}\,m$

OBJECT # 2:

$\vec{r}_{com} = \dfrac{10\binom{0}{0} + 10\binom{10}{0} + 10\binom{0}{10} + 10\binom{0}{-10} + 20\binom{5\cos 30^{\circ}}{5\sin 30^{\circ}} + 20\binom{-10}{0} + 30\binom{-5\cos 40^{\circ}}{-5\sin 40^{\circ}}}{110}$

$= \binom{-2.08}{-0.422}\,m$

3. A. $\vec{r}_{com} = \dfrac{1\binom{-1}{0} + 2\binom{0}{0} + 3\binom{0}{-3} + 4\binom{2\cos30°}{2\sin30°}}{10} = \binom{0.593}{-0.5}\,m$

$I = 1(1)^2 + 2(0)^2 - 3(3)^2 + 4(2)^2 = 44\ Kg\cdot m2$

B. $\vec{r}_{com} = \dfrac{1\binom{-1}{0} + 2\binom{0}{0.5} - 3\binom{0}{-0.5} + 4\binom{1}{0}}{10} = \binom{0.3}{-0.05}\,m$

$I = 1(0)^2 + 2(0.5)^2 + 4(0)^2 + 3(0.5)^2 = 1.25\ Kg\ m^2$

C. $\vec{r}_{com} = \dfrac{1\binom{-1}{0} + 2\binom{-1}{0.5} + 3\binom{0}{0} + 4\binom{1}{0} + 3\binom{1}{0.5}}{13} = \binom{-0.308}{0.192}\,m$

$I = 1(1)^2 + 2(1)^2 + 3(0)^2 - 4(1)^2 + 3(1)^2 = 10\ Kg\ m^2$

D. $\vec{r}_{com} = \dfrac{1\binom{-1}{-1} + 2\binom{-1}{+1} + 4\binom{1}{-1} + 5\binom{+1}{+1}}{12} = \binom{0.5}{0.166}\,m$

$I = 1(\sqrt{2})^2 + 2(\sqrt{2})^2 + 4(\sqrt{2})^2 + 5(\sqrt{2})^2 = 24\ Kg\ m^2$

4. A. $\vec{\tau} = \vec{r}\times\vec{F} = \binom{3}{0}\times\binom{30\cos10°}{30\sin10°} = (3\cdot30\sin10°) - (0) = 15.6\ New\cdot m\ \odot$

B. $\vec{\tau} = \vec{r}\times\vec{F} = \binom{-4\cos5°}{+4\sin5°}\times\binom{0}{-25} = (-4\cos5°\cdot(-25)) - (0) = 99.62\ New\cdot m\ \odot$

C. $\vec{\tau} = \vec{r}\times\vec{F} = \binom{+5}{0}\times\binom{-50\cos30°}{-50\sin30°} = (5\cdot(-50\sin30°)) - (0) = 125\ New\cdot m\ \otimes$

D. $\vec{\tau} = \vec{r}\times\vec{F} = \binom{3\cos80°}{3\sin80°}\times\binom{-10\cos10°}{+10\sin10°} = (3\cos80°\cdot10\sin10°) - (3\sin80°\cdot(-10\cos10°))$
$= 30\ New\cdot m\ \odot$

5. Given: $\omega_0 = +1.5\ ^{RAD}/_{sec}$; $\theta_0 = +^{\pi}/_2$; $\alpha = +0.25\ ^{RAD}/_{sec^2}$

a.) Solve θ at $t = 4$:

$U\alpha M \#2$: $\theta = \theta_0 + \omega_0 t + \frac{1}{2}\alpha t^2 \Rightarrow \theta = \frac{\pi}{2} + 1.5(4) + \frac{1}{2}(0.25)(4)^2$

$= 9.57\ Rads$

b.) Solve ω at $t = 4$:

$U\alpha M \#1$: $\omega = \omega_0 + \alpha t \Rightarrow \omega = 1.5 + 0.25(4)$

$= 2.5\ ^{Rad}/_{sec}$

6. First, apply $\Sigma \vec{F} = m\vec{a}$ to the 2kg block:

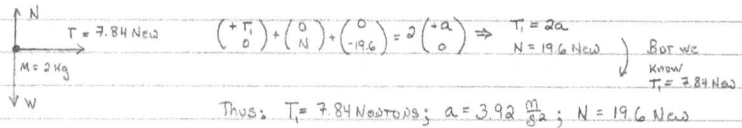

$T = 7.84\ New$ $M = 2kg$ N up, W down

$\begin{pmatrix} +T_1 \\ 0 \end{pmatrix} + \begin{pmatrix} 0 \\ N \end{pmatrix} + \begin{pmatrix} 0 \\ -19.6 \end{pmatrix} = 2\begin{pmatrix} +a \\ 0 \end{pmatrix} \Rightarrow \begin{array}{l} T_1 = 2a \\ N = 19.6\ New \end{array}$) But we know $T_1 = 7.84\ New$

Thus: $T_1 = 7.84\ Newtons$; $a = 3.92\ \frac{m}{s^2}$; $N = 19.6\ New$

Next, apply $\Sigma \vec{F} = m\vec{a}$ to the 4kg block

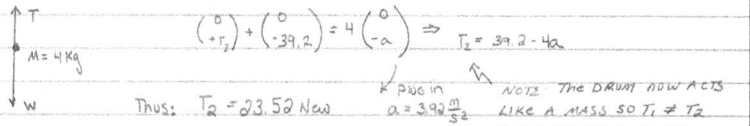

T up, $M = 4kg$, W down

$\begin{pmatrix} 0 \\ +T_2 \end{pmatrix} + \begin{pmatrix} 0 \\ -39.2 \end{pmatrix} = 4\begin{pmatrix} 0 \\ -a \end{pmatrix} \Rightarrow T_2 = 39.2 - 4a$

Thus: $T_2 = 23.52\ New$ $\underset{a = 3.92\ \frac{m}{s^2}}{\text{Plug in}}$ NOTE: The DRUM now acts like a mass so $T_1 \neq T_2$ even though only one rope exists in the problem. TWO TENSIONS EXIST IN THIS PROBLEM

Last, apply $\Sigma \vec{T} = I\vec{\alpha}$ to the drum

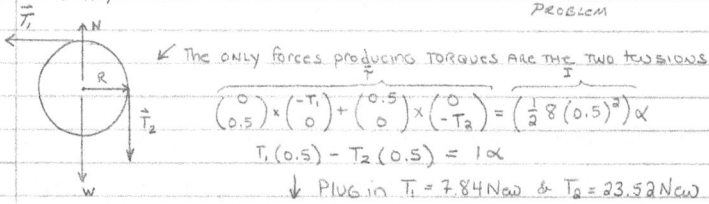

$\overleftarrow{T_1}$ N up R $\overrightarrow{T_2}$ W down

← The only forces producing TORQUES are the two tensions

$\begin{pmatrix} 0 \\ 0.5 \end{pmatrix} \times \begin{pmatrix} -T_1 \\ 0 \end{pmatrix} + \begin{pmatrix} 0.5 \\ 0 \end{pmatrix} \times \begin{pmatrix} 0 \\ -T_2 \end{pmatrix} = \left(\frac{1}{2}8(0.5)^2\right)\alpha$

$T_1(0.5) - T_2(0.5) = I\alpha$

$\downarrow$ Plug in $T_1 = 7.84\ New$ & $T_2 = 23.52\ New$

$\alpha = -7.84\ \frac{RAD}{sec^2}$ OR $7.84\ \frac{Rad}{sec^2}$ ⊗

7. USE LCE

$$mgh = \tfrac{1}{2}mv^2 + \tfrac{1}{2}I\omega^2$$

SUBSTITUTE $\omega = v/R$

$I = f MR^2$, where

$$mgh = \tfrac{1}{2}mv^2 + \tfrac{1}{2}(f\,mR^2)\cdot\frac{v^2}{R^2}$$

$f = \tfrac{2}{5}$ for solid ball

R & m cancel

1 for hoop

$$gh = v^2\left(\tfrac{1}{2} + \tfrac{1}{2}f\right)$$

$\tfrac{1}{2}$ solid cylinder

Solve for v

$$v = \sqrt{\frac{2gh}{1+f}}$$

Thus, THE smaller f has THE greater speed at THE bottom of THE HILL. So THE cylinder finishes in 1ST place, THE solid ball in 2nd, & THE hoop in 3rd.

8. $I_{DRUM} = \tfrac{1}{2}MR^2 = \tfrac{1}{2}\cdot 20 \cdot 0.1^2 = \tfrac{1}{10}$ kg m^2

$I_{Rod-end} = \tfrac{1}{3}m\ell^2 = \tfrac{1}{3}\cdot 3 \cdot 0.25^2 = \tfrac{1}{16}$ kg m^2

USE LCL

$$\Sigma(I\omega)_{before} = \Sigma(I\omega)_{After}$$

$$0 = \left(\tfrac{1}{16}\pi\right)\otimes + \left(\tfrac{1}{10}\omega'_{body}\right)$$

$$\omega'_{body} = -\pi\cdot\tfrac{10}{16}\otimes \text{ or } \pi\cdot\tfrac{10}{16}\odot \text{ (opposite to tail)}$$

The cat spins its tail CW so its body TURNS CCW to conserve angular momentum

Chapter 4: Transition to classical electricity and magnetism

Independent Practice - CH 4

1. $F_c = \frac{K q_1 q_2}{r^2}$

a.) $F_c' = \frac{K q_1 q_2}{(8r)^2} = \frac{1}{64} F_c$ ⇒ So the force decreases by a factor of 64

b.) $F_c' = \frac{K q_1 q_2}{(\frac{1}{4}r)^2} = 16 \cdot F_c$ ⇒ So the force increases by a factor of 16

c.) $F_c' = \frac{K \cdot (2q_1) q_2}{r^2} = 2 \cdot F_c$ ⇒ So the force increases by a factor of 2.

d.) $F_c' = \frac{K q_1 q_2}{(\frac{1}{3}r)^2} = 9 F_c$ ⇒ So the force increases by a factor of 9.

e.) $F_c' = \frac{K (2q_1)(q_2/3)}{r^2} = \frac{2}{3} F_c$ ⇒ So the force is reduced by a factor of $\frac{2}{3}$.

f.) $F_c' = \frac{K(-q_1)(q_2)}{r^2} = -F_c$ ⇒ So the magnitude of F_c does not change but it changes from repulsive to attractive or from attractive to repulsive

g.) $F_c' = \frac{K \cdot (4q_1)(4q_2)}{(4r)^2} = F_c$ ⇒ So the force remains unchanged.

h.) $F_c' = \frac{K \cdot (4q_1)(4q_2)}{(\frac{1}{4}r)^2} = 256 F_c$ ⇒ So the force is increased by a factor of 256.

2.

a.) $F_c' = \frac{K q_1 q_2}{(2r)^2} = \frac{1}{4} F_c$ ⇒ So the force is decreased by a factor of 4

b.) $F_c' = \frac{K \cdot (q_1/2) \cdot (q_2/2)}{(3r)^2} = \frac{1}{36} F_c$ ⇒ So the force is decreased by a factor of 36

c.) $F_c' = \frac{K (3q_1) q_2}{(r/2)^2} = 12 F_c$ ⇒ So the force is increased by a factor of 12.

3.

$q_4 = +10\mu C$

$0.5m$ $0.3m$ $0.5m$

$36°$ $0.4m$

$q_1 = +30\mu C$

$0.3m$

$q_2 = -50\mu C$

$q_3 = -40\mu C$

$\vec{F}_{C\,1\,ON\,4} = \dfrac{9\times10^9 \cdot 30\times10^6 \cdot 10\times10^{-6}}{(0.5)^2} = 10.8\,Newt \Rightarrow (10.8\cos 36.9°,\ 10.8\sin 36.9°)\ New$

$(8.64,\ 6.48)\ Newtons$

$\vec{F}_{C\,2\,ON\,4} = \dfrac{9\times10^9 \cdot 40\times10^6 \cdot 10\times10^{-6}}{(0.6)^2} = 10\,New \Rightarrow (0,\ -10)\ Newtons$

$\vec{F}_{C\,3\,ON\,4} = \dfrac{9\times10^9 \cdot 50\times10^6 \cdot 10\times10^{-6}}{(0.5)^2} = 18\,New \Rightarrow (18\cos 36.9°,\ -18\sin 36.9°)\ New$

$(14.39,\ -10.81)\ Newtons$

$\therefore \vec{F}_{NET} = (23.0,\ -14.33)\ Newtons$

4.

(P)

$0.5m$

$0.3m$

$36°$ $0.4m$

$q_1 = +30\mu C$

$0.3m$

$q_2 = -40\mu C$

To calculate $\vec{E}$, imagine $+1C$ at (P):

a.) $\vec{F}_{C\,1\,ON\,+1C} = \dfrac{9\times10^9 \cdot 30\times10^{-6} \cdot 1}{(0.5)^2} = 1.08\times10^6\,N$

$\Rightarrow (1.08\times10^6 \cos 36.9°,\ 1.08\times10^6 \sin 36.9°)$

$= (8.64\times10^5,\ 6.48\times10^5)\,N$

$\vec{F}_{C\,2\,ON\,+1C} = \dfrac{9\times10^9 \cdot 40\times10^{-6} \cdot 1}{(0.6)^2} = 1\times10^6\,N$

$\Rightarrow (0,\ -1\times10^6)\,N$

$\therefore \vec{E} = (8.64\times10^5,\ -3.52\times10^5)\,\dfrac{N}{C}$

b.) See above - SKETCH HOW a POSITIVE TEST-CHARGE WOULD REACT.

c.) The force on a $+3\mu C$ placed at (P) is: $\vec{F}_C = +3\mu C \cdot \vec{E} = 3\times10^{-6}C * (8.64\times10^5,\ -3.52\times10^5)\dfrac{N}{C}$

$= (2.6,\ -1.056)\,Newtons$

5

a) A fundamental property is a physical property of matter that causes it to experience a force when placed in a field. It is the ingredient that matter needs to experience a particular force. Examples of fundamental properties are mass and electric charge.

b) A field is a model used to explain "action at a distance." It is the mechanism by which masses communicate gravity & by which electric charges communicate the Coulombic force.

c) A field cannot be measured – only forces can be measured

d) We do not know fields really exist – they are models we use to explain "action at a distance."

e) No – electric field lines cannot cross – That would imply a force can exist in two directions at one point. At every point in space, only one well-defined force/field exists

f) The gravitational force is considered the weakest of all the forces in nature

6

(P)

$q_1 = +5C$ (0,0) $q_2 = ?$
$(-2,0)$ m $(+5,0)$ m

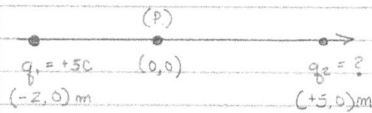

$\vec{E}_{AT\,(P)\,from\,q_1} = \dfrac{K \cdot 5}{(2)^2} = \dfrac{K \cdot 5}{4} \Rightarrow \left(\dfrac{K5}{4}, 0\right)$

$\vec{E}_{AT\,(P)\,from\,q_2} = \dfrac{K \cdot q_2}{(5)^2} = \dfrac{K q_2}{25} \Rightarrow \left(-\dfrac{K q_2}{25}, 0\right)$

$\vec{E}_{AT\,(P)\,TOTAL} = \left(\dfrac{K5}{4} - \dfrac{K q_2}{25}, 0\right)$, To set $\vec{E} = 0$, then $\dfrac{5}{4} - \dfrac{q_2}{25} = 0$ or $q_2 = \dfrac{125}{4} C$

7

$q_1 =$
$+6 \mu C$

2 m $\sqrt{8}$

.2m   2m u90° (P)
              45°

$q_2 =$
$+5 \mu C$

2m $\sqrt{8}$

Imagine placing $+1C$ at (P)

$q_3 =$
$+4 \mu C$

a)

$\vec{F}_{C \text{ on } +1C \text{ from } 1} = \dfrac{9 \times 10^9 \cdot 6 \times 10^{-6}}{(\sqrt{8})^2} = 6750 \, N \Rightarrow (6750 \cos 45°, -6750 \sin 45°) \, N$
$\qquad\qquad (4772.9, -4772.9) \, N$

$\vec{F}_{C \text{ on } +1C \text{ from } 2} = \dfrac{9 \times 10^9 \cdot 5 \times 10^{-6}}{4^2} = 2812.5 \, N \Rightarrow (2812.5, 0) \, N$

$\vec{F}_{C \text{ on } +1C \text{ from } 3} = \dfrac{9 \times 10^9 \cdot 4 \times 10^{-6}}{(\sqrt{8})^2} = 4500 \, N \Rightarrow (-4500 \cos 45°, -4500 \sin 45°) \, N$
$\qquad\qquad (-3182, -3182) \, N$

$\vec{F}_{C \, TOTAL} = (4403.4, -7954.9) \, N \quad , \quad \vec{E} = \dfrac{\vec{F}_C}{+1C}$
$\vec{E} = (4403.4, -7954.9) \dfrac{N}{C}$

b.) $\vec{F}_C \text{ on } -3\mu C = -3\mu C \cdot \vec{E} = -3 \times 10^{-6} C \cdot (4403.4, -7954.9) \dfrac{N}{C}$
$\qquad\qquad\qquad = (-0.0132, +0.0239) \, Newtons$

www.ingramcontent.com/pod-product-compliance
Lightning Source LLC
Chambersburg PA
CBHW082139210326

41599CB00031B/6037